T0215361

PROBABILITY
Theory, Examples, Problems, Simulations

PROBABILITY
Theory, Examples, Problems, Simulations

Hannelore Lisei
Babeş-Bolyai University, Cluj-Napoca, Romania

Wilfried Grecksch
Martin-Luther-University, Halle-Wittenberg, Germany

Mihai Iancu
Babeş-Bolyai University, Cluj-Napoca, Romania

World Scientific

NEW JERSEY · LONDON · SINGAPORE · BEIJING · SHANGHAI · HONG KONG · TAIPEI · CHENNAI · TOKYO

Published by

World Scientific Publishing Co. Pte. Ltd.
5 Toh Tuck Link, Singapore 596224
USA office: 27 Warren Street, Suite 401-402, Hackensack, NJ 07601
UK office: 57 Shelton Street, Covent Garden, London WC2H 9HE

British Library Cataloguing-in-Publication Data
A catalogue record for this book is available from the British Library.

PROBABILITY
Theory, Examples, Problems, Simulations

ISBN 978-981-120-573-6
ISBN 978-981-120-719-8 (pbk)

For any available supplementary material, please visit
https://www.worldscientific.com/worldscibooks/10.1142/11427#t=suppl

Desk Editor: Liu Yumeng

Printed in Singapore

Preface

This textbook gives an accessible approach to probability concepts through examples and problems with solutions, by presenting various techniques from probability theory and its applications. By an appropriate balance between simulations and mathematical results, the reader can experience the excitement of comprehending basic concepts and can develop the intuitive thinking in solving problems. The given Matlab codes and detailed solutions make the textbook useful to students, not only of mathematics but also computer science, engineering, physics, economy and other related fields.

In general, the structure of each section of this textbook is as follows: first, definitions are presented, next, properties and examples are pointed out and then the section ends with a few solved problems.

Many problems and examples are based on the description of an experiment which is simulated in Matlab. Therefore, the reader should be familiar with the basic operations in Matlab. In this textbook we use very often Matlab functions from the Statistics and Machine Learning Toolbox and sometimes from the Symbolic Math Toolbox. However, some of the given Matlab programs work also in GNU Octave with the corresponding Statistics and Symbolic packages. A key role in the simulations is played by the pseudorandom number generators. For simplicity, the expression random numbers is used instead of pseudorandom numbers.

The main purpose of the simulations is to illustrate the intuitive idea that the relative frequency of the occurrence of an event in a large number of repetitions of an experiment is close to the corresponding theoretical probability of the event. We point out that the descriptions of the experiments are adapted for a pure mathematical context without any pretension to model accurately a real situation.

Each chapter ends with a list of problems of varying difficulty. Some of the problems are classic, but rewritten from the authors' point of view, while others are original. In order to solve the mathematical tasks of a problem, the reader should possess basic knowledge of calculus, combinatorics, matrix theory and measure theory (to recall some basic notions and results, see the Appendix) and should use the material presented in the sections. For each problem a complete solution is given. The reader is encouraged to implement the simulations of the described experiments and to test the Matlab codes given in the solutions to get a better understanding of the ideas behind the problems. The programs and solutions in the textbook reflect the authors' perspective, which may inspire the reader to find more elegant ways to write the programs and to solve the problems.

In the following we summarize the contents of the textbook.

Chapter 1 gives a brief introduction into the basic notions of probability, such as events, σ-fields and probability spaces. Some counting examples and simulations are meant to accustom the reader with basic methods and principles in combinatorics.

Chapter 2 deals with random variables and random vectors, their (conditional) distribution functions and (conditional) density functions. Simulation techniques such as the inversion method, the acceptance-rejection method, the Box–Muller algorithm, random shuffling are discussed. At the end of this chapter Bayesian networks are presented.

Chapter 3 deals with numerical characteristics of random variables and random vectors, such as (conditional) expectations, variances, moment generating functions.

Chapter 4 covers results about convergence types for sequences of random variables, laws of large numbers and central limit theorems.

Chapter 5 presents a few applications into the theory of stochastic processes, especially random walks, Markov processes, Markov chains, branching processes, birth and death processes, Poisson processes and Wiener processes.

Finally, the Appendix provides some useful information about the Lebesgue integral, the Lebesgue–Stieltjes integral, the Gamma and Beta Euler functions, it gives some formulae from combinatorics and results from (deterministic) matrix theory.

The textbook is based on the experience of the authors in teaching probability. It is the result of collaborative work spanned over years between the Faculty of Mathematics and Computer Science, Babeş-Bolyai University in

Cluj-Napoca (Romania), and the Institute of Mathematics of the Faculty of Natural Sciences II, Martin-Luther-University in Halle-Wittenberg (Germany).

The authors thank everyone who has helped them to write and publish this textbook. Special thanks to their families for their support and understanding.

H. Lisei, W. Grecksch, M. Iancu

Contents

Chapter 1

Probability Space

1.1 Experiments and Events

The terms **trial** and **experiment** describe any process or action that can be infinitely repeated and has a set of possible outcomes, denoted by Ω, which is called **sample space**. Probability theory deals with experiments that are random, which have more than one possible outcome. The sample space is connected to the notion of probability space, which we consider in Definition 1.6.

The elements of Ω (i.e., the possible outcomes of an experiment) are called **elementary events**. For an elementary event ω we write $\omega \in \Omega$ to denote that ω is an element of Ω. An **event** is a subset of the sample space, i.e., it is a collection of elementary events. If A is an event, we write $A \subseteq \Omega$. An event A is **realized** or **occurs** in an experiment, if the outcome of the experiment is in A. We consider two special events: the **impossible** event, denoted by $\emptyset$, which is the event that does not occur in any trial, and the **sure** (**certain**) event, which is the event that occurs in each trial.

For any event $A \subseteq \Omega$ we may associate its **contrary event** or **complement**, denoted $\bar{A}$, to mean that $\bar{A}$ is realized if and only if A is not realized.

It is said that the event A **implies** the event B if every element of A is also an element of B, written $A \subseteq B$, i.e., if A occurs, then B occurs. If A implies B and B implies A, then the events A and B are **equal** $A = B$.

The **union** of the events A and B is the set of all elementary events that are in A or B, written

$$A \cup B = \{\omega \in \Omega : \omega \in A \text{ or } \omega \in B\},$$

i.e., the event that at least one of the events A and B occurs.

The **intersection** of the events A and B is the set of all elementary events that are both in A and in B, written

$$A \cap B = \{\omega \in \Omega \ : \ \omega \in A \, \text{and} \, \omega \in B\},$$

i.e., the event that both events A and B are realized.

Two events A and B are said to be **disjoint** (or **mutually exclusive**) if A and B have no common outcomes, written $A \cap B = \emptyset$, i.e., the events A and B cannot occur both.

The **difference** of the events A and B is the set of all elementary events that are in A but not in B, written

$$A \setminus B = A \cap \bar{B},$$

i.e., the event that A is realized and B not.

The **symmetric difference** of the events A and B is the set of all elementary events that are in A or B but are not common to both, written

$$A \Delta B = (A \cup B) \setminus (A \cap B),$$

i.e., the event that exactly one of the events A and B is realized.

The operations of union, intersection and symmetric difference are:
- **commutative:**

$$A \cup B = B \cup A, \quad A \cap B = B \cap A, \quad A \Delta B = B \Delta A;$$

- **associative:**

$$(A \cup B) \cup C = A \cup (B \cup C), \quad (A \cap B) \cap C = A \cap (B \cap C),$$

$$(A \Delta B) \Delta C = A \Delta (B \Delta C);$$

- **distributive:**

$$(A \cup B) \cap C = (A \cap C) \cup (B \cap C), \quad (A \cap B) \cup C = (A \cup C) \cap (B \cup C),$$

$$A \cap (B \Delta C) = (A \cap B) \Delta (A \cap C).$$

Example 1.1. Suppose that we make an experiment by tossing a coin two times. Then the sample space is

$$\Omega = \{\omega_1, \omega_2, \omega_3, \omega_4\},$$

where the elementary events (possible outcomes) are:
ω_1, if the first toss gives a head and the second toss gives a tail;
ω_2, if both tosses give tails;
ω_3, if both tosses give heads;
ω_4, if the first toss gives a tail and the second toss gives a head.

Let A be the event that at least one head is obtained during the experiment, B be the event that a head is given by the second toss, C be the event that a tail is given by the first toss and D the event that no heads are obtained. Then we can write

$$A = \{\omega_1, \omega_3, \omega_4\}, \ B = \{\omega_3, \omega_4\}, \ C = \{\omega_2, \omega_4\}, \ D = \{\omega_2\}.$$

We have the following relations for these events

$$D \subseteq C, A \cap D = \emptyset, B \cap C = \{\omega_4\}, A \cup C = \Omega, \bar{D} = A, B \Delta C = \{\omega_2, \omega_3\}. \ \blacktriangle$$

Throughout this book, Ω denotes a sample space, which is always assumed to be nonempty.

Definition 1.1. A collection of events $(A_i)_{i \in I}$, where I is a nonempty set of indices, from Ω is said to be **collectively exhaustive** if

$$\bigcup_{i \in I} A_i = \Omega.$$

A collection $(A_i)_{i \in I}$ of events, where I is a set of indices with $\#I \geq 2$, from Ω is said to be a **partition** of Ω if the events are collectively exhaustive and for all $i, j \in I$, $i \neq j$, the events A_i and A_j are disjoint.

The **lower limit** of a sequence $(A_n)_{n \in \mathbb{N}^*}$ of events from Ω is the event

$$\liminf_{n \to \infty} A_n = \bigcup_{n=1}^{\infty} \bigcap_{i=n}^{\infty} A_i.$$

The **upper limit** of a sequence $(A_n)_{n \in \mathbb{N}^*}$ of events from Ω is the event

$$\limsup_{n \to \infty} A_n = \bigcap_{n=1}^{\infty} \bigcup_{i=n}^{\infty} A_i. \qquad \blacklozenge$$

Remark 1.1. Note that $\liminf\limits_{n \to \infty} A_n$ is the event that occurs if all events of the sequence $(A_n)_{n \in \mathbb{N}^*}$ occur except a finite number of them, while the event $\limsup\limits_{n \to \infty} A_n$ occurs if an infinite number of events of the sequence $(A_n)_{n \in \mathbb{N}^*}$ occur, i.e.,

$$\liminf_{n \to \infty} A_n = \{\omega \in \Omega : \omega \in A_n \text{ for all } n \in \mathbb{N}^* \text{ except a finite number}\},$$

$$\limsup_{n \to \infty} A_n = \{\omega \in \Omega : \omega \in A_n \text{ for infinitely many } n \in \mathbb{N}^*\}. \qquad \blacktriangledown$$

Definition 1.2. A sequence $(A_n)_{n \in \mathbb{N}}$ of events from Ω is said to be **monotonically increasing** if

$$A_n \subseteq A_{n+1} \quad \text{for all } n \in \mathbb{N}$$

and it is **monotonically decreasing** if

$$A_{n+1} \subseteq A_n \quad \text{for all } n \in \mathbb{N}. \qquad \blacklozenge$$

Theorem 1.1.

(1) *A collection $(A_i)_{i \in I}$ of events from Ω satisfies **De Morgan's laws***

$$\overline{\bigcup_{i \in I} A_i} = \bigcap_{i \in I} \bar{A}_i \quad and \quad \overline{\bigcap_{i \in I} A_i} = \bigcup_{i \in I} \bar{A}_i.$$

(2) *For any sequence $(A_n)_{n \in \mathbb{N}^*}$ of events from Ω it holds*

$$\liminf_{n \to \infty} A_n \subseteq \limsup_{n \to \infty} A_n.$$

(3) *For any monotonically increasing sequence $(A_n)_{n \in \mathbb{N}^*}$ of events from Ω it holds*

$$\liminf_{n \to \infty} A_n = \limsup_{n \to \infty} A_n = \bigcup_{n=1}^{\infty} A_n.$$

(4) *For any monotonically decreasing sequence $(A_n)_{n \in \mathbb{N}^*}$ of events from Ω it holds*

$$\liminf_{n \to \infty} A_n = \limsup_{n \to \infty} A_n = \bigcap_{n=1}^{\infty} A_n.$$

1.1.1 Solved Problems

Problem 1.1.1. A point is randomly chosen in $\Omega = [0,8] \times [0,8]$. Let A be the event that the point is in $[0,6] \times [0,5]$ and let B be the event that the point is in $[3,8] \times [2,8]$. Express the following events, using only the events A and B and the operations for events:

1) the point is chosen in the set

$$\{(x,y) \in \Omega : (x,y) \in [3,6] \times [2,5]\};$$

2) the point is chosen in the set

$$\{(x,y) \in \Omega : (x,y) \in [0,3) \times (5,8] \text{ or } (x,y) \in (6,8] \times [0,2)\};$$

3) the point is chosen in the set

$$\{(x,y) \in \Omega : (x,y) \in [0,6] \times [0,2) \text{ or } (x,y) \in [0,3) \times [2,5]\};$$

4) the point is chosen in the set

$$\{(x,y) \in \Omega : (x,y) \notin [3,6] \times [2,5] \text{ and } (x,y) \notin [0,3) \times (5,8]$$
$$\text{and } (x,y) \notin (6,8] \times [0,2)\}.$$

Solution 1.1.1: 1) $A \cap B$; 2) $\bar{A} \cap \bar{B}$; 3) $A \setminus B$; 4) $A \triangle B$.

Problem 1.1.2. 1) Find the number of solutions $(x_1, \ldots, x_k) \in \mathbb{N}^* \times \cdots \times \mathbb{N}^*$ of the equation $x_1 + \ldots + x_k = n$ for $n, k \in \mathbb{N}^*$, $n \geq k$. Write a function in Matlab that returns the solutions of this equation.
2) Find the number of solutions $(x_1, \ldots, x_k) \in \mathbb{N} \times \cdots \times \mathbb{N}$ of the equation $x_1 + \ldots + x_k = n$ for $n, k \in \mathbb{N}^*$. Write a function in Matlab that returns the solutions of this equation.

Solution 1.1.2: 1) Consider the following sequence of n zeros and $n - 1$ spaces: $0\_0\_ \cdots \_0$. Next, replace $k-1$ spaces with ones, delete the remaining spaces and let x_i be the number of consecutive zeros before the ith one, for $i \in \{1, \ldots, k - 1\}$, and x_k be the number of zeros after the last one. Then $(x_1, \ldots, x_k) \in \mathbb{N}^* \times \cdots \times \mathbb{N}^*$ is a solution of our equation. Moreover, every solution of the equation can be represented by a unique binary number constructed as above. Thus, we have $C(n - 1, k - 1)$ solutions, because this is the number of distinct choices of the $k - 1$ spaces to be replaced by ones.

```
function x=solveq1(n,k)
positions2delete=nchoosek(3:2:2*n-1,n-k);
x=zeros(nchoosek(n-1,k-1),k);
for j=1:nchoosek(n-1,k-1)
  v=[1 repmat([0 1],1,n)];
  v(positions2delete(j,:))=[ ];
  x(j,:)=diff(find(v==1))-1;
end
end
```

```
>> solveq1(5,4)
ans = 2 1 1 1
      1 2 1 1
      1 1 2 1
      1 1 1 2
```

2) Every solution $(x_1, \ldots, x_k) \in \mathbb{N} \times \cdots \times \mathbb{N}$ of the equation $x_1 + \ldots + x_k = n$ has a unique corresponding solution $(y_1, \ldots, y_k) \in \mathbb{N}^* \times \cdots \times \mathbb{N}^*$ of the equation $y_1 + \ldots + y_k = n + k$ and vice versa, by letting $y_i = x_i + 1$, for every $i \in \{1, \ldots, k\}$. In view of 1), we have now $C(n+k-1, k-1)$ solutions.

```
function x=solveq2(n,k)
x=diff([zeros(nchoosek(n+k-1,k-1),1) nchoosek(1:n+k-1,k-1)...
    (n+k)*ones(nchoosek(n+k-1,k-1),1)],1,2)-1;
end
```

```
>> solveq2(2,3)
```

ans = 0 0 2
          0 1 1
          0 2 0
          1 0 1
          1 1 0
          2 0 0

Problem 1.1.3. Let $n \in \mathbb{N}^*$. Write a Matlab function that prints the elements of the multisubsets of size n from a given nonempty finite set S (i.e., the sequences of n not necessarily distinct elements of S, whose order is not taken into account: two sequences define the same multiset if one can be obtained from the other by permuting the terms; they are also called n-combinations with repetitions). Find the number of such multisubsets.

Solution 1.1.3: We shall use the function from Problem 1.1.2-2).

```
function multiset(S,n)
k=length(S);
x=solveq2(n,k);
for i=1:nchoosek(n+k-1,k-1)
    for j=1:k
        m=repmat(S(j),1,x(i,j));
        fprintf('%s ',m{:});
    end
    fprintf('\n');
end
end
```

```
>> multiset({'cat','owl','frog'},2)
   frog frog
   owl frog
   owl owl
   cat frog
   cat owl
   cat cat
```

The number of multisubsets of size n from a set S with $k \in \mathbb{N}^*$ elements is equal to $C(k+n-1, n) = C(n+k-1, k-1)$ (see A.2), which is the number of solutions $(x_1, \ldots, x_k) \in \mathbb{N} \times \cdots \times \mathbb{N}$ of the equation $x_1 + \ldots + x_k = n$, where x_i is the number of repetitions of the ith element from S, for $i \in \{1, \ldots, k\}$.

Problem 1.1.4. Using Matlab, print all distinct codes on a circle that contain the characters: $A, A, A, B, B, 0, 0, 0, 1$. Find the number of such codes.

Solution 1.1.4: In the following, we present two ways to generate the desired codes (the second one is faster).

`>> [unique(perms('AAABB000'),'rows') repelem('1',560)')]`

```
function codes=circode
codes=[ ];
code=['AAABB0001'];
posA=nchoosek(1:8,3);
for i=1:nchoosek(8,3)
    code(posA(i,:))='A';
    posB=nchoosek(setdiff(1:8,posA(i,:)),2);
    for j=1:nchoosek(5,2)
        code(posB(j,:))='B';
        pos0=setdiff(1:8,[posA(i,:), posB(j,:)]);
        code(pos0)='0';
        codes=[codes; code];
    end
end
end
```

The number of such codes is equal to the number of permutations with repetitions of the code divided by the number of circular permutations:

$$\frac{1}{9} \cdot \frac{9!}{1!3!3!2!} = 560.$$

1.2   Measurable Spaces

Definition 1.3. A collection $\mathcal{F}$ of events from the sample space Ω is said to be a **σ-field** (or **σ-algebra**) on Ω, if it satisfies the following conditions:

(1) $\mathcal{F} \neq \emptyset$;
(2) if $A \in \mathcal{F}$, then $\bar{A} \in \mathcal{F}$;
(3) if $A_n \in \mathcal{F}$ for all $n \in \mathbb{N}$, then $\bigcup_{n \in \mathbb{N}} A_n \in \mathcal{F}$.

If $\mathcal{F}$ is a σ-field on the sample space Ω, then the pair $(\Omega, \mathcal{F})$ is called a **measurable space.**   ◆

Theorem 1.2. *If $\mathcal{F}$ is a σ-field in Ω, then the following properties hold:*

(1) $\emptyset, \Omega \in \mathcal{F}$.
(2) *If $A, B \in \mathcal{F}$, then $A \cap B$, $A \setminus B$, $A \Delta B \in \mathcal{F}$.*
(3) *If $A_n \in \mathcal{F}$ for all $n \in \mathbb{N}$, then*

$$\bigcap_{n \in \mathbb{N}} A_n \in \mathcal{F}, \quad \liminf_{n \to \infty} A_n \in \mathcal{F}, \quad \limsup_{n \to \infty} A_n \in \mathcal{F}.$$

Example 1.2. (1) The collection of all subsets of a set Ω is a σ-field, denoted by $\mathcal{P}(\Omega)$.
(2) Let Ω be a set and $\emptyset \neq A \subsetneq \Omega$. Then $\{\emptyset, \Omega, A, \bar{A}\}$ is a σ-field on Ω.
(3) Let $(\Omega, \mathcal{F})$ be a measurable space and $\emptyset \neq B \subseteq \Omega$. Then

$$B \cap \mathcal{F} = \{B \cap A : A \in \mathcal{F}\}$$

is a σ-field on the set B.   ▲

Theorem 1.3. *Let $\mathcal{A}$ be a collection of subsets of Ω. Then there exists in $\mathcal{P}(\Omega)$ a smallest σ-field containing $\mathcal{A}$ and denoted by $\sigma(\mathcal{A})$, i.e., $\sigma(\mathcal{A})$ is a σ-field such that it contains $\mathcal{A}$ and for any σ-field $\mathcal{F}$ containing $\mathcal{A}$ we have $\sigma(\mathcal{A}) \subseteq \mathcal{F}$.*

Definition 1.4. The σ-field $\sigma(\mathcal{A})$, whose existence is given by Theorem 1.3, is called **σ-field generated by $\mathcal{A}$**.   ◆

Example 1.3.
(1) Let, as in Figure 1.1: $\Omega = [0,1] \times [0,1]$, $A=[0.2,0.7]\times[0.2,0.6]$, $B=[0.4,0.9]\times[0.3,0.9]$.
Let $\mathcal{F}$ be a σ-field on Ω such that $A, B \in \mathcal{F}$.
The σ-field $\sigma(\mathcal{A})$ generated by $\mathcal{A} = \{A, B\}$
contains the following sets: $\emptyset$, Ω, A, $\bar{A}$, B, $\bar{B}$,
$A \cup B$, $A \cap B$, $\bar{A} \cup \bar{B}$, $\bar{A} \cap \bar{B}$, $A \cap \bar{B}$, $\bar{A} \cap B$, $A \cup \bar{B}$,
$\bar{A} \cup B$, $(A \cap \bar{B}) \cup (\bar{A} \cap B)$, $(A \cap B) \cup (\bar{A} \cap \bar{B})$.

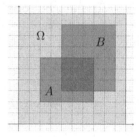

Fig. 1.1: Example 1.3-(1)

(2) Let $(\Omega, \mathcal{F})$ be a measurable space and let $\mathcal{A} = (A_i)_{i \in I}$, $I \subseteq \mathbb{N}$, be a partition of Ω. Then the σ-field generated by $\mathcal{A}$ is

$$\sigma(\mathcal{A}) = \{\emptyset\} \cup \left\{ \bigcup_{i \in J} A_i : \emptyset \neq J \subseteq I \right\}.$$

One proves that the right hand side of the above equality is the smallest σ-field which contains $\mathcal{A}$. We conclude that the collection of all unions of sets in a (countable) partition of Ω is a σ-field.   ▲

Notation: Let $(\mathbb{R}, \mathcal{B})$ be the measurable space $\mathbb{R}$ endowed with the σ-field $\mathcal{B} = \sigma(\mathcal{S}_{\text{open}})$ generated by

$$\mathcal{S}_{\text{open}} = \{A \subseteq \mathbb{R} : A \text{ is an open set}\}.$$

Let $(\mathbb{R}^n, \mathcal{B}^n)$ be the measurable space $\mathbb{R}^n$ endowed with the σ-field $\mathcal{B}^n = \sigma(\mathcal{S}^n_{\text{open}})$ generated by

$$\mathcal{S}^n_{\text{open}} = \{A \subseteq \mathbb{R}^n : A \text{ is an open set}\}.$$

Let $([0,1], \mathcal{B}([0,1]))$ be the measurable space on $[0,1]$ endowed with the σ-field $\mathcal{B}([0,1])$ generated by the open sets in $[0,1]$.

For more details and properties of the above spaces, see [Gut (2005), Chapter 1, Sections 3.3–3.4].

1.2.1 *Solved Problems*

Problem 1.2.1. Let Ω be a set and $(\Omega^*, \mathcal{F}^*)$ be a measurable space. Consider the function $g : \Omega \to \Omega^*$. Then

$$g^{-1}(\mathcal{F}^*) = \{g^{-1}(A^*) : A^* \in \mathcal{F}^*\}$$

is a σ-field on Ω.

Solution 1.2.1: We check Definition 1.3: Observe that $g^{-1}(\mathcal{F}^*) \neq \emptyset$, since $\Omega = g^{-1}(\Omega^*)$ and $\Omega^* \in \mathcal{F}^*$. Further, consider $A \in g^{-1}(\mathcal{F}^*)$. Then there exists $A^* \in \mathcal{F}^*$ such that $A = g^{-1}(A^*)$. This implies $\bar{A} = \overline{g^{-1}(A^*)}$. But $\overline{g^{-1}(A^*)} = g^{-1}(\overline{A^*})$, because
$$x \in \overline{g^{-1}(A^*)} \Leftrightarrow x \notin g^{-1}(A^*) \Leftrightarrow g(x) \notin A^* \Leftrightarrow g(x) \in \overline{A^*} \Leftrightarrow x \in g^{-1}(\overline{A^*}).$$
But $\overline{A^*} \in \mathcal{F}^*$, therefore $\bar{A} = g^{-1}(\overline{A^*}) \in g^{-1}(\mathcal{F}^*)$.
Let $A_n \in g^{-1}(\mathcal{F}^*)$ for $n \in \mathbb{N}$. Then there exists $A_n^* \in \mathcal{F}^*$ such that $A_n = g^{-1}(A_n^*)$ for all $n \in \mathbb{N}$. But $\bigcup\limits_{n \in \mathbb{N}} A_n^* \in \mathcal{F}^*$ and

$$x \in \bigcup_{n \in \mathbb{N}} g^{-1}(A_n^*) \Leftrightarrow \exists i \in \mathbb{N} \text{ such that } x \in g^{-1}(A_i^*)$$

$$\Leftrightarrow \exists i \in \mathbb{N} \text{ such that } g(x) \in A_i^*$$

$$\Leftrightarrow g(x) \in \bigcup_{n \in \mathbb{N}} A_n^* \Leftrightarrow x \in g^{-1}\Big(\bigcup_{n \in \mathbb{N}} A_n^* \Big).$$

Then we write

$$\bigcup_{n \in \mathbb{N}} A_n = \bigcup_{n \in \mathbb{N}} g^{-1}(A_n^*) = g^{-1}\Big(\bigcup_{n \in \mathbb{N}} A_n^* \Big) \in g^{-1}(\mathcal{F}^*).$$

Problem 1.2.2. Give an example of a sample space Ω and a nonempty collection $\mathcal{F}$ of events from Ω such that:
- for each $A \in \mathcal{F}$ we have $\bar{A} \in \mathcal{F}$;
- for each $A_1, \dots, A_n \in \mathcal{F}$, $n \in \mathbb{N}^*$, we have $\bigcup\limits_{k=1}^{n} A_k \in \mathcal{F}$;
- $\mathcal{F}$ is not a σ-field.

Solution 1.2.2: The main idea of the following example is to consider the experiment of tossing a coin infinitely many times and associate to it the events regarding only the output of the first finitely many tosses. Let

$$\Omega = \{(x_1, x_2, \ldots) : x_k \in \{0,1\}, \ k \in \mathbb{N}^*\}$$

and for every $n \in \mathbb{N}^*$ let

$$\Omega_n = \{(x_1, \ldots, x_n) : x_k \in \{0,1\}, \ k \in \{1, \ldots, n\}\}.$$

For every $n \in \mathbb{N}^*$ and $S \subseteq \Omega_n$ let

$$A_{n,S} = \{(x_1, x_2, \ldots) \in \Omega : (x_1, \ldots, x_n) \in S\}.$$

We have $A_{n,\emptyset} = \emptyset$ and $A_{n,\Omega_n} = \Omega$. Let

$$\mathcal{F} = \{A_{n,S} : n \in \mathbb{N}^*, S \subseteq \Omega_n\}.$$

Let $A_{n,S} \in \mathcal{F}$, where $n \in \mathbb{N}$ and $S \subseteq \Omega_n$. Then $\bar{A}_{n,S} = A_{n,\bar{S}}$, where $\bar{S}$ is the complement of S in Ω_n, and thus $\bar{A}_{n,S} \in \mathcal{F}$. Next, let $A_{n,S}, A_{m,T} \in \mathcal{F}$, where $m, n \in \mathbb{N}^*$, $m \le n$, $S \subseteq \Omega_n$, $T \subseteq \Omega_m$. To prove that the second condition holds, it suffices to prove that $A_{n,S} \cup A_{m,T} \in \mathcal{F}$. Let $R = \{(x_1, \ldots, x_n) \in \Omega_n : (x_1, \ldots, x_m) \in T\}$. Then $A_{m,T} = A_{n,R}$. Since $A_{n,S} \cup A_{n,R} = A_{n,S\cup R}$, we deduce that $A_{n,S} \cup A_{m,T} \in \mathcal{F}$.

Let $O = \{(0,0,\ldots,0,\ldots)\}$. For every $n \in \mathbb{N}^*$ let $O_n = \{(0,0,\ldots,0)\} \subset \Omega_n$. Because $O \ne A_{n,S}$ for all $n \in \mathbb{N}^*$ and $S \subseteq \Omega_n$, we have $O \notin \mathcal{F}$. But $\bigcap_{n=1}^{\infty} A_{n,O_n} = O$ and thus $\bigcup_{n=1}^{\infty} A_{n,\bar{O}_n} = \bar{O}$. We have $\bar{O} \notin \mathcal{F}$, because, otherwise, $\bar{O} \in \mathcal{F}$ implies $O \in \mathcal{F}$, which would be a contradiction. We obtain $\bigcup_{n=1}^{\infty} A_{n,\bar{O}_n} \notin \mathcal{F}$, but $A_{n,\bar{O}_n} \in \mathcal{F}$ for each $n \in \mathbb{N}^*$, hence $\mathcal{F}$ is not a σ–field. ♦

1.3   Probabilities

The **classical definition of probability** is:

Definition 1.5. We consider an experiment whose outcomes are finite and equally likely. The **probability** that an event A occurs is the number

$$P(A) = \frac{\text{number of outcomes favorable for the occurrence of } A}{\text{number of all possible outcomes of the experiment}},$$

i.e., $P(A) = \dfrac{\#A}{\#\Omega}$, where Ω is a corresponding sample space such that $A \subseteq \Omega$. The **relative frequency** $r_n(A)$ of the occurrences of the event A, with $n \in \mathbb{N}^*$, is defined as $r_n(A) = \dfrac{k}{n}$, when A appears k times after repeating the experiment n times, where $k \in \{1, \ldots, n\}$.   ♦

In practice, by repeating the experiment for a large number of times, one observes that the relative frequency $r_n(A)$ depends on the outcomes of the n experiments and it oscillates around the number $P(A)$ (compare with Example 4.3).

In the following, we shall simulate very often experiments based on randomly drawing (sampling) objects, with or without replacement, from a given family of objects. To do this in Matlab, we shall use the functions `randsample` or `randi`.

Example 1.4.

(1) Let us implement in Matlab the following Monte Carlo algorithm (i.e., an algorithm whose output is random):

```
input: a vector v of 100 distinct real numbers
  choose randomly 20 elements from v without replacement
    in order to form a subvector w of v
output: the minimum of w
```

```
function m=minimum(v)
m=min(randsample(v,20));
end
```

(2) Next, we estimate the probability that the output of the algorithm is the minimum of the input v.

```
function minimum_sim(v,N)
count=0;
for i=1:N
    if min(v)==minimum(v)
        count=count+1;
    end
end
fprintf('The estimated probability is %4.3f.\n',count/N);
end
end
```

```
>>minimum_sim(1:100,10000)
The estimated probability is 0.204.
```

(3) In what follows we compute the probability that the output of the algorithm is the minimum of the input v. The output of the algorithm is the minimum of v if and only if w contains the minimum of v. Thus the desired probability is $p = \dfrac{C(99, 19)}{C(100, 20)} = 0.2$. ▲

Definition 1.6. Let $\mathcal{F}$ be a σ-field in the sample space Ω. A mapping $P : \mathcal{F} \to \mathbb{R}$ is called **probability** if it satisfies the following axioms:
(1) $P(\Omega) = 1$;
(2) $P(A) \geq 0$ for every $A \in \mathcal{F}$;
(3) for any sequence $(A_n)_{n \in \mathbb{N}^*}$ of mutually exclusive events from $\mathcal{F}$, the following relation holds

$$P\Big(\bigcup_{n=1}^{\infty} A_n \Big) = \sum_{n=1}^{\infty} P(A_n).$$

The triplet $(\Omega, \mathcal{F}, P)$ consisting of a measurable space $(\Omega, \mathcal{F})$ and a probability $P : \mathcal{F} \to \mathbb{R}$ is called **probability space.** ◆

Example 1.5. Let Ω be a finite sample space, let $\mathcal{F} = \mathcal{P}(\Omega)$ (see Example 1.2-(1)) and P be given as in Definition 1.5. Then $(\Omega, \mathcal{F}, P)$ is a probability space.

Theorem 1.4. *Let $(\Omega, \mathcal{F}, P)$ be a probability space and $A, B \in \mathcal{F}$. Then the following properties are true:*
(1) $P(\bar{A}) = 1 - P(A)$ *and* $0 \leq P(A) \leq 1$.
(2) $P(\emptyset) = 0$.
(3) $P(A \setminus B) = P(A) - P(A \cap B)$.
(4) *If* $A \subseteq B$, *then* $P(A) \leq P(B)$, *i.e.,* P *is monotonically increasing.*
(5) $P(A \cup B) = P(A) + P(B) - P(A \cap B)$.

Notation: Let $\lambda_{\mathbb{R}^n}$ denote the Lebesgue measure on $\mathbb{R}^n$ (see [Nelson (2015)]). For $n \in \{1, 2, 3\}$ the Lebesgue measure coincides with the standard measure of length, area, respectively volume.

Definition 1.7. Let $\Omega \subset \mathbb{R}^n$ be Lebesgue measurable with $\lambda_{\mathbb{R}^n}(\Omega) \in (0, \infty)$ and let $\mathcal{F} = \{B \cap \Omega : B \in \mathcal{B}^n\}$. The **geometric probability** of $A \in \mathcal{F}$ is given by

$$P(A) = \frac{\lambda_{\mathbb{R}^n}(A)}{\lambda_{\mathbb{R}^n}(\Omega)}. \tag{1.1}$$

Note that $(\Omega, \mathcal{F}, P)$ is a probability space. ◆

Remark 1.2. (1) Let $\Omega \subset \mathbb{R}^n$ be Lebesgue measurable with $\lambda_{\mathbb{R}^n}(\Omega) \in (0, \infty)$ and let $\mathcal{F} = \{B \cap \Omega : B \in \mathcal{B}^n\}$. A random point is **generated uniformly** in Ω, if all sets in $\mathcal{F}$ of the same measure are equally likely to contain the random point. Let the probability of the event "a uniformly generated random point belongs to A" be the geometric probability of A for each $A \in \mathcal{F}$.

(2) In Matlab, a random point is generated uniformly in the interval $[0, 1]$ by the function **rand**.

(3) Using the **rand** function in Matlab, we can generate uniformly a random point (x, y) in the rectangle $[0, a] \times [0, b]$, where $a, b > 0$, as follows:

$>>$ x=a$*$**rand**, y=b$*$**rand**

▼

Example 1.6. Consider a uniformly generated random point in the square $[0, 2] \times [0, 2]$. The probability that the point belongs to the gray star shaped region (the union of the two congruent gray squares) of Figure 1.2 is $2 - \sqrt{2} \approx 0.5858$, since the area of this gray region is $4(2 - \sqrt{2}) \approx 2.3431$.

We estimate the geometric probability in Matlab, by generating random points in the square with the function **rand** and counting those that belong to the gray region.

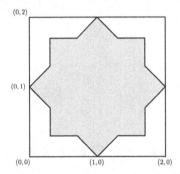 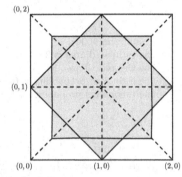

Fig. 1.2: Example 1.6

```
function star_sim(N)
close all; figure; hold on; axis square; axis off;
rectangle('Position',[0 0 2 2],'FaceColor','w');
count=0;
for i=1:N
  x=2*rand;y=2*rand;
  if (pdist([x y;1 1])<pdist([x y; 0 0])&&pdist([x y;1 1])<pdist([x y; 0 2])...
      &&pdist([x y;1 1])<pdist([x y; 2 0])&&pdist([x y;1 1])<pdist([x y; 2 2]))...
      ||(abs(x-1)<sqrt(2)/2&&abs(y-1)<sqrt(2)/2)
    plot(x,y,'hb','MarkerSize',3,'MarkerFaceColor','k');
    count=count+1;
  end
end
fprintf('The estimated area is %6.5f.\n',count/N*4);
area=4*(2-sqrt(2));
fprintf('The theoretical area is %6.5f.\n',area);
```

```
fprintf('The estimated probability is %7.6f.\n',count/N);
fprintf('The theoretical probability is %7.6f.\n',area/4);
end
```

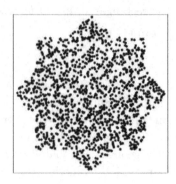

Fig. 1.3: Random points for Example 1.6

```
>>star_sim(5000)
The estimated area is 2.34320.
The theoretical area is 2.34315.
The estimated probability is 0.585800.
The theoretical probability is 0.585786.
```

▲

Throughout the rest of this section, $(\Omega, \mathcal{F}, P)$ is a probability space and all results are with respect to it.

Theorem 1.5. *If $(A_n)_{n \in \mathbb{N}^*}$ is a sequence of events from $\mathcal{F}$, then the following formula of the inclusion-exclusion principle (also known as Poincaré's formula) holds*

$$P\Big(\bigcup_{i=1}^{n} A_i\Big) = \sum_{i=1}^{n} P(A_i) - \sum_{1 \le i < j \le n} P(A_i \cap A_j) + \cdots + (-1)^{n-1} P\Big(\bigcap_{i=1}^{n} A_i\Big)$$

for all $n \in \mathbb{N}^*$.

Theorem 1.6. *The following properties are true:*

(1) *If $(A_n)_{n \in \mathbb{N}^*}$ is an increasing sequence of events from $\mathcal{F}$, then*

$$\lim_{n \to \infty} P(A_n) = P\Big(\bigcup_{n=1}^{\infty} A_n\Big).$$

(2) *If $(A_n)_{n\in\mathbb{N}^*}$ is a decreasing sequence of events from $\mathcal{F}$, then*

$$\lim_{n\to\infty} P(A_n) = P\left(\bigcap_{n=1}^{\infty} A_n\right).$$

Example 1.7. Let $\Omega = [0,1]$, $\mathcal{F} = \mathcal{B}([0,1])$ (see the notations from Section 1.2) and $P = \lambda_{\mathbb{R}}$ the Lebesgue measure on $[0,1]$. Then $P([a,b]) = P((a,b)) = P((a,b]) = P([a,b)) = b - a$ for $0 \le a < b \le 1$. Consider for $n \in \mathbb{N}, n \ge 5$,

$$A_n = \left[\frac{1}{n}, \frac{1}{2} - \frac{1}{n}\right] \in \mathcal{F} \quad \text{and} \quad B_n = \left[\frac{1}{4} - \frac{1}{n}, \frac{3}{4} + \frac{1}{n}\right] \in \mathcal{F}.$$

By Theorem 1.6, we have

$$\lim_{n\to\infty} P(A_n) = P\left(\bigcup_{n=5}^{\infty} A_n\right) = \frac{1}{2},$$

$$\lim_{n\to\infty} P(B_n) = P\left(\bigcap_{n=5}^{\infty} B_n\right) = \frac{1}{2}. \qquad \blacktriangle$$

Definition 1.8. Let $A, B \in \mathcal{F}$. The **conditional probability** of A given B is $P(\cdot|B) : \mathcal{F} \to \mathbb{R}$ defined by

$$P(A|B) = \frac{P(A \cap B)}{P(B)},$$

provided $P(B) > 0$. $\qquad\qquad\blacklozenge$

Remark 1.3. For $A, B \in \mathcal{F}$ with $P(B) \in (0,1)$ it holds:
$P(\bar{A}|B) = 1 - P(A|B)$,
$P(\bar{A}|\bar{B}) = 1 - P(A|\bar{B})$,
$P(A \cap B) = P(B)P(A|B)$,
$P(\bar{A} \cap B) = P(B)P(\bar{A}|B)$,
$P(A \cap \bar{B}) = P(\bar{B})P(A|\bar{B})$,
$P(\bar{A} \cap \bar{B}) = P(\bar{B})P(\bar{A}|\bar{B})$.

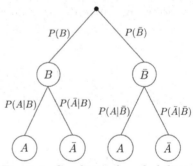

Fig. 1.4: Conditional probabilities $\qquad\blacktriangledown$

Theorem 1.7. (Chain rule; multiplication rule) *Let $A_1, \ldots, A_n \in \mathcal{F}$, $n \in \mathbb{N}$, $n \ge 2$, be such that $P(A_1 \cap \cdots \cap A_{n-1}) > 0$. Then*

$$P(A_1 \cap \cdots \cap A_n) = P(A_1)P(A_2|A_1)\ldots P(A_n|A_1 \cap \cdots \cap A_{n-1}).$$

Theorem 1.8. (Law of total probability) *Consider* $(A_i)_{i \in I}$ *to be a partition of* Ω *with* $A_i \in \mathcal{F}$ *and* $P(A_i) > 0$ *for all* $i \in I$. *Then*

$$P(A) = \sum_{i \in I} P(A_i)P(A|A_i) \quad \text{for all } A \in \mathcal{F}.$$

Theorem 1.9. (Bayes' formula) *Consider* $(A_i)_{i \in I}$ *to be a partition of* Ω *with* $A_i \in \mathcal{F}$ *and* $P(A_i) > 0$ *for all* $i \in I$ *and let* $A \in \mathcal{F}$ *be such that* $P(A) > 0$. *Then*

$$P(A_j|A) = \frac{P(A_j)P(A|A_j)}{P(A)} = \frac{P(A_j)P(A|A_j)}{\sum_{i \in I} P(A_i)P(A|A_i)} \quad \text{for all } j \in I.$$

Remark 1.4. In Bayes' formula, we have a set of **prior probabilities** $P(A_i)$ for the **hypothesis** A_i, where $i \in I$, an event A called **evidence** and we can calculate the probabilities for the hypothesis given the evidence $P(A_i|A)$, where $i \in I$, called **posterior probability**. ▼

Example 1.8. An urn contains 2 red marbles and 3 blue marbles. Three marbles are successively drawn without replacement. We compute the probability that:
(1) the first marble is red and the second blue;
(2) the first two marbles have the same colour ;
(3) the second marble is blue;
(4) the first marble was red, given that the second marble is blue?
(5) the first marble is red, while the second and third marble are both blue.
Solution: In order to compute the probabilities, we consider for $i \in \{1, 2, 3\}$ the events (see Figure 1.5):

$$B_i : \text{ the } i\text{th drawn marble is blue}$$

$$R_i : \text{ the } i\text{th drawn marble is red.}$$

Note that $R_i = \bar{B}_i$ for $i \in \{1, 2, 3\}$.
(1) We compute $P(R_1 \cap B_2) = P(R_1)P(B_2|R_1) = \dfrac{2}{5} \cdot \dfrac{3}{4} = 0.3$.
(2) We compute

$$P\big((R_1 \cap R_2) \cup (B_1 \cap B_2)\big) = P(R_1 \cap R_2) + P(B_1 \cap B_2)$$

$$= P(R_1)P(R_2|R_1) + P(B_1)P(B_2|B_1) = \frac{2}{5} \cdot \frac{1}{4} + \frac{3}{5} \cdot \frac{2}{4} = 0.4.$$

(3) By the law of total probability (see Theorem 1.8) we have

$$P(B_2) = P(R_1)P(B_2|R_1) + P(B_1)P(B_2|B_1) = \frac{2}{5} \cdot \frac{3}{4} + \frac{3}{5} \cdot \frac{2}{4} = 0.6.$$

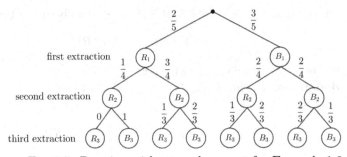

Fig. 1.5: Drawing without replacement for Example 1.8

(4) We use Bayes' formula from Theorem 1.9

$$P(R_1|B_2) = \frac{P(R_1)P(B_2|R_1)}{P(B_2)} = \frac{\frac{2}{5} \cdot \frac{3}{4}}{\frac{2}{5} \cdot \frac{3}{4} + \frac{3}{5} \cdot \frac{2}{4}} = 0.5.$$

(5) By the multiplication rule (see Theorem 1.7) we write

$$P(R_1 \cap B_2 \cap B_3) = P(R_1) \cdot P(B_2|R_1) \cdot P(B_3|R_1 \cap B_2) = \frac{2}{5} \cdot \frac{3}{4} \cdot \frac{2}{3} = 0.2.$$

Next, we estimate the above probabilities by some simulations in Matlab, using the function **randsample**.

```
function p=marbles(N)
count=zeros(1,5);
for i=1:N
  m=randsample({'red','red','blue','blue','blue'},3);
  if strcmp(m{1},'red')&&strcmp(m{2},'blue')
     count(1)=count(1)+1;
  end
  if strcmp(m{1},m{2})
     count(2)=count(2)+1;
  end
  if strcmp(m{2},'blue')
     count(3)=count(3)+1;
  end
  if strcmp(m{1},'red')&&strcmp(m{2},'blue')&&strcmp(m{3},'blue')
     count(5)=count(5)+1;
  end
end
p=count/N;
p(4)=count(1)/count(3);
end

>> marbles(10000)
ans = 0.2975 0.4045 0.6048 0.4919 0.1972
```

Definition 1.9. The events $A, B \in \mathcal{F}$ are said to be **independent** if

$$P(A \cap B) = P(A)P(B).$$

Let I be a set of indices such that $\#I \geq 2$. $(A_i)_{i \in I}$ is called a **family of independent events** from $\mathcal{F}$, if for each finite subset $\{i_1, \ldots, i_m\} \subseteq I$, $m \in \mathbb{N}$, $m \geq 2$, we have

$$P(A_{i_1} \cap \ldots \cap A_{i_m}) = P(A_{i_1}) \cdot \ldots \cdot P(A_{i_m}).$$

$(A_i)_{i \in I}$ is a **family of pairwise independent events** from $\mathcal{F}$, if for each $i, j \in I$ with $i \neq j$ the events A_i and A_j are independent. ◆

Example 1.9. We give an example of three events A, B and C such that each one is independent of the intersection of the other two (i.e., A and $B \cap C$ are independent, B and $A \cap C$ are independent, C and $A \cap B$ are independent), but A, B and C are not independent.

Solution: Let $\Omega = \{1, \ldots, 10\}$, $\mathcal{F} = \mathcal{P}(\Omega)$ and $P(S) = \dfrac{\#S}{10}$ for $S \subseteq \Omega$. Let $A = \{1, 2, 3, 5, 6\}$, $B = \{1, 2, 4, 7, 8\}$, $C = \{1, 3, 4, 9, 10\}$ (see Figure 1.6). We have

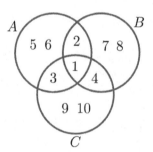

Fig. 1.6: Example 1.9

$$P(A \cap B \cap C) = P(A)P(B \cap C) = P(B)P(A \cap C) = P(C)P(A \cap B) = \frac{1}{10},$$

but

$$P(A \cap B \cap C) = \frac{1}{10} \neq P(A)P(B)P(C) = \frac{1}{8}. \qquad \blacktriangle$$

Example 1.10. Using Matlab, we want to find the minimum number of independent calls with the same input **v** of the algorithm given in Example 1.4:

```
input: a vector v of 100 distinct real numbers
  choose randomly 20 elements from v without replacement
    in order to form a subvector w of v
output: the minimum of w
```

such that the probability that the minimum of the outputs equals the minimum of v is at least 99%.

Solution: If we call independently the algorithm $n \in \mathbb{N}^*$ times with the same input v, then the probability that none of the outputs equals the minimum of v is $(1-p)^n = 0.8^n$, where $p = 0.2$ is the probability that the output of the algorithm is the minimum of the input v. Thus, we have to find the minimum value of $n \in \mathbb{N}^*$ such that:

$$1 - 0.8^n \geq 0.99 \Leftrightarrow 0.8^n \leq 0.01 \Leftrightarrow n \geq \log_{0.8}(0.01).$$

```
>> n=ceil(log(0.01)/log(0.8))
n = 21
```

▲

The following theorem is usually called the Borel–Cantelli Lemma.

Theorem 1.10. (1) Let $(A_n)_{n \in \mathbb{N}^*}$ be a sequence of events from $\mathcal{F}$ such that

$$\sum_{n=1}^{\infty} P(A_n) < \infty.$$

Then

$$P\left(\limsup_{n \to \infty} A_n\right) = 0.$$

(2) Let $(A_n)_{n \in \mathbb{N}^*}$ be sequence of independent events from $\mathcal{F}$ with

$$\sum_{n=1}^{\infty} P(A_n) = \infty.$$

Then

$$P\left(\limsup_{n \to \infty} A_n\right) = 1.$$

Example 1.11. A monkey hits randomly and independently the keys of a certain keyboard having the 26 capital letters of the English alphabet and the space key, such that each key is hit every time with the same probability. The monkey thus creates an infinite string of characters. Using the Borel–Cantelli Lemma (see Theorem 1.10), we shall deduce that the probability of the event that this string contains infinitely many times the string ⌞COMPUTER⌟ is 1.

Proof: The string has 10 characters. For every $n \in \mathbb{N}^*$ let A_n be the event that the nth block of 10 consecutive characters in the string typed by the monkey is ⌞COMPUTER⌟. Then $(A_n)_{n \in \mathbb{N}^*}$ is a sequence of independent events, because the monkey hits the keys independently, and $P(A_n) = \left(\dfrac{1}{27}\right)^{10}$ for all $n \in \mathbb{N}^*$. Hence, $\displaystyle\sum_{n=1}^{\infty} P(A_n) = \infty$ and thus, by the Borel–Cantelli Lemma (see Theorem 1.10-(2)), $P(\limsup_{n \to \infty} A_n) = 1$, which is the desired conclusion, in view of Remark 1.1. ▲

1.3.1 Solved Problems

Problem 1.3.1. Let $(A_n)_{n \in \mathbb{N}^*}$ be a sequence of events in a probability space $(\Omega, \mathcal{F}, P)$. Define the events

$$B_1 = A_1, \quad B_n = A_n \setminus \left(\bigcup_{i=1}^{n-1} A_i\right), \quad n \in \mathbb{N}, n \geq 2.$$

Prove that:

1) $\displaystyle\bigcup_{n=1}^{\infty} A_n = \bigcup_{n=1}^{\infty} B_n$ and $B_m \cap B_n = \emptyset$, if $m, n \in \mathbb{N}^*, m \neq n$.

2) $P\left(\displaystyle\bigcup_{n=1}^{\infty} A_n\right) \leq \displaystyle\sum_{n=1}^{\infty} P(A_n)$.

Solution 1.3.1: 1) Obviously, $\displaystyle\bigcup_{n=1}^{\infty} B_n \subseteq \bigcup_{n=1}^{\infty} A_n$. Consider $a \in \displaystyle\bigcup_{n=1}^{\infty} A_n$. Let $j \in \mathbb{N}^*$ be the smallest index for which $a \in A_j$. Then $a \in A_j$ and $a \notin \displaystyle\bigcup_{i=1}^{j-1} A_i$, so $a \in B_j \subseteq \displaystyle\bigcup_{n=1}^{\infty} B_n$. Thus $\displaystyle\bigcup_{n=1}^{\infty} A_n \subseteq \bigcup_{n=1}^{\infty} B_n$. The double inclusion implies the stated equality.

Let $m, n \in \mathbb{N}^*$ with $m < n$. We prove $B_m \cap B_n = \emptyset$ by assuming the opposite, i.e., there exists $b \in B_m \cap B_n$. We have $b \in B_m \subseteq A_m$ and $b \in B_n = A_n \setminus \left(\displaystyle\bigcup_{i=1}^{n-1} A_i\right)$. Then $b \notin A_i$ for all $i \in \{1, \ldots, n-1\}$, hence $b \notin A_m$. This is in contradiction with our above reasoning.

2) We have $B_n \subseteq A_n$ for all $n \in \mathbb{N}^*$, which implies $P(B_n) \leq P(A_n)$, by Theorem 1.4-(4). Then it follows from 1) that

$$P\Big(\bigcup_{n=1}^{\infty} A_n \Big) = P\Big(\bigcup_{n=1}^{\infty} B_n \Big) = \sum_{n=1}^{\infty} P(B_n) \leq \sum_{n=1}^{\infty} P(A_n).$$

Problem 1.3.2. A gambler has 200\$ and plays roulette with the following strategy:
• he bets 100\$ on the red color, which has a 47.4% chance of win;
• if the red color comes up, he gets his 100\$ back and the casino pays him an extra 100\$;
• if the red color does not come up, he does not get his 100\$ back.
He plays with this strategy until: either he has no money left or he doubles his initial money.
1) Simulate $N \in \{5000, 10000, 20000\}$ times the play of the gambler and estimate the probability that he will double his initial money.
2) Compute the probability that the gambler will double his initial money.
3) Is it better for the gambler to make only one bet, by putting all his money on the red color?

Solution 1.3.2: 1)

```
function gambler(N)
sums=200*ones(1,N);
for i=1:N
    while sums(i)>0 && sums(i)<400
        sums(i)=sums(i)+randsample([-100,100],1,true,[52.6 47.4]);
    end
end
fprintf('The estimated probability is %4.3f.\n',mean(sums==400));
end
```

```
>> gambler(20000)
The estimated probability is 0.449.
```

2) Consider the following events, regarding the first two bets of the gambler, which must necessarily occur in order to decide what the gambler should do next:

A: "The gambler wins the first two bets".

B: "The gambler loses the first two bets".

C: "The gambler wins one bet and loses the other one".

Let W be the event that the gambler doubles his initial money (after some repetitions of his strategy).

Clearly, $\{A, B, C\}$ is a partition of the certain event (i.e., the gambler makes the first two bets). The law of total probability (see Theorem 1.8) implies

$$P(W) = P(W|A)P(A) + P(W|B)P(B) + P(W|C)P(C).$$

Since the roulette trials are independent, the probability of the event W is equal to the probability of the event W given C (i.e., the gambler has 200\$ again and doubles his inital money by playing the next bets with his strategy), hence $P(W) = P(W|C)$. Then

$$P(W) = 1 \cdot 0.474^2 + 0 \cdot 0.526^2 + P(W) \cdot (2 \cdot 0.474 \cdot 0.526),$$

and thus

$$P(W) = \frac{0.474^2}{1 - 2 \cdot 0.474 \cdot 0.526} \approx 44.8\%.$$

3) Yes, because the probability to double his money by betting all his money on the red color is 47.4% (also, he saves some time).

Problem 1.3.3. In an urn there are 20% red marbles, 30% blue marbles and 50% yellow marbles. Three marbles are successively drawn with replacement. Using the **randsample** function in some simulations in Matlab, estimate the probabilities of the following events:

1) the marbles have different colors;

2) the marbles have the same color;

3) at least two marbles have the same color;

4) the marbles have the same color, knowing that at least two marbles have the same color.

Find the probabilities of the above events and compare them with the obtained estimations.

Solution 1.3.3:

```
function p=marbles2(N)
p=zeros(1,4);
for i=1:N
 m=randsample({'red','blue','yellow'},3,true,[20,30,50]);
 l=length(unique(m));
 if l==3 p(1)=p(1)+1;
  elseif l==2 p(3)=p(3)+1;
   else p(2)=p(2)+1; p(3)=p(3)+1;
  end
end
p=p/N; p(4)=p(2)/p(3);
end

>> marbles2(10000)
ans = 0.1846 0.1614 0.8154 0.1979
```

For every $i \in \{1, 2, 3\}$ let R_i, B_i, respectively Y_i, be the event that the ith marble is red, blue, respectively yellow.

The probability for 1) is

$$3! \cdot P(R_1 \cap B_2 \cap Y_3) = 6 \cdot 0.2 \cdot 0.3 \cdot 0.5 = 0.18.$$

The probability for 2) is

$$P(R_1 \cap R_2 \cap R_3) + P(B_1 \cap B_2 \cap B_3) + P(Y_1 \cap Y_2 \cap Y_3)$$
$$= 0.2^3 + 0.3^3 + 0.5^3 = 0.16.$$

The probability for 3) is $1 - 0.18 = 0.82$, in view of 1).

The probability for 4) is $\dfrac{0.16}{0.82} = \dfrac{8}{41} \approx 0.1951$, in view of 2) and 3).

1.4 Problems for Chapter 1

Problem 1.4.1. Suppose a 10-bit message is sent on a binary symmetric channel with bit error probability $p = 0.02$, i.e., if the bit $X \in \{0, 1\}$ was sent, then the bit $Y = X \oplus E$ is received, where $\oplus$ is the xor operation between bits and E is the random error such that

$$P(E = 1) = p \text{ (error occurred)}, \quad P(E = 0) = 1 - p \text{ (no error occurred)}.$$

Errors occur independently.

1) What is the probability of getting exactly three errors during the transmission?

2) The message 0100101101 was sent. What is the probability of receiving the message 1100101000?

Solution 1.4.1: 1) $C(10, 3)(0.02)^3(0.98)^7$. 2) $(0.02)^3(0.98)^7$.

Problem 1.4.2.

1) Implement in Matlab the following Monte Carlo algorithm:

```
input: a vector v of 100 distinct real numbers
  choose randomly a number k from the set {1,2,...,100}
  choose randomly k elements from v without replacement
    in order to form a subvector w of v
output: the maximum of w
```

2) Estimate the probability that the output of the algorithm is the maximum of the input v.

3) Find the probability that the output of the algorithm is the maximum of the input v.

Solution 1.4.2: 1)

```
function m=maximum(v)
k=randi(100);
m=max(randsample(v,k));
end
```

2)

```
function maximum_sim(v,N)
count=0;
for i=1:N
    if max(v)==maximum(v)
        count=count+1;
    end
end
fprintf('The estimated probability is %4.3f.\n',count/N);
end
```

```
>>maximum_sim(1:100,10000)
The estimated probability is 0.504.
```

3) The output of the algorithm is the maximum of v if and only if w contains the maximum of v. Thus, by the law of total probability (see Theorem 1.8), the desired probability is

$$\sum_{k=1}^{100} \frac{C(99, k-1)}{C(100, k)} \cdot \frac{1}{100} = \sum_{k=1}^{100} \frac{k}{100^2} = \frac{101}{200} = 0.505.$$

Problem 1.4.3. Suppose that four new computer models M_1, M_2, M_3, M_4 are being tested for their reliability. The probability that a model fits the latest market standards is $p_1 = 0.8$ for the model M_1, $p_2 = 0.7$ for model the M_2, $p_3 = 0.9$ for model the M_3, and $p_4 = 0.6$ for model the M_4. Determine the probability p that at least three models match the profile.

Solution 1.4.3: We give a more general setting. In an experiment only two possible outcomes can occur and their probabilities do not remain the same throughout the trials. We call the two possible outcomes: "success" and "failure". The probability that success occurs in the ith trial is $p_i \in [0,1]$ and the probability of failure is $q_i = 1 - p_i$ for $i \in \mathbb{N}^*$. The probability of the event that in n trials we obtain k successes and $n - k$ failures is

$$b(k, n; p_1, \ldots, p_n) = \sum_{1 \le i_1 < \cdots < i_k \le n} p_{i_1} \cdots p_{i_k} q_{i_{k+1}} \cdots q_{i_n}$$

for $k \in \mathbb{N}$ and $n \in \mathbb{N}^*$ with $k \le n$. We note that the number $b(k, n; p_1, \ldots, p_n)$ for $k \in \mathbb{N}$ and $n \in \mathbb{N}^*$ with $k \le n$ represents the coefficient of x^k in the polynomial $(p_1 x + q_1) \ldots (p_n x + q_n)$.

We apply the above reasoning for $p_1 = 0.8$, $p_2 = 0.7$, $p_3 = 0.9$ and $p_4 = 0.6$. At least three models match the profile if exactly three models or exactly four models match, so $p = b(3, 4; p_1, p_2, p_3, p_4) + b(4, 4; p_1, p_2, p_3, p_4)$, representing the sum of the coefficients of x^3 and x^4 in the polynomial

$$(0.8x + 0.2)(0.7x + 0.3)(0.9x + 0.1)(0.6x + 0.4).$$

Using the Symbolic Math Toolbox in Matlab, we find the coefficients of the above polynomial and the desired probability.

```
>> syms x
>> c=coeffs((0.8*x+0.2)*(0.7*x+0.3)*(0.9*x+0.1)*(0.6*x+0.4));
>> p=c(4)+c(5)
```

We obtain $p = \dfrac{1857}{2500} = 0.7428$.

Problem 1.4.4. A gambler has 300\$ and plays roulette with the following strategy:
• he bets 100\$ on the red color, which has a 47.4% chance of win;
• if the red color comes up, he gets his 100\$ back and the casino pays him an extra 100\$;
• if the red color does not come up, he does not get his 100\$ back.
He plays with this strategy until: either he has no money left or he doubles his initial money.
1) Simulate $N \in \{5000, 10000, 20000\}$ times the play of the gambler and estimate the probability that he will double his initial money.
2) Compute the probability that the gambler will double his initial money (one can use Matlab for some computations).

Solution 1.4.4: 1)

```
function gambler(N)
sums=300*ones(1,N);
for i=1:N
    while sums(i)>0 && sums(i)<600
        sums(i)=sums(i)+randsample([-100,100],1,true,[52.6 47.4]);
    end
end
fprintf('The estimated probability is %4.3f.\n',mean(sums==600));
end

>> gambler(20000)
The estimated probability is 0.423.
```

2) For $S \in \{0, 100, 200, \ldots, 600\}$ let W_S be the event that the gambler has S\$ and, by playing with his strategy the next bets, he reaches the sum of

600\$ and stops playing. Note that W_0 is the impossible event, while W_{600} is the sure event. We have to compute W_{300}.

For $n \in \{1,2,3\}$ and $k \in \{0,\ldots,n\}$ let $B_{k,n}$ be the event that the gambler wins k bets out of the first n bets. Since the roulette trials are independent, note that the probability of the event W_S given $B_{k,n}$ (i.e., $P(W_S|B_{k,n})$) is equal to the probability of the event that the gambler has $S+100k-100(n-k)\$$ and, by playing with his strategy the next bets, he reaches the sum of 600\$ (i.e., $P(W_{S+100(2k-n)})$), for $S \in \{0,100,200,\ldots,600\}$, $n \in \{1,2,3\}$ and $k \in \{1,\ldots,n\}$ such that $0 \le S + 100(2k-n) \le 600$.

Let $p = 0.474$. The law of total probability (see Theorem 1.8) implies

$$
\begin{cases}
P(W_{300}) = P(W_{300}|B_{3,3})P(B_{3,3}) + P(W_{300}|B_{2,3})P(B_{2,3}) \\
\qquad + P(W_{300}|B_{1,3})P(B_{1,3}) + P(W_{300}|B_{0,3})P(B_{0,3}) \\
P(W_{400}) = P(W_{400}|B_{2,2})P(B_{2,2}) + P(W_{400}|B_{1,2})P(B_{1,2}) \\
\qquad + P(W_{400}|B_{0,2})P(B_{0,2}) \\
P(W_{200}) = P(W_{200}|B_{2,2})P(B_{2,2}) + P(W_{200}|B_{1,2})P(B_{1,2}) \\
\qquad + P(W_{200}|B_{0,2})P(B_{0,2})
\end{cases}
$$

$$
\iff
\begin{cases}
P(W_{300}) = 1 \cdot p^3 + P(W_{400}) \cdot 3p^2(1-p) \\
\qquad + P(W_{200}) \cdot 3p(1-p)^2 + 0 \cdot (1-p)^3 \\
P(W_{400}) = 1 \cdot p^2 + P(W_{400}) \cdot 2p(1-p) \\
\qquad + P(W_{200}) \cdot (1-p)^2 \\
P(W_{200}) = P(W_{400}) \cdot p^2 + P(W_{200}) \cdot 2p(1-p) \\
\qquad + 0 \cdot (1-p)^2.
\end{cases}
$$

Let $x = P(W_{300})$, $y = P(W_{400})$ and $z = P(W_{200})$. Then (x,y,z) is the solution of the linear system

$$
\begin{cases}
x - 3p^2(1-p) \cdot y - 3p(1-p)^2 \cdot z = p^3 \\
(1 - 2p(1-p)) \cdot y - (1-p)^2 \cdot z = p^2 \\
-p^2 \cdot y + (1 - 2p(1-p)) \cdot z = 0.
\end{cases}
$$

We can solve this linear system in Matlab:

```
>> A=[1, -3*p^2*(1-p), -3*p*(1-p)^2;...
      0, 1-2*p*(1-p), -(1-p)^2; 0, -p^2, 1-2*p*(1-p)];
>> X=[p^3;p^2;0];
>> A\X
ans = 0.4226
      0.5954
      0.2668
```

Hence, the probability that the gambler will double his initial money is $x \approx 42.2\%$.

Problem 1.4.5. A shop buys a certain product from three suppliers S_1, S_2 and S_3. 40% of the products are acquired from S_1, 35% from S_2 and 25% from S_3. Suppose that 2% of the products coming from S_1 are defective, 1% of those from S_2 are defective and 3% of those from S_3 are defective. Furthermore, suppose that a customer randomly buys such a product and then requires a refund because the product was defective. What is the probability that this was a product acquired from S_2? What is the probability that a randomly chosen product is defective?

Solution 1.4.5: We consider the following events:

D: the product is defective

A_i: the product was acquired from S_i, where $i = 1, 2, 3$.

We then have $P(A_1) = 0.4$, $P(A_2) = 0.35$, $P(A_3) = 0.25$, $P(D|A_1) = 0.02$, $P(D|A_2) = 0.01$, $P(D|A_3) = 0.03$. By Bayes' formula (see Theorem 1.9), we compute the probability that the defective product was acquired from S_2

$$ P(A_2|D) = \frac{P(A_2)P(D|A_2)}{\sum\limits_{i=1}^{3} P(A_i)P(D|A_i)} = \frac{7}{38} \approx 0.1842 \,. $$

A randomly chosen product is defective with probability

$$ P(D) = \sum_{i=1}^{3} P(A_i)P(D|A_i) = 0.019 \,. $$

Problem 1.4.6. 1) Write a program in Matlab that, for a given $n \in \{1, \ldots, 10\}$, simulates the following game on a (standard) chessboard:
• first, place randomly a black queen on a square of the chessboard;
• then, place randomly n white rooks on the remaining free squares of the chessboard;
• count how many white rooks does the black queen attack;
• count how many white rooks attack the black queen.
2) Simulate $N \in \{1000, 5000, 10000\}$ times the above game and estimate the probabilities that the black queen attacks no white rooks, respectively the black queen is attacked by no white rook. Compute in Matlab the corresponding (theoretical) probabilities.

Solution 1.4.6: 1)

```
function chess(n) %n<=10
clf; axis equal; hold on; axis([0.5 8.5 0.5 8.5]);
%we draw the chessboard
for i=1:8
  for j=1:8
    if mod(i,2)==mod(j,2)
      rectangle('Position',[i-0.5 j-0.5 1 1],'FaceColor', [0.8 0.8 0.8]);
    else
      rectangle('Position',[i-0.5 j-0.5 1 1],'FaceColor', [0.6 0.6 0.6]);
    end
  end
end
%we choose the squares, first, for the queen and, next, for the rooks
pieces=randsample(0:63,n+1);
%we establish the coordinates of the squares of the pieces
X=mod(pieces,8)+1; Y=fix(pieces./8)+1;
%we draw the queen
rectangle('Position',[X(1)-0.40 Y(1)-0.40 0.8 0.8], 'Curvature',[1 1],'FaceColor','k');
count1=0; count2=0; %the counters for the attacked pieces
for i=1:n
  %we draw the ith rook
  rectangle('Position',[X(i+1)-0.40 Y(i+1)-0.40 0.8 0.8],...
            'Curvature',[1 1],'FaceColor','w');
  if X(i+1)==X(1)||Y(i+1)==Y(1) count1=count1+1; count2=count2+1;
  elseif abs(X(i+1)-X(1))==abs(Y(i+1)-Y(1)) count1=count1+1;
  end
end
fprintf('The black queen attacks %d white rook(s).\n',count1);
fprintf('The black queen is attacked by %d white rook(s).\n', count2);
end
```

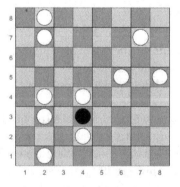

Fig. 1.7: Problem 1.4.6

```
>> chess(10)
The black queen attacks 5 white rook(s).
The black queen is attacked by 3 white rook(s).
```

2)

```
function chess_simul(n,N)
count1=0; count2=0; %the counters for the occurrences of the desired events
for j=1:N
    %we choose the squares, first, for the queen and, next, for the rooks
    pieces=randsample(0:63,n+1);
    %we establish the coordinates of the squares of the pieces
    X=mod(pieces,8)+1; Y=fix(pieces./8)+1;
    ok1=0; ok2=0;
    for i=1:n
        %ok1=1 when the queen attacks the ith rook
        %ok2=1 when the queen is attacked by the ith rook
        if X(i+1)==X(1)||Y(i+1)==Y(1) ok1=1; ok2=1; break;
        elseif abs(X(i+1)-X(1))==abs(Y(i+1)-Y(1))&&ok1==0 ok1=1;
        end
    end
    count1=count1+ok1; count2=count2+ok2;
end
fprintf(['The black queen attacks no white rook with'...
        'estimated probability %4.3f.\n'],1-count1/N);
fprintf(['The black queen is attacked by no white rook with'...
        'estimated probability %4.3f.\n'],1-count2/N);
end
```

```
>> chess_simul(4,5000)
The black queen attacks no white rook with estimated probability 0.159.
The black queen is attacked by no white rook with estimated probability 0.352.
```

Next, we apply the law of total probability (see Theorem 1.8) to compute the probability p_1 that the black queen attacks no white rook. If the black queen is placed on one of the squares on the edges of the chessboard (there are $8^2 - 6^2 = 28$ such squares), then there are 42 squares that are not attacked by the black queen. Thus, the probability that the black queen attacks no white rook given that the black queen is placed on such a square is $\dfrac{C(42, n)}{C(63, n)}$ for $n \in \{1, \ldots, 10\}$. Moving (towards the center of the chessboard) to the next frame of squares and reasoning similarly, we compute p_1 in Matlab as follows:

```
function chess_probability1(n)
p1=0;
for i=0:3
    p1=p1+((8-2*i)^2-(6-2*i)^2)/64*nchoosek(42-2*i,n)/nchoosek(63,n);
end
```

fprintf('The black queen attacks no white rook with probability %4.3f.\n',p1);
end

>>chess_probability1(4)
The black queen attacks no white rook with probability 0.160.

Since the black queen is attacked by no white rook if and only if none of the white rooks is on the same line or on the same column as the black queen, the second probability, denoted by p_2, is computed in Matlab as follows:

function chess_probability2(n)
p2=**nchoosek**(49,n)/**nchoosek**(63,n);
fprintf('The black queen is attacked by no white rook with probability %4.3f.\n',p2);
end

>>chess_probability2(4)
The black queen is attacked by no white rook with probability 0.356.

Problem 1.4.7. (An application of the symmetry principle in probability – the bridges problem) A traveler approaches the North bank of a sea and needs to reach the South bank, using a network of islands and bridges, as represented below. The islands are distributed on n lines and m columns, $n, m \in \mathbb{N}^*$.

North bank

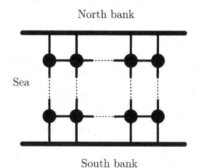

South bank

Fig. 1.8: Problem 1.4.7

The day before the arrival of the traveler, there has been a great storm and each bridge, independent of the others, was as likely as not to be washed away. In the next figure we have a network of bridges and islands after the storm, for $n = 1$ and $m = 2$, which allows the traveler to cross the sea, respectively a network of bridges and islands after the storm, for $n = 2$ and $m = 3$, which does not allow the traveler to cross the sea.

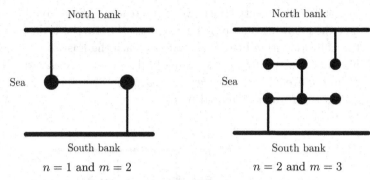

$n = 1$ and $m = 2$ $n = 2$ and $m = 3$

1) Simulate in Matlab a graphical representation of the network of islands and bridges after the storm and print a message regarding the possibility for the traveler to cross the sea.

2) Prove that: if $m = n + 1$, then the traveler has a 50% chance to cross the sea.

3) Prove that: the traveler has less than 50% chance to cross the sea if and only if $m \leq n$.

Solution 1.4.7: 1)

```
function storm(n,m)
clf; axis equal; axis off;hold on; axis([-1 m+2 -2 n+3]);
plot(repmat(1:n,m,1),'ok','MarkerFaceColor','k','MarkerSize',10);
line([0 m+1],[0 0],'Color','k','LineWidth',3); text(m/2-0.5,-0.5,'South bank');
line([0 m+1],[n+1 n+1],'Color','k','LineWidth',3); text(m/2-0.5,n+1.5,'North bank');
text(-1,n/2+0.5,'Sea','Color','b');
G=graph; G=addnode(G,m*(n+2));
for i=1:n+1
  for j=1:m
   if randsample([0,1],1)
    line([j j],[i-1 i],'Color','k','LineWidth',1.5); G=addedge(G,j+(i-1)*m,j+i*m);
   end
  end
end
for i=1:n
  for j=1:m-1
   if randsample([0,1],1)
    line([j j+1],[i i],'Color','k','LineWidth',1.5); G=addedge(G,j+i*m,j+1+i*m);
   end
  end
end
if any(any(distances(G,1:m,(n+1)*m+(1:m))~=Inf))
    disp('The traveler can cross the sea.');
 else disp('The traveler can not cross the sea.');
 end
end
```

2) Let A be the event: "the traveler can reach the South bank". We consider now the following dual-problem: suppose the traveler is on the West side of the network with a boat and wants to reach the East side. A bridge blocks his way through if and only if the bridge has not been washed away. Let B be the event: "the traveler can reach the East side". Next, we identify each area of water delimited by bridges, with a node, as it is shown in the figure below.

North bank

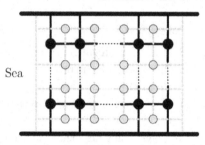

South bank

We note that the network of nodes on the sea has the same structure as the network of islands (n lines and $n + 1$ columns, from the point of view of the traveler on sea, respectively on land) and any standing, respectively destroyed, bridge between two islands represents a path, respectively a blockage, between two nodes on the sea. Since every bridge is independently destroyed with probability 0.5, we conclude that $P(A) = P(B)$. Since the traveler can reach the North bank if and only if the traveler can not reach the East side with a boat on sea, we have $P(A) = P(\bar{B})$. Thus $P(A) = 0.5$.

North bank

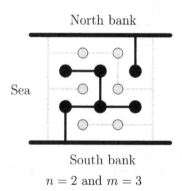

South bank

$n = 2$ and $m = 3$

We mention that the problem (respectively the solution) is based on the so-called symmetry principle (see [Beekman (2017), Puzzle 37]).

3) Suppose that $m \leq n$. If we connect the network of islands and bridges with a network of islands and bridges which has n lines and $n - m + 1$ columns, then the probability that the traveler crosses the sea using this new network after the storm is 0.5, according to 2). Hence the probability that the traveler crosses the sea using the old network is less than 0.5.

Similarly, we can prove that: if $m > n + 1$, then the probability that the traveler crosses the sea is more than 0.5. Therefore, we have: the probability that the traveler crosses the sea is less than 0.5 if and only if $m \leq n$.

Problem 1.4.8. 1) Simulate graphically in Matlab the following procedure: Generate randomly and independently two numbers in the interval $[0, 1]$ by using the function **rand** and denote them by X_1 and X_2 such that $X_1 \leq X_2$. Verify whether the segments generated by X_1 and X_2: $[0, X_1]$, $[X_1, X_2]$, $[X_2, 1]$, form a triangle or not. If they do, draw the triangle, then specify the type of the triangle: acute, obtuse or right.

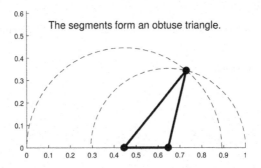

Fig. 1.9: Problem 1.4.8

2) Estimate the probability that the segments generated by X_1 and X_2 form an acute triangle given that the segments form a triangle, using simulations in Matlab, without a graphical representation.

3) Generate independently and uniformly $N \in \{1000, 5000, 10000\}$ random points in the square $[0, 1] \times [0, 1]$. Plot with blue the points (X, Y) for which the segments $[0, X_1]$, $[X_1, X_2]$ and $[X_2, 1]$ generate a triangle, where $X_1 = \min\{X, Y\}$, $X_2 = \max\{X, Y\}$, and with red those that generate an acute triangle. Return the ratio of the number of red points and the number of blue points.

4) Find the probability that the segments generated by X_1 and X_2 form an acute triangle given that the segments form a triangle.

Solution 1.4.8: 1)

```
function triangle
close all; fig = figure;
while ishghandle(fig)
 clf; hold on; axis equal; axis([0 1 0 0.61]);
 x=sort(rand([1 2]));
 plot(x,[0 0],'or','MarkerSize',10,'MarkerFaceColor','r');
 a=x(1);b=x(2)-x(1);c=1-x(2);
 if (a<b+c)&&(b<a+c)&&(c<a+b)
   X=((a^2-c^2)/b+x(1)+x(2))/2; Y=sqrt(a^2-(X-x(1))^2);
   u=0:0.1:pi; plot(x(1)+a*cos(u),a*sin(u),'--k');
   plot(x(2)+c*cos(u),c*sin(u),'--k');
   plot([x(1) x(2) X x(1)], [0 0 Y 0],'k','LineWidth',3);
   plot([x(1) x(2) X], [0 0 Y],'or','MarkerSize',10,'MarkerFaceColor','r');
   if (a^2<b^2+c^2)&&(b^2<a^2+c^2)&&(c^2<a^2+b^2)
     text(0.1,0.55,'The segments form an acute triangle.','FontSize',15);
   elseif (a^2==b^2+c^2)||(b^2==a^2+c^2)||(c^2==a^2+b^2)
     text(0.1,0.55,'The segments form a right triangle.','FontSize',15);
   else
     text(0.1,0.55,'The segments form an obtuse triangle.','FontSize',15);
   end
 else
   text(0.1,0.55,'The segments cannot form a triangle.','FontSize',15);
 end
 try
       waitforbuttonpress;
 catch
 end
end
end
```

2)

```
function p=triangle_sim(N)
count1=0; count2=0;
for i=1:N
 x=sort(rand(1,2));
 a=x(1);b=x(2)-x(1);c=1-x(2);
 if (a<b+c)&&(b<a+c)&&(c<a+b)
   count1=count1+1;
   if (a^2<b^2+c^2)&&(b^2<a^2+c^2)&&(c^2<a^2+b^2)
     count2=count2+1;
   end
 end
end
p=count2/count1;
end
```

```
>> triangle_sim(10000)
ans = 0.3196
```

3)

```
function p=triangle_sim2(N)
clf;axis equal; axis([0 1 0 1]); hold on;
countA=0; countO=0;
for i=1:N
  P=rand([1 2]);
  if P(1)<=P(2)
    a=P(1);b=P(2)-P(1);c=1-P(2);
    else a=P(2);b=P(1)-P(2);c=1-P(1);
  end
  if (a<b+c)&&(b<a+c)&&(c<a+b)
   if (a^2<b^2+c^2)&&(b^2<a^2+c^2)&&(c^2<a^2+b^2)
    plot(P(1),P(2),'ob','MarkerSize',4,'MarkerFaceColor','b');
    countA=countA+1;
   else
    plot(P(1),P(2),'or','MarkerSize',4,'MarkerFaceColor','r');
    countO=countO+1;
   end
  end
end
p=countA/(countA+countO);
end
end
```

>> triangle_sim2(10000)
ans = 0.3194

4) Let $(x_1, x_2) \in [0,1] \times [0,1]$. Then $x_1, x_2 \in [0,1]$ divide the segment $[0,1]$ in three segments that form a triangle if and only if

$$\begin{cases} x_1 \in \left(0, \dfrac{1}{2}\right) \\ \dfrac{1}{2} < x_2 < x_1 + \dfrac{1}{2} \end{cases} \quad \text{or} \quad \begin{cases} x_2 \in \left(0, \dfrac{1}{2}\right) \\ \dfrac{1}{2} < x_1 < x_2 + \dfrac{1}{2}. \end{cases}$$

Hence, the (geometric) probability that the segments form a triangle is

$$2 \int_0^{\frac{1}{2}} \int_{\frac{1}{2}}^{x_1+\frac{1}{2}} 1 dx_1 dx_2 = 2 \int_0^{\frac{1}{2}} \left(x_1 + \frac{1}{2}\right) - \frac{1}{2} \, dx_1 = \frac{1}{4}.$$

Next, we note that x_1 and x_2 divide the segment $[0,1]$ in three segments that form an obtuse triangle if and only if

$$\begin{cases} x_2 \in \left(\dfrac{1}{2}, 1\right) \\ x_2 - 1 + \dfrac{1}{2x_2} < x_1 < \dfrac{1}{2} \end{cases} \quad \text{or} \quad \begin{cases} x_2 \in \left(\dfrac{1}{2}, 1\right) \\ x_2 - \dfrac{1}{2} < x_1 < 1 - \dfrac{1}{2x_2} \end{cases}$$

$$\text{or} \quad \begin{cases} x_1 \in \left(0, \dfrac{1}{2}\right) \\ \dfrac{1}{2} < x_2 < 1 + x_1 - \dfrac{1}{2(1 - x_1)} \end{cases}$$

or any of the previous cases with x_1 and x_2 interchanged. Hence, the (geometric) probability that the segments form an obtuse triangle is

$$
2\left(\int_{\frac{1}{2}}^{1} \int_{x_2-1+\frac{1}{2x_2}}^{\frac{1}{2}} 1 dx_2 dx_1 + \int_{\frac{1}{2}}^{1} \int_{x_2-\frac{1}{2}}^{1-\frac{1}{2x_2}} 1 dx_2 dx_1 \right.
$$

$$
\left. + \int_{0}^{\frac{1}{2}} \int_{\frac{1}{2}}^{1+x_1-\frac{1}{2(1-x_1)}} 1 dx_1 dx_2 \right) = 2\left(\int_{\frac{1}{2}}^{1} \frac{1}{2} - \left(x_2 - 1 + \frac{1}{2x_2} \right) dx_2 \right.
$$

$$
\left. + \int_{\frac{1}{2}}^{1} \left(1 - \frac{1}{2x_2} \right) - \left(x_2 - \frac{1}{2} \right) dx_2 + \int_{0}^{\frac{1}{2}} \left(1 + x_1 - \frac{1}{2(1-x_1)} \right) - \frac{1}{2} dx_1 \right)
$$

$$
= \frac{9}{4} - 3 \log 2.
$$

Hence the desired conditional probability is

$$
1 - \frac{\frac{9}{4} - 3 \log 2}{\frac{1}{4}} = 12 \log 2 - 8 \approx 0.3178.
$$

Problem 1.4.9. (Cayley's problem) Let $n, k \in \mathbb{N}$, $n \geq 2k$ and $k \geq 3$.
1) Find the number of convex k-gons whose vertices are vertices of a given regular convex n-gon and all of whose sides are diagonals of the n-gon.
2) Represent graphically in Matlab all convex k-gons whose vertices are vertices of a given regular convex n-gon and all of whose sides are diagonals of the n-gon.
3) Estimate in Matlab the probability that, choosing randomly k distinct vertices of a given regular convex n-gon, all the sides of the k-gon are diagonals of the n-gon. Compare the estimated probability with the exact value of the probability, for some values of n and k.

Solution 1.4.9: 1) In the following, we consider binary numbers on a circle, which contain k ones and $n - k$ zeros. To each of such binary numbers on the circle we associate a cyclic permutation of the vertices of the n-gon and we obtain a k-gon whose vertices are the vertices of the n-gon corresponding to the ones.

Let x_i be the number of consecutive zeros after the ith one in the binary cyclic number, for $i \in \{1, \ldots, k\}$. Then $(x_1, \ldots, x_k) \in \mathbb{N}^* \times \ldots \times \mathbb{N}^*$ is a solution of the equation $x_1 + \ldots + x_k = n - k$. By Problem 1.1.2-1), we have $C(n - k - 1, k - 1)$ such solutions. Hence, by the above procedure we obtain $n \cdot C(n - k - 1, k - 1)$ desired k-gons. If $x_1 = \ldots = x_k$, then we obtain k repetitions of the same k-gon, when we permute cyclically the vertices of the n-gon (i.e., we obtain k cyclic permutations of the vertices of the same k-gon, which is regular in this case). If $x_1, \ldots, x_k$ are not all equal, then we obtain n distinct k-gons, when we permute cyclically the vertices of the n-gon. However, note that any cyclic permutation of the solution $x_1, \ldots, x_k$, when they are not all equal, is a distinct solution of the above equation, but which generates the same set of k-gons, when we permute cyclically the vertices of the n-gon. We conclude that the total number of desired k-gons is

$$\frac{n}{k} \cdot C(n - k - 1, k - 1).$$

For other related problems and details, see [Yaglom and Yaglom (1964), Chapter IV].

2)

```
function cayley_g(V,k)
%example of call: >>cayley_g('ABCDEFG',3)
n=length(V);
v=cayley(n,k);
v=[v v(:,1)];
x=linspace(0,2*pi,n+1);
clf; hold on; axis off;
plot(cos(x),sin(x),'-k',"linewidth",3);
plot(cos(x),sin(x),'ok',"markerfacecolor",'k',"markersize",10);
for j=1:n
   text(1.1*cos(x(j)),1.1*sin(x(j)),V(j));
end
fprintf('The %d-gons whose sides are diagonals of the regular %d-gon:\n',k,n);
for i=1:size(v,1)
   pause(1.5);
   plot(cos(x(v(i,:))),sin(x(v(i,:))),"linewidth",3);
   fprintf('%s\n',V(v(i,1:k)));
end
end


function v=cayley(n,k) %n>=2k>=6
v=[ ];
postions2delete=nchoosek(3:2:2*(n-k)-1,n-2*k);
for i=1:nchoosek(n-k-1,k-1)
   b=repmat([1 0],1,n-k);
   if size(postions2delete)
      b(postions2delete(i,:))=[ ];
   end
   X=1:n;
   v=[v; X(b==1)];
   for j=1:n-1
      X=circshift(X',1)';
      v=[v; X(b==1)];
   end
```

```
end
v=unique(sort(v,2),'rows');
end
```

3)

```
function p=cayley_sim(n,k,N_sim)
count=0;
for i=1:N_sim
    w=sort(randsample(1:n,k));
    if sum(diff(w)==1)==0 && w(k)-w(1)~=n-1
        count=count+1;
    end
end
p=count/N_sim;
end
```

```
>> cayley_sim(10,4,10000)
ans = 0.1196
>> 10/4*nchoosek(5,3)/nchoosek(10,4)
ans = 0.1190
```

Problem 1.4.10. (The minimum cut problem — Karger's algorithm)
Consider a graph $G = (V, E)$ (V is the set of vertices and E is the set of
edges) which is undirected (i.e., every edge is associated with an unordered
pair of two vertices), connected (there is a path between any two vertices).
$C \subseteq E$ is called a cut-set for G if the graph $G_C = (V, E \backslash C)$ is not connected.
The minimum cut problem is to find a cut-set C for G such that $\#C$ is
minimum. Such a cut-set is said to be minimum.

The main idea of Karger's algorithm Random Min-Cut is to contract,
in each iteration, a random edge, until there are only two vertices left. To
contract an edge $e = (u, v)$, which connects the vertices u and v, means:
i) to eliminate every edge that connects u and v;
ii) to replace the vertices u and v with a new vertex w;
iii) to keep the rest of the edges, including the ones that were connected
with either u or v — they are now connected with w.

The pseudocode of Karger's algorithm:

```
RndMinCut(G)
while |V|>2
    randomly pick an edge and contract it
return the set of remaining edges
```

1) Implement Karger's algorithm in Matlab with a graphic representation
for each iteration and test it on the following graph:

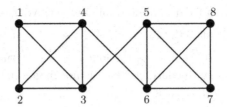

Use the function **graph** to generate a graph object and then use the corresponding object functions and properties.

2) Let $|V| = n$, $n \in \mathbb{N}$, $n > 2$. Prove that: if we run **RndMinCut(G)** at least $n^2 \log n$ times and we choose from all outputs the one with the minimum cardinality, the probability that the chosen one is not a minimum cut-set is at most $\dfrac{1}{n^2}$.

Solution 1.4.10: 1)

```
%Code for Matlab 2018a or newer versions
%E.g. how to generate a graph object and how to call RndMinCut:
%>>E=[1 2;1 3;1 4;2 3;2 4;3 4;3 5;4 6;5 6;5 7;5 8;6 7;6 8;7 8];
%>>G=graph(E(:,1),E(:,2));
%>>H=RndMinCut(G);
%>>H.Nodes, H.Edges
function G=RndMinCut(G)
n=numnodes(G); G.Nodes.Name(:,1)=cellstr(string(1:n));
subplot(1,n-1,1); plot(G,'-k');
axis square; axis off;
for i=2:n-1
 e=randi(numedges(G));
 u=findnode(G,G.Edges.EndNodes(e,1));
 v=findnode(G,G.Edges.EndNodes(e,2));
 G=rmedge(G,u,v);
 G=addnode(G,strcat(G.Nodes.Name(u,1),',',G.Nodes.Name(v,1)));
 G_old=G;
 G=rmnode(G,G.Nodes.Name([u v],1));
  for j=1:numedges(G_old)
   if ismember(G_old.Edges.EndNodes(j,1),G_old.Nodes.Name([u v],1))
      G=addedge(G,G_old.Edges.EndNodes(j,2),...
              G_old.Nodes.Name(numnodes(G_old),1));
    elseif ismember(G_old.Edges.EndNodes(j,2),G_old.Nodes.Name([u v],1))
      G=addedge(G,G_old.Edges.EndNodes(j,1),...
              G_old.Nodes.Name(numnodes(G_old),1))
   end
  end
  subplot(1,n-1,i); plot(G,'-k'); axis square; axis off;
end
end
```

2) Let C be a fixed minimum min-cut set. We shall prove that the probability that RndMinCut(G) returns the set C is at least $\dfrac{2}{n(n-1)}$.

Let $k = \#C$. Let A_i be the following event: "in iteration i we randomly choose an edge to contract which is not in C" and let $G_i = (V_i, E_i)$ be the graph after i iterations, $i = \overline{1, n-2}$.

We note that:

• if in each iteration we pick an edge that is not in C, then $C \subseteq E_i$ and C is a cut-set for G_i, $i = \overline{1, n-2}$.

• RndMinCut(G) returns the set C if and only if $A_1 \cap \ldots \cap A_{n-2}$ occurs. Indeed, if $A_1 \cap \ldots \cap A_{n-2}$ occurs, then C is contained in the output of RndMinCut(G) and there are no other edges in the output, because otherwise C would not be a cut-set (the output of RndMinCut(G) is always a cut-set). The other implication is clear.

Since $\#C = k$, for every vertex there are at least k edges passing through it, hence, $\#E \geq \dfrac{nk}{2}$. The probability that in the first iteration we choose an edge from C is at most $\dfrac{k}{\frac{nk}{2}} = \dfrac{2}{n}$. Therefore,

$$P(A_1) \geq 1 - \frac{2}{n} = \frac{n-2}{n}.$$

Using the same arguments for G_1, we deduce that if in the first iteration we have picked an edge that is not in C (i.e., A_1 has occurred), then: $\#V_1 = n - 1$, $\#E_1 \geq \dfrac{(n-1)k}{2}$ and the probability that in the second iteration we pick an edge from C is at least $\dfrac{k}{\frac{(n-1)k}{2}} = \dfrac{2}{n-1}$. Therefore,

$$P(A_2|A_1) \geq 1 - \frac{2}{n-1} = \frac{n-3}{n-1}.$$

By induction, we deduce that

$$P(A_i|A_1 \cap \ldots \cap A_{i-1}) \geq 1 - \frac{2}{n-(i-1)} = \frac{n-i-1}{n-i+1}, \quad i = \overline{2, n-2}.$$

Using the chain rule from Theorem 1.7, we get

$$P(A_1 \cap \ldots \cap A_{n-2}) = P(A_1) \cdot P(A_2|A_1) \cdot \ldots \cdot P(A_{n-2}|A_1 \cap \ldots \cap A_{n-3})$$
$$\geq \frac{n-2}{n} \cdot \frac{n-3}{n-1} \cdot \ldots \cdot \frac{1}{3} = \frac{2}{n(n-1)}.$$

If we run RndMinCut(V,E) once, then the probability that the output is not a minimum cut-set is at most $1 - \dfrac{2}{n(n-1)}$. If we run RndMinCut(V,E) at least $n^2 \log n$ times, then the probability that every output is not a minimum cut-set is at most

$$\left(1 - \frac{2}{n(n-1)}\right)^{n^2 \log n} \leq \left(\exp\left\{-\frac{2}{n(n-1)}\right\}\right)^{n^2 \log n} \leq \frac{1}{n^2},$$

where we use the following inequality: $1 - x \leq e^{-x}$, $x \geq 0$.

For this and other random algorithms regarding graph theory, we refer to [Mitzenmacher and Upfal (2005)].

Problem 1.4.11. Let $p \in (0, 1)$. A certain type of cell either divides itself with probability p, or it dies with probability $1 - p$. If the initial cell succeeds to divide, then, in the next generation, we have two cells of the same type, which behave independently in the same way as the initial cell does. And so on, until a possible extinction of the population occurs.

1) Simulate in Matlab the above process, with a graphical representation of the population of cells, for at most 5 generations and some $p \in (0, 1)$.

2) Write a function in Matlab that returns the probability that the population of cells goes extinct in at most $n \in \mathbb{N}$, $n \geq 2$ generations (consider that the initial cell represents the first generation).

3) Determine the probability that the population of cells goes extinct.

Solution 1.4.11: 1)

```
function celldiv
clf;axis([-5 4 -5.5 0.5]);
axis equal; axis off; hold on;
celldiv_g(0,0.7,1);
end

function celldiv_g(cel,p,n)
%cel=the vector of abscissas of the centers of the "cells"
%n=number of the generation
%p=probability of division
if n>5||isempty(cel)
    return;
elseif n==5
    str=['Generation no. ' num2str(n)]; text(-5,-n,str,'fontsize',12);
    for i=1:length(cel)
        circle(cel(i),-n,'r');
    end
    fprintf('The size of the population for generation no. %d is %d.\n',n,length(cel));
else
    cel_new=[ ];
    str=['Generation no. ' num2str(n)]; text(-5,-n,str,'fontsize',12)
    for i=1:length(cel)
        circle(cel(i),-n,'r');
```

```
      if randsample([0,1],1,true,[1-p p])
            cel_new=[cel_new cel(i)-1/2^(n-1) cel(i)+1/2^(n-1)];
      else
            circle(cel(i),-n,'b');
      end
  end
  fprintf('The size of the population for generation no. %d is %d.\n',n,length(cel));
  pause(0.5);
  celldiv_g(cel_new,p,n+1);
end
end

function circle(x,y,col)
 t=linspace(0,2*pi,100);
 fill(x+1/8*cos(t),y+1/8*sin(t),col,'FaceColor',col);
end
```

2) Let D be the event that the initial cell divides itself. For $n \in \mathbb{N}^*$ let A_n be the event that the population of cells goes extinct by the nth generation and denote $x_n = P(A_n)$. For $n \in \mathbb{N}$, $n \geq 2$, and $i \in \{1,2\}$ let $B_{n,i}$ be the event that the population of cells generated by the ith cell in the second generation goes extinct by the nth generation, given that the initial cell divides itself. Note that $B_{n,1}$ and $B_{n,2}$ are independent and that $P(B_{n,1}) = P(B_{n,2}) = x_{n-1}$ for $n \in \mathbb{N}$, $n \geq 2$, because the cells in the first generation behave independently like the initial cell. By the law of total probability (see Theorem 1.8), we have for $n \in \mathbb{N}$, $n \geq 2$,

$$
\begin{aligned}
x_n = P(A_n) &= P(A_n|D)P(D) + P(A_n|\overline{D})P(\overline{D}) \\
&= P(B_{n,1} \cap B_{n,2})p + 1 - p = P(B_{n,1})P(B_{n,2})p + 1 - p \\
&= px_{n-1}^2 + 1 - p,
\end{aligned}
$$

where we use the fact that the extinction of the population generated by the initial cell, given that the cell divides itself, is equivalent with the extinction of the two populations generated by the cells resulted from the initial division. So, $x_1 = 0$ and $x_n = px_{n-1}^2 + 1 - p$, $n \in \mathbb{N}$, $n \geq 2$.

```
function x=celldiv_prob(n,p)
 x=0;
 for i=2:n
    x=p*x^2+1-p;
 end
end
```

3) Let A be the event that the population generated by one cell goes extinct

and denote $x = P(A)$. We have $A = \bigcup_{n=1}^{\infty} A_n$, and thus

$$P(A) = P\left(\bigcup_{n=1}^{\infty} A_n\right) = \lim_{n \to \infty} P(A_n),$$

because $(A_n)_{n \in \mathbb{N}^*}$ is an increasing family of events (see Theorem 1.6). So,

$$\lim_{n \to \infty} x_n = x.$$

From the recurrence relation, given in the proof of 2), we deduce that x satisfies the following equation

$$px^2 - x + 1 - p = 0.$$

Hence, $x \in \left\{1, \dfrac{1-p}{p}\right\}$. Using the recurrence relation again, we can easily prove, by induction, that $x_n < \dfrac{1-p}{p}$ for $n \in \mathbb{N}^*$. So, $x \leq \dfrac{1-p}{p}$.

If $p \in (0, 0.5]$, then $\dfrac{1-p}{p} \geq 1$, and thus $x = 1$. If $p \in (0.5, 1)$, then $\dfrac{1-p}{p} < 1$, and thus $x \leq \dfrac{1-p}{p} < 1$, therefore, $x = \dfrac{1-p}{p}$. We conclude that $P(A) = 1$ if $p \in (0, 0.5]$ and $P(A) = \dfrac{1}{p} - 1$ if $p \in (0.5, 1)$.

Chapter 2

Random Variables and Vectors

2.1 Definitions and Properties

Definition 2.1. Let $(E_1, \mathcal{E}_1)$ and $(E_2, \mathcal{E}_2)$ be measurable spaces. A function $G : E_1 \to E_2$ is said to be $\boldsymbol{\mathcal{E}_1/\mathcal{E}_2}$ **measurable** if

$$G^{-1}(B) = \{e \in E_1 \: : \: G(e) \in B\} \in \mathcal{E}_1 \quad \text{for all } B \in \mathcal{E}_2. \qquad \blacklozenge$$

In this chapter, $(\Omega, \mathcal{F}, P)$ denotes a probability space. We consider the special cases $E_1 = \Omega$, $\mathcal{E}_1 = \mathcal{F}$, and $E_2 = \mathbb{R}$, $\mathcal{E}_2 = \mathcal{B}$, respectively $E_2 = \mathbb{R}^n$, $\mathcal{E}_2 = \mathcal{B}^n$ (see Section 1.2).

Definition 2.2. A function $X : \Omega \to \mathbb{R}$ is called a **random variable** if it is $\mathcal{F}/\mathcal{B}$ measurable, i.e.,

$$X^{-1}(B) = \{\omega \in \Omega \: : \: X(\omega) \in B\} \in \mathcal{F} \text{ for all } B \in \mathcal{B}.$$

A function $\mathbb{X} : \Omega \to \mathbb{R}^n$ is called a **random vector** if it is $\mathcal{F}/\mathcal{B}^n$ measurable, i.e.,

$$\mathbb{X}^{-1}(B) = \{\omega \in \Omega \: : \: \mathbb{X}(\omega) \in B\} \in \mathcal{F} \quad \text{for all } B \in \mathcal{B}^n.$$

In fact, $\mathbb{X} = (X_1, \ldots, X_n)$ with the components $X_1, \ldots, X_n$ being random variables; $(\cdot)^T$ represents the transpose operation of a matrix, in particular, a vector:

$$(X_1, \ldots, X_n)^T = \begin{pmatrix} X_1 \\ \vdots \\ X_n \end{pmatrix}.$$

$\blacklozenge$

Notation: Let $\mathbb{X} = (X_1, \ldots, X_n)$ be a random vector. Then $\{\mathbb{X} \in B\}$ denotes the event $\mathbb{X}^{-1}(B)$ for $B \in \mathcal{B}^n$ and $\{X_1 \in B_1, \ldots, X_n \in B_n\}$ denotes the event $\mathbb{X}^{-1}(B_1 \times \ldots \times B_n)$ for $B_1, \ldots, B_n \in \mathcal{B}$. Also, $\{\mathbb{X} = \mathbf{x}\}$ and $\{X_1 = x_1, \ldots, X_n = x_n\}$ denote the event $\mathbb{X}^{-1}(\{\mathbf{x}\})$ for $\mathbf{x} = (x_1, \ldots, x_n) \in \mathbb{R}^n$. We consider the same notations for random variables.

Moreover, if X and Y are random variables, then $\{X \geq Y\}$ denotes the event $(X - Y)^{-1}([0, \infty))$ and, similarly, we consider the notations for the other standard comparison signs.

Note that in the above notations (and also in other corresponding situations in this book) the comma represents the intersection operation. For example, $\{X_1 \leq x_1, \ldots, X_n \leq x_n\}$ denotes $\{X_1 \leq x_1\} \cap \ldots \cap \{X_n \leq x_n\}$ for $(x_1, \ldots, x_n) \in \mathbb{R}^n$.

Remark 2.1. Let $\mathbb{X}$ be a random vector and $A \in \mathcal{B}^n$. It is said that

$$\mathbb{X} \in A \text{ almost surely,}$$

if $P(\{\omega \in \Omega : \mathbb{X}(\omega) \in A\}) = 1$, and we write

$$\mathbb{X} \in A \text{ a.s.} \quad \text{or} \quad \mathbb{X}(\omega) \in A \text{ for a.e. } \omega \in \Omega.$$

Obviously, we consider the same notations for random variables.   ▼

Example 2.1. (1) Let $\mathbb{X}$ and $\mathbb{Y}$ be random vectors. We write $\mathbb{X} = \mathbb{Y}$ a.s. or $\mathbb{X}(\omega) = \mathbb{Y}(\omega)$ for a.e. $\omega \in \Omega$, if $\mathbb{X} - \mathbb{Y} \in \{0_n\}$ a.s.
(2) Let X and Y be random variables. We write

$$X < Y \text{ a.s. or } X(\omega) < Y(\omega) \text{ for a.e. } \omega \in \Omega,$$

if $X - Y \in (0, \infty)$ a.s.   ▲

Example 2.2. Let $A \in \mathcal{F}$. An example of a random variable is the **indicator function of the event** A

$$\mathbb{I}_A : \Omega \to \mathbb{R} \text{ defined by } \mathbb{I}_A(\omega) = \begin{cases} 1, & \text{if } \omega \in A \\ 0, & \text{if } \omega \in \bar{A}. \end{cases}$$

   ▲

Definition 2.3. X is said to be a **discrete random variable** if its range $X(\Omega)$ is a countable set, i.e.,

$$X(\Omega) = \{x_i : i \in I\}, \text{ where } \emptyset \neq I \subseteq \mathbb{N} \text{ and } P(X = x_i) > 0 \text{ for each } i \in I.$$

Similarly, $\mathbb{X}$ is said to be a **discrete random vector** if its range $\mathbb{X}(\Omega)$ is a countable set, i.e.,

$\mathbb{X}(\Omega) = \{\mathbb{x}_i : i \in I\}$, where $\emptyset \neq I \subseteq \mathbb{N}$ and $P(\mathbb{X} = \mathbb{x}_i) > 0$ for each $i \in I$. ♦

Note that a discrete random vector $\mathbb{X}$ with the range as in Definition 2.3 has the representation

$$\mathbb{X}(\omega) = \sum_{i \in I} \mathbb{x}_i \mathbb{I}_{A_i}(\omega) \quad \text{for all } \omega \in \Omega,$$

where $A_i = \{\mathbb{X} = \mathbb{x}_i\} \in \mathcal{F}$, $i \in I$, and $(A_i)_{i \in I}$ is a partition of Ω with $P(A_i) > 0$ for each $i \in I$. Similarly, one can reason for a discrete random variable.

Theorem 2.1. *The following statements are equivalent:*
(1) X *is a random variable.*
(2) $\{X < x\} \in \mathcal{F}$ *for all* $x \in \mathbb{R}$.
(3) $\{X \leq x\} \in \mathcal{F}$ *for all* $x \in \mathbb{R}$.
(4) $\{X > x\} \in \mathcal{F}$ *for all* $x \in \mathbb{R}$.
(5) $\{X \geq x\} \in \mathcal{F}$ *for all* $x \in \mathbb{R}$.

Definition 2.4. For a random variable X we consider

$$\mathcal{F}_X = X^{-1}(\mathcal{B}) = \{X^{-1}(B) : B \in \mathcal{B}\}.$$

$\mathcal{F}_X$ is called the **σ-field generated by** X. Let $X_1, \ldots, X_n$ be random variables, denote $\mathbb{X} = (X_1, \ldots, X_n)$ and consider

$$\mathcal{F}_{X_1, \ldots, X_n} = \{\mathbb{X}^{-1}(B) : B \in \mathcal{B}^n\}.$$

$\mathcal{F}_{X_1, \ldots, X_n}$ is called the **σ-field generated by** $X_1, \ldots, X_n$. ♦

Remark 2.2. (1) Note that $\mathcal{F}_X$ verifies Definition 1.3 of a σ-field (see Problem 1.2.1 for details). It is the smallest σ-field such that $X : \Omega \to \mathbb{R}$ is $\mathcal{F}_X/\mathcal{B}$ measurable.

(2) We have $\mathcal{F}_{X_1, \ldots, X_n} = \sigma \left(\bigcup_{i=1}^{n} \mathcal{F}_{X_i} \right) = \sigma \left(\bigcup_{i=1}^{n} X_i^{-1}(\mathcal{B}) \right)$. ▼

Example 2.3. (1) Let $A \in \mathcal{F}$ and consider the random variable $\mathbb{I}_A$ (see Example 2.2). The σ-field generated by $\mathbb{I}_A$ is

$$\mathcal{F}_{\mathbb{I}_A} = \{\emptyset, \Omega, A, \bar{A}\}.$$

(2) Let X be a discrete random variable with $X(\Omega) = \{x_i : i \in I\}$. Denote $\mathcal{A} = (A_i)_{i \in I}$, $A_i = \{X = x_i\}, i \in I$. By Example 1.3-(2) we have

$$\mathcal{F}_X = \sigma(\mathcal{A}) = \{\emptyset\} \cup \left\{ \bigcup_{i \in J} \{X = x_i\} : \emptyset \neq J \subseteq I \right\}.$$ ▲

Theorem 2.2. *Let* $\mathbb{X}$ *be a random vector. Then the mapping*

$$P_{\mathbb{X}} : \mathcal{B}^n \to \mathbb{R} \quad \text{defined by} \quad P_{\mathbb{X}}(B) = P(\mathbb{X} \in B), \ B \in \mathcal{B}^n,$$

is a probability over $\mathcal{B}^n$.

Definition 2.5. (1) The probability $P_{\mathbb{X}} : \mathcal{B}^n \to \mathbb{R}$ given by Theorem 2.2 is called the **distribution** of $\mathbb{X}$.
(2) A family of random variables $(X_k)_{k \in K}$ (where K is a set of indices) is said to be **identically distributed**, if the random variables of this family have the same distribution.

Definition 2.6. (1) If X is a discrete random variable with range $X(\Omega) = \{x_i : i \in I\}$, then P_X is completely determined by the values

$$P(X = x_i) = P(\{\omega \in \Omega : X(\omega) = x_i\}), \ i \in I,$$

the so-called **probability mass function of the discrete random variable** X. We say that X has a **discrete distribution**.
(2) If $\mathbb{X}$ is a discrete random vector with range $\mathbb{X}(\Omega) = \{\mathbb{x}_i : i \in I\}$, then $P_{\mathbb{X}}$ is completely determined by the values

$$P(\mathbb{X} = \mathbb{x}_i) = P(\{\omega \in \Omega : \mathbb{X}(\omega) = \mathbb{x}_i\}), \ i \in I,$$

the so-called **joint probability mass function of the discrete random vector** $\mathbb{X}$. We say that $\mathbb{X}$ has a **discrete distribution**.   ♦

Example 2.4. A discrete random vector (X, Y) taking the values $(x_i, y_j), (i, j) \in I \times J$ ($\emptyset \neq I, J \subseteq \mathbb{N}$ are sets of indices) is characterized by

$$p_{ij} = P((X, Y) = (x_i, y_j)) = P(\{X = x_i\} \cap \{Y = y_j\}),$$

where $p_{ij} > 0$ for all $i \in I, j \in J$, and $\displaystyle\sum_{(i,j) \in I \times J} p_{ij} = 1$.

The joint probability mass function of the discrete random vector (X, Y)

is often given by a **contingency table**:

$\diagdown$	$\cdots$	y_j	$\cdots$
$\vdots$	$\vdots$	$\vdots$	$\vdots$
x_i	$\cdots$	p_{ij}	$\cdots$
$\vdots$	$\vdots$	$\vdots$	$\vdots$

Observe that if the joint probability mass function of the discrete random vector (X, Y) is known, then the probability mass functions of X and of Y are respectively given by the **marginal probability mass functions**:

$$P(X = x_i) = \sum_{j \in J} p_{ij} \quad \text{for all } i \in I \,,$$

$$P(Y = y_j) = \sum_{i \in I} p_{ij} \quad \text{for all } j \in J. \qquad \blacktriangle$$

Theorem 2.3. *Let $G : \mathbb{R}^n \to \mathbb{R}$ be a $\mathcal{B}^n/\mathcal{B}$ measurable function and let $\mathbb{X}$ be a random vector. Then $G(\mathbb{X})$ is a random variable.*

Example 2.5. Let X and Y be random variables. Then aX (where $a \in \mathbb{R}$), $|X|$, $\min\{X, Y\}$, $\max\{X, Y\}$, $X + Y$, $X - Y$ and $X \cdot Y$ are random variables. Moreover, if $Y(\omega) \neq 0$ for all $\omega \in \Omega$, then also $\dfrac{X}{Y}$ is a random variable. $\blacktriangle$

Example 2.6. Let X and Y be random variables. Then

$$\{X > Y\} \in \mathcal{F}, \quad \{X \geq Y\} \in \mathcal{F} \quad \text{and} \quad \{X = Y\} \in \mathcal{F}.$$

Proof: We observe that

$$\{\omega \in \Omega : X(\omega) > Y(\omega)\} = \bigcup_{r \in \mathbb{Q}} \Big(\{\omega \in \Omega : X(\omega) > r\} \cap \{\omega \in \Omega : r > Y(\omega)\} \Big).$$

By using Theorem 2.1 (applied to X, respectively Y) and the properties of a σ-field, it follows that $\{X > Y\} \in \mathcal{F}$. By the same argument, we also obtain $\{Y > X\} \in \mathcal{F}$. Finally, we write

$$\{X \geq Y\} = \overline{\{Y > X\}} \in \mathcal{F}$$

and

$$\{X = Y\} = \{X \geq Y\} \setminus \{X > Y\} \in \mathcal{F}. \qquad \blacktriangle$$

2.1.1   *Solved Problems*

Problem 2.1.1. Let $(X_n)_{n \in \mathbb{N}}$ be a sequence of random variables. Prove that $\sup\limits_{n \in \mathbb{N}} X_n$, $\inf\limits_{n \in \mathbb{N}} X_n$, $\limsup\limits_{n \to \infty} X_n$, $\liminf\limits_{n \to \infty} X_n$ are random variables.

Solution 2.1.1: We start with the relation

$$\{\omega \in \Omega : \sup_{n \in \mathbb{N}} X_n(\omega) > x\} = \bigcup_{n=1}^{\infty} \{\omega \in \Omega : X_n(\omega) > x\} \in \mathcal{F},$$

and conclude that $\sup\limits_{n \in \mathbb{N}} X_n$ is a random variable. Using the identity

$$\inf_{n \in \mathbb{N}} X_n = -\sup_{n \in \mathbb{N}}(-X_n),$$

it follows that $\inf\limits_{n \in \mathbb{N}} X_n$ is also a random variable. The rest follows from the first two results and the identities

$$\limsup_{n \to \infty} X_n = \inf_{n \in \mathbb{N}}\left(\sup_{k \geq n} X_k\right), \quad \liminf_{n \to \infty} X_n = \sup_{n \in \mathbb{N}}\left(\inf_{k \geq n} X_k\right).$$

Problem 2.1.2. Find a probability space $(\Omega, \mathcal{F}, P)$ and a sequence of random variables $(X_n)_{n \in \mathbb{N}^*}$ such that for every $n \in \mathbb{N}^*$ and $x_1, \ldots, x_n \in \{1, \ldots, 6\}$

$$P(X_1 = x_1, \ldots, X_n = x_n) = \frac{1}{6^n},$$

i.e., for $i \in \mathbb{N}^*$ the random variable X_i indicates the result of the ith roll of a fair die.

Solution 2.1.2: Let $\Omega = [0,1]$, $\mathcal{F} = \mathcal{B}([0,1])$ and P be the Lebesgue measure $\lambda_{\mathbb{R}}$ restricted to $[0,1]$. Next, we consider the senary (base-6) representation of the numbers in $[0,1]$. Note that some numbers in $[0,1]$ have a double senary representation, e.g., $0.01 = 0.00(5) = 0.0055\ldots5\ldots$ In the following, we shall omit the representation $0.\omega_1\omega_2\ldots\omega_n(5) = 0.\omega_1\omega_2\ldots\omega_n\,5\,5\ldots5\ldots$, for $\omega_1, \ldots, \omega_n \in \{0,1,\ldots,5\}$, $n \in \mathbb{N}^*$, and always use instead the corresponding representation $0.\omega_1\omega_2\ldots\omega_n + 0.\underbrace{0\ldots0}_{n-1}1$ (note that the set of these numbers, which have a finite number of nonzero senary digits, has measure zero, because it is countable). For every $i \in \mathbb{N}^*$, let $X_i(\omega) = \omega_i + 1$, where ω_i is the ith senary digit of $\omega \in [0,1]$. For every $n \in \mathbb{N}^*$ and $x_1, \ldots, x_n \in \{1, \ldots, 6\}$ we denote $d_i = x_i - 1$, $i \in \{1, \ldots, n\}$, and we deduce that

$$P(X_1 = x_1, \ldots, X_n = x_n) = P\left(\{\omega \in [0,1] : \omega = 0.d_1\ldots d_n\ldots\}\right)$$

$$= \lambda_{\mathbb{R}}\left([0.d_1\ldots d_n\,,\,0.d_1\ldots d_n + 0.\underbrace{0\ldots0}_{n-1}1)\right) = 0.\underbrace{0\ldots0}_{n-1}1 = \frac{1}{6^n}.$$

2.2 Distribution Functions and Density Functions

Definition 2.7. The function $F : \mathbb{R} \to \mathbb{R}$, defined by

$$F(x) = P(X \leq x), \ x \in \mathbb{R},$$

is called the **distribution function** or **cumulative distribution function** of the random variable X. ◆

Example 2.7. (1) If X is a discrete random variable and takes the values $\{x_i : i \in I\}$, then its distribution function is

$$F(x) = \sum_{\substack{i \in I \\ x_i \leq x}} P(X \leq x_i), \ x \in \mathbb{R}.$$

(2) Let $0 < a < b < c$ be given and consider the discrete random variable X such that

$$P(X = x) = \frac{x}{a+b+c} \quad \text{for } x \in \{a, b, c\}.$$

Then the distribution function of X is $F : \mathbb{R} \to \mathbb{R}$ given by

$$F(x) = \begin{cases} 0, & \text{if } x < a \\ \dfrac{a}{a+b+c}, & \text{if } a \leq x < b \\ \dfrac{a+b}{a+b+c}, & \text{if } b \leq x < c \\ 1, & \text{if } c \leq x. \end{cases}$$

▲

Theorem 2.4. *The distribution function* $F : \mathbb{R} \to \mathbb{R}$ *of a random variable* X *has the following properties:*

(1) $P(a < X \leq b) = F(b) - F(a)$ *for* $a < b$;
(2) $P(X = b) = F(b) - F(b - 0)$ *for* $b \in \mathbb{R}$;
(3) $P(X < b) = F(b - 0)$ *for* $b \in \mathbb{R}$;
(4) F *is monotonically increasing;*
(5) F *is right-continuous, i.e.,* $F(x + 0) = F(x)$ *for* $x \in \mathbb{R}$;
(6) $\lim\limits_{x \to -\infty} F(x) = 0$ *and* $\lim\limits_{x \to \infty} F(x) = 1$;
(7) *the set of discontinuity points of* F *is countable.*

Example 2.8. The properties (4), (5) and (6) from Theorem 2.4 characterize a distribution function, i.e., if a function $F : \mathbb{R} \to [0, 1]$ has these properties, then there exists a probability space and a random variable X

which has F as its distribution function.

Proof: To prove this property, consider the function $G : (0, 1) \to \mathbb{R}$ defined by $G(y) = \inf\{x \in \mathbb{R} : F(x) \geq y\}$ if $y \in (0, 1)$. Let $\Omega = [0, 1]$, $\mathcal{F} = \mathcal{B}([0, 1])$ and let $P = \lambda_{\mathbb{R}}$ be the Lebesgue measure restricted to $[0, 1]$. We define the random variable $X : \Omega \to \mathbb{R}$ by $X(\omega) = G(\omega)$, if $\omega \in (0, 1)$, and $X(\omega) = \omega$, if $\omega \in \{0, 1\}$. Then for $x \in \mathbb{R}$

$$P(X \leq x)$$
$$= P\big(\{\omega \in (0, 1) : G(\omega) \leq x\} \cup \{\omega \in \{0\} : 0 \leq x\} \cup \{\omega \in \{1\} : 1 \leq x\}\big)$$
$$= \lambda_{\mathbb{R}}\big(\{y \in (0, 1) : y \leq F(x)\}\big) = F(x).$$
▲

Definition 2.8. Let X be a random variable and let $F : \mathbb{R} \to \mathbb{R}$ be its distribution function. If there exists a function $f : \mathbb{R} \to \mathbb{R}$ such that

$$F(x) = \int_{-\infty}^{x} f(t)dt \text{ for all } x \in \mathbb{R}, \tag{2.1}$$

then f is called (probability) **density function** of X. If the random variable X admits a density function, then X is said to be a **continuous random variable**. We say it has a **continuous distribution**. ♦

Remark 2.3. Let X be a continuous random variable. If f_1 is a density function of X and f_2 is another density function of X, then $f_1(x) = f_2(x)$ for a.e. $x \in \mathbb{R}$, i.e.,

$$\lambda_{\mathbb{R}}\big(\{x \in \mathbb{R} : f_1(x) \neq f_2(x)\}\big) = 0.$$
▼

Theorem 2.5. *Let $a, b \in \mathbb{R}, a < b$. If X is a continuous random variable having the distribution function F and a density function f, then the following properties hold:*

(1) *F is absolutely continuous and $F'(x) = f(x)$ for a.e. $x \in \mathbb{R}$;*
(2) *$f(x) \geq 0$ for a.e. $x \in \mathbb{R}$;*
(3) *$\int_{\mathbb{R}} f(t)dt = 1$;*
(4) *for $a, b \in \mathbb{R}$ with $a < b$ we have $P(X = b) = 0$ and*
$$P(a < X < b) = P(a \leq X < b) = P(a \leq X \leq b)$$
$$= P(a < X \leq b) = \int_{a}^{b} f(t)dt;$$

(5) *For $B \in \mathcal{B}$ we have $P(X \in B) = \int_{B} f(x)dx.$*

Example 2.9. Let $F : \mathbb{R} \to \mathbb{R}$ be given by

$$F(x) = \begin{cases} d, & \text{if } x < 0 \\ ax^2 + bx + c, & \text{if } 0 \le x < 2 \\ e, & \text{if } x \ge 2, \end{cases}$$

where $a, b, c, d, e \in \mathbb{R}$ are fixed. We shall find a, b, c, d, e such that F is the distribution function of a continuous random variable X with

$$P(1 < X < 2) = \frac{3}{4}.$$

Solution: From Theorem 2.4-(6), we have $0 = \lim\limits_{x \to -\infty} F(x) = d$ and $1 = \lim\limits_{x \to \infty} F(x) = e$. By Theorem 2.5-(1) F is continuous, since X is continuous, and thus $c = F(0) = F(0 - 0) = d = 0$ and $1 = e = F(2) = F(2 - 0) = 4a + 2b + c = 4a + 2b$. By Theorems 2.4-(1) and 2.5-(4), we have

$$P(1 < X < 2) = F(2) - F(1) = e - a - b - c = 1 - a - b = \frac{3}{4}.$$

Hence, we deduce that $a = \dfrac{1}{4}, b = 0, c = 0, d = 0, e = 1$.

We point out that $f : \mathbb{R} \to \mathbb{R}$ defined by

$$f(x) = \begin{cases} \dfrac{x}{2}, & \text{if } x \in (0, 2) \\ 0, & \text{otherwise} \end{cases}$$

is a density function of X. ▲

Definition 2.9. Let F be a distribution function. The function $Q : (0, 1) \to \mathbb{R}$ defined by

$$Q(p) = \inf\{x \in \mathbb{R} : F(x) \ge p\} \text{ for each } p \in (0, 1)$$

is called **quantile function.** ◆

The above function is also called the generalized inverse or pseudo-inverse function of F. For various results concerning quantile functions, see [Pfeiffer (1990), Section 11.4 and Chapter 11a].

Theorem 2.6. *Let F be a distribution function and Q its corresponding quantile function. Then the following properties hold:*

(1) Q is monotonically increasing.
(2) Let $u \in (0, 1)$ and $t \in \mathbb{R}$. Then $Q(u) \le t$ if and only if $u \le F(t)$.

(3) *If F is continuous, strictly increasing on the interval (a, b), then Q is continuous, strictly increasing on $\big(F(a), F(b)\big)$ and $Q(u) = t$ if and only if $F(t) = u$ for $u \in \big(F(a), F(b)\big)$ and $t \in (a, b)$.*

(4) *Q is a left-continuous function.*

(5) *$F(Q(u) - 0) \leq u \leq F\big(Q(u)\big)$ for each $u \in (0, 1)$; if F is continuous at $Q(u)$, then $F\big(Q(u)\big) = u$.*

Example 2.10. Consider the distribution function $F : \mathbb{R} \to [0, 1]$ and the corresponding quantile function $Q : (0, 1) \to \mathbb{R}$

$$F(x) = \begin{cases} 0, & \text{if } x < 0 \\ \dfrac{x}{2}, & \text{if } 0 \leq x < 1 \\ 0.5, & \text{if } 1 \leq x < 2 \\ \dfrac{x-1}{2}, & \text{if } 2 \leq x < 3 \\ 1, & \text{if } 3 \leq x \end{cases} \Rightarrow Q(p) = \begin{cases} 2p, & \text{if } 0 < p \leq 0.5 \\ 2p + 1, & \text{if } 0.5 < p < 1. \end{cases}$$

Observe that F is a continuous function, which is strictly increasing on $[0, 1)$ and on $[2, 3)$. The function Q is discontinuous at 0.5 and strictly increasing on $(0, 1)$. ▲

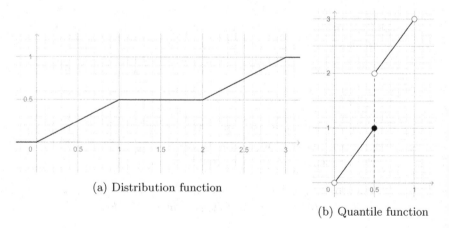

(a) Distribution function

(b) Quantile function

Fig. 2.1: Example 2.10

2.2.1 Solved Problems

Problem 2.2.1. Let $f : \mathbb{R} \to \mathbb{R}$ be given by

$$f(x) = \begin{cases} (ax + b)e^{-x}, & \text{if } x > 0 \\ 0, & \text{if } x \le 0, \end{cases}$$

where $a, b \in \mathbb{R}$ are fixed.

1) Find a and b such that f is a density function of a continuous random variable X with $P(X < 1) = \dfrac{e - 2}{e}$.

2) Find the distribution function F of X.

3) Plot in Matlab the quantile function of X on the interval $[F(1), \ F(9)]$.

Solution 2.2.1: 1) Since $\displaystyle\int_{-\infty}^{\infty} f(x)dx = 1$, we have

$$1 = \int_{0}^{\infty} (ax + b)e^{-x}dx = a + b.$$

Since $P(X < 1) = \displaystyle\int_{-\infty}^{1} f(x)dx$, we have

$$1 - 2e^{-1} = \int_{0}^{1} (ax + b)e^{-x}dx = a + b - (2a + b)e^{-1}.$$

Therefore, $a = 1$ and $b = 0$.

2) $F(x) = \displaystyle\int_{0}^{x} f(t)dt = 1 - (x + 1)e^{-x}$, if $x > 0$, and $F(x) = 0$, if $x \le 0$.

3)

```
>> F=@(x) 1-(x+1).*exp(-x);
>> Q=linspace(1,9,1000);
>> plot(F(Q),Q);
```

Problem 2.2.2. The distribution function $F : \mathbb{R} \to [0, 1]$ is defined by

$$F(x) = \begin{cases} e^{x}, & \text{if } x < -1 \\ 0.5, & \text{if } -1 \le x < 2 \\ 1 - e^{-x}, & \text{if } 2 \le x. \end{cases}$$

Determine the corresponding quantile function Q.

Solution 2.2.2: The quantile function $Q : (0, 1) \to \mathbb{R}$ has the expression

$$Q(p) = \begin{cases} \log p, & \text{if } 0 < p \le e^{-1} \\ 2, & \text{if } e^{-1} < p \le 1 - e^{-2} \\ -\log(1 - p), & \text{if } 1 - e^{-2} < p < 1. \end{cases}$$

Note that F is strictly monotonically increasing on $(-\infty, -1) \cup [2, \infty)$, but is discontinuous at -1 and 2. The function Q is strictly increasing on $(0, e^{-1}] \cup [1 - e^{-2}, 1)$ and discontinuous at e^{-1}.

2.3   Joint Distributions

Definition 2.10. The function $F : \mathbb{R}^n \to \mathbb{R}$ defined by

$$F(x_1, \ldots, x_n) = P(X_1 \leq x_1, \ldots, X_n \leq x_n) \text{ for each } x_1, \ldots, x_n \in \mathbb{R}$$

is called the **joint distribution function** of the random vector $(X_1, \ldots, X_n)$. ◆

Remark 2.4. Two random vectors have the same distribution (see Definition 2.5) if and only if they have the same joint distribution function. ▼

Example 2.11.

Consider the discrete random vector (X_1, X_2) with joint probability mass function given by the contingency table:

X_1 \ X_2	0	1
-1	0.4	0.3
1	0.2	0.1

The joint distribution function of the random vector (X_1, X_2) is given by $F_{X_1,X_2} : \mathbb{R} \times \mathbb{R} \to [0,1]$

$$F_{X_1,X_2}(x_1, x_2) = \begin{cases} 0, & \text{if } x_1 < -1 \text{ or } x_2 < 0 \\ 0.4, & \text{if } -1 \leq x_1 < 1 \text{ and } 0 \leq x_2 < 1 \\ 0.7, & \text{if } -1 \leq x_1 < 1 \text{ and } 1 \leq x_2 \\ 0.6, & \text{if } 1 \leq x_1 \text{ and } 0 \leq x_2 < 1 \\ 1, & \text{if } 1 \leq x_1 \text{ and } 1 \leq x_2. \end{cases}$$

▲

Theorem 2.7. *If $(X_1, \ldots, X_n)$ is a random vector having the joint distribution function F, then the following properties hold:*
(1)

$$P(a_1 < X_1 \leq b_1, \ldots, a_n < X_n \leq b_n)$$

$$= F(b_1, \ldots, b_n) - \sum_{i=1}^{n} F(b_1, \ldots, a_i, \ldots, b_n)$$

$$+ \sum_{\substack{i,j=1 \\ i<j}}^{n} F(b_1, \ldots, a_i, \ldots, a_j, \ldots, b_n) - \cdots + (-1)^n F(a_1, \ldots, a_n),$$

whenever $a_i, b_i \in \mathbb{R}$ and $a_i < b_i$ for all $i \in \{1, \ldots, n\}$.
(2) F is monotonically increasing in each argument.
(3) F is right-continuous in each argument.

(4) $\lim\limits_{x_1,\ldots,x_n \to \infty} F(x_1,\ldots,x_n) = 1.$

(5) *For each* $k \in \{1,\ldots,n\}$ *we have*

$$\lim_{x_k \to -\infty} F(x_1,\ldots,x_{k-1},x_k,x_{k+1},\ldots,x_n) = 0$$

for all $x_1,\ldots,x_{k-1},x_{k+1},\ldots,x_n \in \mathbb{R}.$

Definition 2.11. Let $(X_1,\ldots,X_n)$ be a random vector and let F be its joint distribution function. If there exists a function $f : \mathbb{R}^n \to \mathbb{R}$ such that

$$F(x_1,\ldots,x_n) = \int\limits_{-\infty}^{x_1} \cdots \int\limits_{-\infty}^{x_n} f(t_1,\ldots,t_n)dt_1\ldots dt_n, \ (x_1,\ldots,x_n) \in \mathbb{R}^n,$$

then f is called **joint density function** of the random vector $(X_1,\ldots,X_n)$. If the random vector $(X_1,\ldots,X_n)$ has a joint density function, then it is said to be a **continuous random vector**. We say that it has a **continuous distribution**. ♦

Remark 2.5. Let $\mathbb{X}$ be a random vector. If f_1 is a joint density function of $\mathbb{X}$ and f_2 is another joint density function of $\mathbb{X}$, then $f_1(\mathbb{x}) = f_2(\mathbb{x})$ for a.e. $\mathbb{x} \in \mathbb{R}^n$, i.e.,

$$\lambda_{\mathbb{R}^n}\big(\{\mathbb{x} \in \mathbb{R}^n : f_1(\mathbb{x}) \neq f_2(\mathbb{x})\}\big) = 0. \qquad \blacktriangledown$$

Theorem 2.8. *If* $(X_1,\ldots,X_n)$ *is a continuous random vector having the joint distribution function* F *and a joint density function* f, *then the following properties hold.*

(1) F *is absolutely continuous function and*

$$\frac{\partial^n F(x_1,\ldots,x_n)}{\partial x_1 \partial x_2 \ldots \partial x_n} = f(x_1,\ldots,x_n)$$

for a.e. $(x_1,\ldots,x_n) \in \mathbb{R}^n.$

(2) $f(x_1,\ldots,x_n) \geq 0$ *for a.e.* $(x_1,\ldots,x_n) \in \mathbb{R}^n.$

(3) $\underbrace{\int \cdots \int}_{\mathbb{R}^n} f(t_1,\ldots,t_n)dt_1\ldots dt_n = 1.$

(4) *For any* $B \in \mathcal{B}^n$

$$P\big((X_1,\ldots,X_n) \in B\big) = \underbrace{\int \cdots \int}_{B} f(t_1,\ldots,t_n)dt_1\ldots dt_n.$$

Theorem 2.9. *Let X and Y be random variables.*
(1) If $F_{X,Y}$ is the joint distribution function of (X,Y), then
$$F_X(x) = \lim_{y \to \infty} F_{X,Y}(x,y) \quad for\ x \in \mathbb{R}$$
and
$$F_Y(y) = \lim_{x \to \infty} F_{X,Y}(x,y) \quad for\ y \in \mathbb{R},$$
where F_X and F_Y are the distribution functions of X, respectively Y.
(2) If (X,Y) has a joint density function $f_{X,Y}$, then the functions given by
$$f_X(x) = \int_{\mathbb{R}} f_{X,Y}(x,y)dy \quad for\ x \in \mathbb{R}$$
and
$$f_Y(y) = \int_{\mathbb{R}} f_{X,Y}(x,y)dx \quad for\ y \in \mathbb{R}$$
are density functions of X, respectively Y.

Theorem 2.10. *Let $(X_1, \ldots, X_n)$ be a continuous random vector having the joint distribution function F and a joint density function f. If $i_1, \ldots, i_k \in \{1, \ldots, n\}$ and $i_1 < \cdots < i_k$, then the following properties hold:*

(1) The function $F_{i_1,\ldots,i_k} : \mathbb{R}^k \to \mathbb{R}$ defined for $(x_{i_1}, \ldots, x_{i_k}) \in \mathbb{R}^k$ by
$$F_{i_1,\ldots,i_k}(x_{i_1}, \ldots, x_{i_k}) = \lim_{\substack{x_j \to \infty \\ j \neq i_1, \ldots, i_k}} F(x_1, \ldots, x_n)$$
is the joint distribution function of the random vector $(X_{i_1}, \ldots, X_{i_k})$.
(2) For a.e. $(x_{i_1}, \ldots, x_{i_k}) \in \mathbb{R}^k$
$$f_{i_1,\ldots,i_k}(x_{i_1}, \ldots, x_{i_k}) = \underbrace{\int \ldots \int}_{\mathbb{R}^{n-k}} f(x_1, \ldots, x_n)dx_{j_1} \cdots dx_{j_{n-k}},$$
where $j_1, \ldots, j_{n-k} \in \{1, \ldots, n\} \backslash \{i_1, \ldots, i_k\}$, $j_1 < \ldots < j_{n-k}$ and $f_{i_1,\ldots,i_k}$ is a joint density function of the random vector $(X_{i_1}, \ldots, X_{i_k})$.

Definition 2.12. (1) $X_1, \ldots, X_n$, where $n \in \mathbb{N}$, $n \geq 2$, are called **independent random variables**, if
$$F_{X_1,\ldots,X_n}(x_1, \ldots, x_n) = F_{X_1}(x_1) \cdot \ldots \cdot F_{X_n}(x_n)$$
for all $x_1, \ldots, x_n \in \mathbb{R}$.
(2) Let I be a set of indices such that $\#I \geq 2$. $(X_i)_{i \in I}$ is a **family of independent random variables**, if for each finite subset $\{i_1, \ldots, i_m\} \subseteq I$, $m \in \mathbb{N}$, $m \geq 2$, the random variables $X_{i_1}, \ldots, X_{i_m}$ are independent. $(X_i)_{i \in I}$ is a **family of pairwise independent random variables**, if for each $i, j \in I$ with $i \neq j$ the random variables X_i and X_j are independent. ♦

Remark 2.6. Note that in Definition 2.12, if I is finite, then (2) is equivalent with (1). ▼

Theorem 2.11. *The random variables $X_1, \ldots, X_n$ are independent if and only if*

$$P(X_1 \in B_1, \ldots, X_n \in B_n) = P(X_1 \in B_1) \cdot \ldots \cdot P(X_n \in B_n)$$

for every $B_1, \ldots, B_n \in \mathcal{B}$.

Definition 2.13. $\mathcal{F}_1, \ldots, \mathcal{F}_n \subseteq \mathcal{F}$ are **independent σ-fields**, if

$$P(A_1 \cap \ldots \cap A_n) = P(A_1) \cdot \ldots \cdot P(A_n)$$

for every $A_1 \in \mathcal{F}_1, \ldots, A_n \in \mathcal{F}_n$. ◆

Remark 2.7. By Theorem 2.11, the random variables $X_1, \ldots, X_n$ are independent if and only if the σ-fields $\mathcal{F}_{X_1}, \ldots, \mathcal{F}_{X_n}$ are independent. ▼

Theorem 2.12. (see [Breiman (1992), Proposition 3.5, p. 37]) *Let $X_1, \ldots, X_m, Y_1, \ldots Y_n$ be independent random variables, where $m, n \in \mathbb{N}^*$. Then $\mathcal{F}_{X_1, \ldots, X_m}$ and $\mathcal{F}_{Y_1, \ldots, Y_n}$ are independent. In particular, if g is a $\mathcal{B}^m/\mathcal{B}$ measurable function and h is a $\mathcal{B}^n/\mathcal{B}$ measurable function, then $g(X_1, \ldots, X_m)$ and $h(Y_1, \ldots, Y_n)$ are independent random variables.*

Definition 2.14. We say that a **random variable** X is **independent of a σ-field** $\mathcal{G} \subseteq \mathcal{F}$, if $\mathcal{F}_X$ and $\mathcal{G}$ are independent. ◆

Example 2.12. A random variable X is independent of a σ-field $\mathcal{G} \subseteq \mathcal{F}$ if and only if X and $\mathbb{I}_A$ are independent for all $A \in \mathcal{G}$.
Proof: Let $\mathcal{F}_X$ be the σ-field generated by X. If $\mathcal{F}_X$ and $\mathcal{G}$ are independent, then for each $B \in \mathcal{B}$ and each $C \in \mathcal{G}$

$$P(\{X \in B\} \cap C) = P(X \in B)P(C). \qquad (2.2)$$

Let $A \in \mathcal{G}$, $B_1 \in \mathcal{B}$ and $B_2 \in \mathcal{B}$. We denote

$$C = \{\mathbb{I}_A \in B_2\} = \begin{cases} A, & \text{if } 1 \in B_2, \, 0 \notin B_2 \\ \bar{A}, & \text{if } 0 \in B_2, \, 1 \notin B_2 \\ \Omega, & \text{if } 0, 1 \in B_2 \\ \emptyset, & \text{if } 0, 1 \notin B_2. \end{cases}$$

Obviously, $C \in \mathcal{G}$. Then by (2.2) it follows

$$\begin{aligned} P(X \in B_1, \mathbb{I}_A \in B_2) &= P(\{X \in B_1\} \cap C) \\ &= P(X \in B_1)P(C) = P(X \in B_1)P(\mathbb{I}_A \in B_2), \end{aligned}$$

and thus X and $\mathbb{I}_A$ are independent.

For the converse implication let $A \in \mathcal{G}$, $B_1 \in \mathcal{B}$, and $B_2 \in \mathcal{B}$ such that $1 \in B_2$, $0 \notin B_2$. Since X and $\mathbb{I}_A$ are independent, by Theorem 2.11 we can write

$$P(X \in B_1, A) = P(\{X \in B_1\} \cap \{\mathbb{I}_A \in B_2\})$$
$$= P(X \in B_1)P(\mathbb{I}_A \in B_2) = P(X \in B_1)P(A).$$

It follows that $\mathcal{F}_X$ and $\mathcal{G}$ are independent.   ▲

Remark 2.8. Let X be a random variable and $\mathbb{Y} = (Y_1, \ldots Y_n)$ be a random vector, where $n \in \mathbb{N}^*$.
(1) If X is independent of $\mathcal{F}_{Y_1,\ldots,Y_n}$ and $g = (g_1, \ldots, g_m)$ is a $\mathcal{B}^n/\mathcal{B}^m$ measurable function, where $m \in \mathbb{N}^*$, then X is independent of $\mathcal{F}_{g_1(\mathbb{Y}),\ldots,g_m(\mathbb{Y})}$. This follows by observing that $\mathcal{F}_{g_1(\mathbb{Y}),\ldots,g_m(\mathbb{Y})} \subseteq \mathcal{F}_{Y_1,\ldots,Y_n}$. Note that, in particular, if X is independent of $\mathcal{F}_{Y_1,\ldots,Y_n}$ and g is a $\mathcal{B}^n/\mathcal{B}$ measurable function, then X and $g(Y_1, \ldots, Y_n)$ are independent random variables.
(2) If $X, Y_1, \ldots Y_n$ are independent, then X is independent of $\mathcal{F}_{Y_1,\ldots,Y_n}$, by Theorem 2.12.   ▼

Theorem 2.13. (1) *Let $X_1, \ldots, X_n$ be discrete random variables. They are independent if and only if*

$$P(X_1 = x_1) \cdot \ldots \cdot P(X_n = x_n) = P(X = x_1, \ldots, X_n = x_n)$$

for all $x_1, \ldots, x_n \in \mathbb{R}$.
(2) *Let $(X_1, \ldots, X_n)$ be a continuous random vector. Then $X_1, \ldots, X_n$ are independent if and only if*

$$f_{X_1}(x_1) \cdot \ldots \cdot f_{X_n}(x_n) = f_{X_1,\ldots,X_n}(x_1, \ldots, x_n)$$

for a.e. $(x_1, \ldots, x_n) \in \mathbb{R}^n$, where $f_{X_1,\ldots,X_n}$ is a joint density function of $(X_1, \ldots, X_n)$ and f_{X_i} is a density function of X_i for $i \in \{1, \ldots, n\}$.

Example 2.13. Let $f : \mathbb{R}^n \to \mathbb{R}^n$ be given by

$$f(x_1, \ldots, x_n) = \begin{cases} \alpha \exp\{-x_1 - \ldots - x_n\}, & \text{if } x_1, \ldots, x_n > 0 \\ 0, & \text{otherwise}, \end{cases}$$

where $\alpha > 0$ is fixed. First, we find α such that f is a joint density function of a continuous random vector $\mathbb{X} = (X_1, \ldots, X_n)$. Next, we determine the joint distribution function of $\mathbb{X}$. Finally, we compute the probability

$$P\left(\bigcap_{i=1}^{n} \{1 \leq X_i \leq 2\}\right).$$

Solution: We have

$$1 = \int_{-\infty}^{\infty} \cdots \int_{-\infty}^{\infty} f(t_1, \ldots, t_n) dt_1 \ldots dt_n$$

$$= \alpha \int_{0}^{\infty} \cdots \int_{0}^{\infty} \exp\{-t_1 - \ldots - t_n\} dt_1 \ldots dt_n = \alpha.$$

The joint distribution function of $\mathbb{X}$ is

$$F(x_1, \ldots, x_n) = \int_{-\infty}^{x_1} \cdots \int_{-\infty}^{x_n} f(t_1, \ldots, t_n) dt_1 \ldots dt_n$$

$$= \int_{0}^{x_1} \cdots \int_{0}^{x_n} \exp\{-t_1 - \ldots - t_n\} dt_1 \ldots dt_n$$

$$= (1 - e^{-x_1}) \cdot \ldots \cdot (1 - e^{-x_n}),$$

if $x_1, \ldots, x_n > 0$, and $F(x_1, \ldots, x_n) = 0$, otherwise.

Next, we present two methods to compute $P\left(\bigcap_{i=1}^{n}\{1 \le X_i \le 2\}\right)$. For the first one, by Theorem 2.7-(1), we have

$$P\left(\bigcap_{i=1}^{n}\{1 \le X_i \le 2\}\right) = \sum_{k=0}^{n}(-1)^k C(n, k)\left(1 - e^{-1}\right)^k\left(1 - e^{-2}\right)^{n-k}$$

$$= (e^{-1} - e^{-2})^n.$$

For the second one, we note that $X_1, \ldots, X_n$ are independent. Indeed, for every $i \in \{1, \ldots, n\}$ the function $F_i : \mathbb{R} \to \mathbb{R}$ given by

$$F_i(x_i) = \lim_{\substack{x_j \to \infty \\ j=\overline{1,n}, j \neq i}} F(x_1, \ldots, x_n) = \begin{cases} 1 - e^{-x_i}, & x_i > 0 \\ 0, & \text{otherwise,} \end{cases}$$

is, in view of Theorem 2.10-(1), the distribution function of X_i, and thus

$$F(x_1, \ldots, x_n) = F_1(x_1) \cdot \ldots \cdot F_n(x_n) \text{ for all } x_1, \ldots, x_n \in \mathbb{R}.$$

Hence, by Theorem 2.11, we have

$$P\left(\bigcap_{i=1}^{n}\{1 \le X_i \le 2\}\right) = \prod_{i=1}^{n} P(1 \le X_i \le 2) = (e^{-1} - e^{-2})^n. \quad \blacktriangle$$

Definition 2.15. The random variables $X_1, \ldots, X_n$ are called **random sample** corresponding to the random variable X, if $X_1, \ldots, X_n$ are independent random variables and they are identically distributed, having the same distribution as X. $\qquad \blacklozenge$

Definition 2.16. The **empirical cumulative distribution function** corresponding to a random sample $X_1, \ldots, X_n$ of the random variable X is the function $\hat{F}_n : \mathbb{R} \to [0, 1]$ defined as

$$\hat{F}_n(x) = \frac{1}{n} \sum_{i=1}^{n} \mathbb{I}_{\{X_i \leq x\}}. \qquad \blacklozenge$$

Remark 2.9. (1) Let $X_1, \ldots, X_n$ be a random sample corresponding to a random variable X, which has the distribution function F. The empirical distribution function $\hat{F}_n$ has the following property due to Glivenko–Cantelli (see [Gut (2005), Theorem 9.2, p. 306])

$$P\left(\lim_{n \to \infty} \sup_{x \in \mathbb{R}} \left| \hat{F}_n(x) - F(x) \right| = 0 \right) = 1.$$

Note that the empirical distribution function approximates the theoretical (cumulative) distribution function, if the amount of data is large enough. The plot of the empirical distribution function can be visually compared to known distribution functions to verify if the data eventually came from one of the frequently used distributions. The knowing of the underlying distribution is helpful for statistical inferences.

(2) Very often, in examples and problems, we shall generate arrays of random numbers to simulate a random sample of a certain random variable. To represent graphically these arrays we shall use the function `ecdf`, which approximates the empirical cumulative function corresponding to our sample, and the function `histogram`, which approximates (by using relative frequencies) the probability mass function, if the random variable of our sample is discrete, respectively the density function, if the random variable of our sample is continuous. For more details, see the documentations of the Matlab functions in `help`. $\qquad \blacktriangledown$

In what follows we give the following basic theorem regarding the transformation of continuous random vectors by using a bijective (one-to-one) function (see also [Hesse (2003), Satz 2.7.4, p. 71], [Rudin (1987), Theorem 7.26, p. 153]).

Theorem 2.14. (Transforming Random Vectors)
Consider a bijective function $g : S \to g(S)$, where $S \subseteq \mathbb{R}^n$ and $g(S) \subseteq \mathbb{R}^n$ is open. Assume that the inverse $g^{-1} = h = (h_1, \ldots, h_n)$ has continuous

partial derivatives on $g(S)$ and denote by J_h the Jacobian matrix:

$$J_h = \begin{pmatrix} \dfrac{\partial h_1}{\partial y_1} & \dfrac{\partial h_1}{\partial y_2} & \cdots & \dfrac{\partial h_1}{\partial y_n} \\ \dfrac{\partial h_2}{\partial y_1} & \dfrac{\partial h_2}{\partial y_2} & \cdots & \dfrac{\partial h_2}{\partial y_n} \\ \vdots & \vdots & \ddots & \vdots \\ \dfrac{\partial h_n}{\partial y_1} & \dfrac{\partial h_n}{\partial y_2} & \cdots & \dfrac{\partial h_n}{\partial y_n} \end{pmatrix}.$$

If $\mathbb{X}$ is a continuous random vector, having a joint density function $f_{\mathbb{X}}$, such that $\mathbb{X} \in S$ a.s. and $\mathbb{Y}$ is random vector where $\mathbb{Y} = g(\mathbb{X})$ a.s., then $\mathbb{Y}$ is a continuous random vector and the function $f_{\mathbb{Y}} : \mathbb{R}^n \to \mathbb{R}$ defined by

$$f_{\mathbb{Y}}(\mathbb{y}) = \begin{cases} f_{\mathbb{X}}(h_1(\mathbb{y}), ..., h_n(\mathbb{y}))|\det(J_h(\mathbb{y}))|, & \text{if } \mathbb{y} \in g(S) \\ 0, & \text{if } \mathbb{y} \notin g(S) \end{cases}$$

is a joint density function of $\mathbb{Y}$.

Example 2.14. If $\mathbb{X}$ is a continuous random vector and $B \in \mathcal{M}_{n \times n}(\mathbb{R})$ is an invertible matrix and $\mathbb{b} \in \mathbb{R}^n$, then for the random vector $\mathbb{Y} = \mathbb{X}B + \mathbb{b}$ the function

$$f_{\mathbb{Y}}(\mathbb{y}) = \frac{1}{|\det(B)|} f_{\mathbb{X}}\big((\mathbb{y} - \mathbb{b})B^{-1}\big), \ \mathbb{y} \in \mathbb{R}^n$$

is a joint density function, where $f_{\mathbb{X}}$ is a joint density function of $\mathbb{X}$. This is deduced by using Theorem 2.14 for the affine function $g : \mathbb{R}^n \to \mathbb{R}^n$ $g(\mathbb{x}) = \mathbb{x}B + \mathbb{b}$ with $\mathbb{x} \in \mathbb{R}^n$ which has the inverse function $h : \mathbb{R}^n \to \mathbb{R}^n$ defined by $h(\mathbb{y}) = (\mathbb{y} - \mathbb{b})B^{-1}$ for $\mathbb{y} \in \mathbb{R}^n$. In this case, $J_h = (B^{-1})^T$, hence

$$\det(J_h) = \det\left((B^{-1})^T\right) = \frac{1}{\det(B)}. \qquad \blacktriangle$$

Example 2.15. Let $\mathbb{X} = (X_1, X_2)$ be a continuous random vector and let f_{X_1, X_2} be a joint density function of $\mathbb{X}$. Density functions of the sum $X_1 + X_2$, the product $X_1 \cdot X_2$ and the quotient $\dfrac{X_1}{X_2}$ (when $X_2(\omega) \neq 0$ for all $\omega \in \Omega$) of X_1 and X_2 are:

$$f_{X_1+X_2}(z) = \int_{\mathbb{R}} f_{X_1,X_2}(u, z - u)du, \ z \in \mathbb{R},$$

$$f_{X_1 \cdot X_2}(z) = \int_{\mathbb{R}^*} \frac{1}{|u|} f_{X_1,X_2}\left(u, \frac{z}{u}\right)du, \ z \in \mathbb{R},$$

$$f_{\frac{X_1}{X_2}}(z) = \int_{\mathbb{R}} |v| f_{X_1,X_2}(vz,v)dv, \quad z \in \mathbb{R}.$$

Proof: Denote $\mathbb{X} = (X_1, X_2)$ and $B = \begin{pmatrix} 1 & 1 \\ 1 & 0 \end{pmatrix}$. Then $B^{-1} = \begin{pmatrix} 0 & 1 \\ 1 & -1 \end{pmatrix}$ and $\det(B) = -1$. Consider the affine transformation $g : \mathbb{R}^2 \to \mathbb{R}^2$ defined by

$$g(x_1, x_2) = (x_1 + x_2, x_1) = (x_1, x_2)B, \quad (x_1, x_2) \in \mathbb{R}^2,$$

and let $h = g^{-1}$, i.e.,

$$h(y_1, y_2) = (y_2, y_1 - y_2) = (y_1, y_2)B^{-1}, \quad (y_1, y_2) \in \mathbb{R}^2.$$

It follows by Example 2.14 that

$$f_{X_1+X_2,X_1}(y_1, y_2) = f_{X_1,X_2}(y_2, y_1 - y_2), \quad (y_1, y_2) \in \mathbb{R}^n$$

is a density function of $\mathbb{Y} = (X_1 + X_2, X_1)$. By Theorem 2.9, the following function

$$f_{X_1+X_2}(y_1) = \int_{\mathbb{R}} f_{X_1,X_2}(y_2, y_1 - y_2)dy_2, \quad y_1 \in \mathbb{R}$$

is a density function of $X_1 + X_2$. Moreover, if X_1 and X_2 are independent, then we have

$$f_{X_1+X_2}(y_1) = \int_{\mathbb{R}} f_{X_1}(y_2) f_{X_2}(y_1 - y_2)dy_2, \quad y_1 \in \mathbb{R},$$

where f_{X_i} is a density function of X_i for $i \in \{1, 2\}$.

For the product of X_1 and X_2, consider $g : \mathbb{R}^* \times \mathbb{R} \to \mathbb{R} \times \mathbb{R}^*$ defined by

$$g(x_1, x_2) = (x_1 \cdot x_2, x_1), \quad (x_1, x_2) \in \mathbb{R}^* \times \mathbb{R},$$

and let $h = g^{-1}$, i.e.,

$$h(y_1, y_2) = \left(y_2, \frac{y_1}{y_2} \right), \quad (y_1, y_2) \in \mathbb{R} \times \mathbb{R}^*.$$

Since $P(X_1 = 0) = 0$, we have $(X_1, X_2) \in \mathbb{R}^* \times \mathbb{R}$ a.s. and $(X_1 \cdot X_2, X_1) = g(X_1, X_2)$ a.s. Hence, by Theorem 2.14, the following is a density function of $\mathbb{Y} = (X_1 \cdot X_2, X_1)$

$$f_{X_1 \cdot X_2, X_1}(y_1, y_2) = \begin{cases} \frac{1}{|y_2|} f_{X_1,X_2}\left(y_2, \frac{y_1}{y_2} \right), & \text{if } (y_1, y_2) \in \mathbb{R} \times \mathbb{R}^* \\ 0, & \text{otherwise.} \end{cases}$$

By Theorem 2.9

$$f_{X_1 \cdot X_2}(y_1) = \int_{\mathbb{R}^*} \frac{1}{|y_2|} f_{X_1,X_2}\left(y_2, \frac{y_1}{y_2} \right) dy_2, \quad y_1 \in \mathbb{R}$$

is a density function of $X_1 \cdot X_2$. Moreover, if X_1 and X_2 are independent, then we have

$$f_{X_1 \cdot X_2}(y_1) = \int_{\mathbb{R}^*} \frac{1}{|y_2|} f_{X_1}(y_2) f_{X_2}\left(\frac{y_1}{y_2}\right) dy_2, \quad y_1 \in \mathbb{R}.$$

For the quotient of X_1 and X_2, consider $g : \mathbb{R} \times \mathbb{R}^* \to \mathbb{R} \times \mathbb{R}^*$ defined by

$$g(x_1, x_2) = \left(\frac{x_1}{x_2}, x_2\right), \quad (x_1, x_2) \in \mathbb{R} \times \mathbb{R}^*,$$

and let $h = g^{-1}$, i.e.,

$$h(y_1, y_2) = (y_1 \cdot y_2, y_2), \quad (y_1, y_2) \in \mathbb{R} \times \mathbb{R}^*.$$

We use Theorem 2.14 to obtain that

$$f_{\frac{X_1}{X_2}, X_2}(y_1, y_2) = |y_2| f_{X_1, X_2}(y_1 \cdot y_2, y_2), \quad (y_1, y_2) \in \mathbb{R} \times \mathbb{R}^*$$

is a density function of $\mathbb{Y} = \left(\frac{X_1}{X_2}, X_2\right)$. By Theorem 2.9

$$f_{\frac{X_1}{X_2}}(y_1) = \int_{\mathbb{R}} |y_2| f_{X_1, X_2}(y_1 \cdot y_2, y_2) dy_2, \quad y_1 \in \mathbb{R}$$

is a density function of $\dfrac{X_1}{X_2}$. Moreover, if X_1 and X_2 are independent, then we have

$$f_{\frac{X_1}{X_2}}(y_1) = \int_{\mathbb{R}} |y_2| f_{X_1}(y_1 \cdot y_2) f_{X_2}(y_2) dy_2, \quad y_1 \in \mathbb{R}. \qquad \blacktriangle$$

2.3.1 Solved Problems

Problem 2.3.1. Choose uniformly and independently two random numbers in the interval $[0, 1]$. Compute the density functions of their sum and of their product.

Solution 2.3.1: Let X and Y be independent random variables having uniform distribution on the interval $[0, 1]$ (see Section 2.5.1). Then

$$f_X(x) = f_Y(x) = \begin{cases} 1, & \text{if } x \in [0, 1] \\ 0, & \text{if } x \notin [0, 1] \end{cases}$$

are density funtions of X, respectively Y. By Example 2.15

$$f_{X+Y}(z) = \int_{-\infty}^{\infty} f_X(u) f_Y(z - u) du = \int_{0}^{1} f_Y(z - u) du, \quad z \in \mathbb{R}$$

is a density function of $X + Y$.

For $0 \le z \le 1$ we have $f_{X+Y}(z) = \int_0^z 1 dy = z$, for $1 < z \le 2$ we have

$$f_{X+Y}(z) = \int_{z-1}^1 1 dy = 2 - z,$$ while for $z < 0$ or $z > 2$ we have $f_{X+Y}(z) = 0$.

Hence, a density function of $X + Y$ is

$$f_{X+Y}(z) = \begin{cases} z, & \text{if } 0 \le z \le 1 \\ 2 - z, & \text{if } 1 < z \le 2 \\ 0, & \text{otherwise.} \end{cases}$$

A density function of $X \cdot Y$ is obtained by using Example 2.15:

$$f_{X \cdot Y}(z) = \int_{\mathbb{R}^*} \frac{1}{|u|} f_X(u) f_Y\left(\frac{z}{u}\right) du = \int_0^1 \frac{1}{u} f_Y\left(\frac{z}{u}\right) du \quad \text{for } z \in \mathbb{R}.$$

So,

$$f_{X \cdot Y}(z) = \begin{cases} -\log z, & \text{if } 0 < z < 1 \\ 0, & \text{otherwise.} \end{cases}$$

Problem 2.3.2.

Let $X_i = (X_{i1}, X_{i2})$, $i \in \{1, 2\}$, be independent vectors with the same joint probability mass function given by the contingency table:

X_{i1} \ X_{i2}	-1	1
-1	0.5	0.2
1	0.2	0.1

1) Using the **randsample** function in some simulations in Matlab, estimate the value of $P(X_{11} + X_{22} < X_{12} + X_{21})$.

2) Find the exact value of $P(X_{11} + X_{22} < X_{12} + X_{21})$.

Solution 2.3.2: 1)

```
function p=contingency(N)
count=0;
for i=1:N
  x=randsample({[-1;-1],[-1;1],[1;-1],[1;1]},2,true,[0.5,0.2,0.2,0.1]);
  if x{1}(1)+x{2}(2)<x{1}(2)+x{2}(1)
    count=count+1;
  end
end
p=count/N;

>> contingency(10000)
ans = 0.2823
```

2) We note that $X_{11} + X_{22} < X_{12} + X_{21}$ if and only if

$$\begin{pmatrix} X_1 \\ X_2 \end{pmatrix} \in \left\{ \begin{pmatrix} -1 & 1 \\ 1 & -1 \end{pmatrix}, \begin{pmatrix} -1 & 1 \\ -1 & -1 \end{pmatrix}, \begin{pmatrix} -1 & -1 \\ 1 & -1 \end{pmatrix}, \begin{pmatrix} 1 & 1 \\ 1 & -1 \end{pmatrix}, \begin{pmatrix} -1 & 1 \\ 1 & 1 \end{pmatrix} \right\}.$$

Hence,

$$P(X_{11} + X_{22} < X_{12} + X_{21})$$
$$= 0.2 \cdot 0.2 + 0.2 \cdot 0.5 + 0.5 \cdot 0.2 + 0.1 \cdot 0.2 + 0.2 \cdot 0.1 = 0.28 .$$

2.4 Discrete Distributions

In this section, some classic discrete distributions are given by their (joint) probability mass functions. Also, we point out the corresponding Matlab functions (whenever they are available in Statistics and Machine Learning Toolbox) for:

- (joint) probability mass functions, which have the extension pdf
- (joint) cumulative distribution functions, which have the extension cdf
- quantile functions, which have the extension inv
- random number generators, which have the extension rnd.

2.4.1 *Discrete Uniform Distribution*

A random variable X has **discrete uniform distribution** with parameter $n \in \mathbb{N}^*$, if its probability mass function is

$$P(X = k) = \frac{1}{n} \quad \text{for } k \in \{1, \ldots, n\} .$$

Matlab functions: unidpdf, unidcdf, unidinv, unidrnd.
Notation: $X \sim Unid(n)$, $n \in \mathbb{N}^*$.

2.4.2 *Bernoulli Distribution*

Consider an experiment that has only two possible outcomes, which we classify as either success or failure and whose probabilities remain the same throughout the independent repetitions of the experiment, called **Bernoulli trials**. Let X be a random variable such that $X = 1$ when the outcome is success, and $X = 0$ when it is failure. If the probability of success is $p \in (0, 1)$, then X has **Bernoulli distribution** with parameter p. Its probability mass function is

$$P(X = 0) = 1 - p \quad \text{and} \quad P(X = 1) = p.$$

Notation: $X \sim Ber(p)$, $p \in (0, 1)$.

2.4.3   Binomial Distribution

Given n Bernoulli trials with probability of success p and probability of failure $1 - p$, let X denote the total number of successes in the n trials, where $n \in \mathbb{N}^*$ and $p \in (0, 1)$. Then X has **binomial distribution** with parameters n and p. Its probability mass function is

$$P(X = k) = C(n, k)p^k(1 - p)^{n-k} \quad \text{for } k \in \{0, 1, \ldots, n\}.$$

Matlab functions: `binopdf`, `binocdf`, `binoinv`, `binornd`.
Notation: $X \sim Bino(n, p)$, $n \in \mathbb{N}^*, p \in (0, 1)$.

Remark 2.10. (1) The number $C(n, k)p^k(1 - p)^{n-k}$ represents the coefficient of x^k in the binomial expansion $(px + 1 - p)^n$ for $k \in \{0, 1, \ldots, n\}$.
(2) If $X \sim Bino(1, p)$, then $X \sim Ber(p)$. In particular, we note that we can generate random numbers that follow the $Ber(p)$ distribution by using the `binornd` function, e.g., the following call generates a 2×3 matrix of random numbers following the $Ber(0.5)$ distribution:

>> **binornd**(1,0.5,2,3)
ans = 0 1 1
     1 0 0

▼

Example 2.16. In each scanning cycle a radar registers the presence of an object with probability 0.9. The probability that the radar detects an object at least once in 4 scanning cycles is

$$\sum_{k=1}^{4} C(4, k)0.9^k 0.1^{4-k} = 1 - 0.1^4 = 0.9999. \qquad \blacktriangle$$

2.4.4   Multinomial Distribution

Consider $n \in \mathbb{N}^*$ independent trials such that each trial can have several possible mutually exclusive outcomes $O_1, \ldots, O_k$ ($k \in \mathbb{N}^*$) with $P(O_i) = p_i \in (0, 1)$, $i \in \{1, \ldots, k\}$. Obviously, $p_1 + \cdots + p_k = 1$. Denote by X_i the number occurrences of O_i, $i \in \{1, \ldots, k\}$. Then $(X_1, \ldots, X_k)$ has **multinomial distribution** with parameters n, $p_1, \ldots, p_k$. Its joint probability mass function is given by

$$P(X_1 = n_1, \ldots, X_k = n_k) = \frac{n!}{n_1! n_2! \ldots n_k!} p_1^{n_1} p_2^{n_2} \ldots p_k^{n_k}$$

which is the probability that O_i occurs n_i times in n trials for $n_i \in \mathbb{N}$, $i \in \{1, \ldots, k\}$, $n_1 + \cdots + n_k = n$.

Matlab functions: `mnpdf`, `mnrnd`.
Notation: $(X_1, \ldots, X_k) \sim Multino(n, p_1, \ldots, p_k)$, $n \in \mathbb{N}^*$, $p_1, \ldots, p_k \in (0, 1)$, $p_1 + \ldots + p_k = 1$.

Remark 2.11. The number $\dfrac{n!}{n_1! n_2! \ldots n_k!} p_1^{n_1} p_2^{n_2} \ldots p_k^{n_k}$ represents the coefficient of $x_1^{n_1} \cdot \ldots \cdot x_k^{n_k}$ in the expansion of $(p_1 x_1 + \ldots + p_k x_k)^n$. ▼

Example 2.17. Suppose that an urn contains two red marbles, one yellow marble and three blue marbles. Six marbles are drawn randomly with replacement from the urn (i.e., before the next marble is drawn, each drawn marble is put back into the urn). The probability that there are three red marbles, two yellow marbles and one blue marble drawn is $\dfrac{6!}{3!2!1!} \cdot \dfrac{1}{3^3} \cdot \dfrac{1}{6^2} \cdot \dfrac{1}{2}$.

▲

2.4.5 *Hypergeometric Distribution*

A random variable X has **hypergeometric distribution** with parameters $n, n_1, n_2 \in \mathbb{N}^*$ and $n \leq n_1 + n_2$, if its probability mass function is

$$P(X = k) = \frac{C(n_1, k)C(n_2, n - k)}{C(n_1 + n_2, n)} \quad \text{for } k \in \{0, 1, \ldots, \min\{n, n_1\}\}.$$

Matlab functions: `hygepdf`, `hygecdf`, `hygeinv`, `hygernd`.
Notation: $X \sim Hyge(n, n_1, n_2)$, $n, n_1, n_2 \in \mathbb{N}^*, n \leq n_1 + n_2$.

Remark 2.12. Let $n, n_1, n_2 \in \mathbb{N}^*$ be such that $n \leq n_1 + n_2$. In a box there are n_1 white balls and n_2 black balls of the same size. We randomly draw a ball from the box and do not place the ball back into the box (i.e., we sample without replacement). If X denotes the number of white balls in n repetitions of this experiment, then $X \sim Hyge(n, n_1, n_2)$.

Example 2.18. A team of 5 persons is randomly selected from a group of 10 women and 11 men. The probability that more women than men are selected is

$$\frac{C(10, 5)C(11, 0) + C(10, 4)C(11, 1) + C(10, 3)C(11, 2)}{C(21, 5)}.$$

▲

2.4.6 *Multivariate Hypergeometric Distribution*

Consider k classes $\mathcal{C}_1, \ldots, \mathcal{C}_k$ of items, $k \in \mathbb{N}^*$, such that class $\mathcal{C}_i$ has n_i elements, $n_i \in \mathbb{N}^*$, $i \in \{1, \ldots, k\}$. We draw randomly without replacement n items, $n \in \mathbb{N}^*$, $n \leq n_1 + \ldots + n_k$. Denote by X_i the number of items drawn from class C_i, $i \in \{1, \ldots, k\}$. Then $(X_1, \ldots, X_n)$ has **multivariate hypergeometric distribution** with parameters $n, n_1, \ldots, n_k$. Its joint probability mass function is given by

$$P(X_1 = m_1, \ldots, X_k = m_k) = \frac{C(n_1, m_1) \cdot \ldots \cdot C(n_k, m_k)}{C(n_1 + \cdots + n_k, m_1 + \cdots + m_k)},$$

which is the probability that there are m_i drawn items belonging to class C_i for $m_i \in \{0, 1, \ldots, n_i\}$, $i \in \{1, \ldots, k\}$, $m_1 + \ldots + m_k = n$.

Notation: $(X_1, \ldots, X_k) \sim MVHyge(n, n_1, \ldots, n_k)$, $n, n_1, \ldots, n_k \in \mathbb{N}^*, n \leq n_1 + \ldots + n_k$.

Example 2.19. A standard 52-card deck is shuffled and a hand of 5 cards is dealt to a poker player. We can simulate in Matlab this experiment as follows:

```
>> deck=[repmat({'spades'},1,13),repmat({'diamond'},1,13),...
         repmat({'clubs'},1,13),repmat({'hearts'},1,13)];
>> randsample(deck,5)
ans = 1x5 cell array
   {'spades'} {'diamond'} {'spades'} {'hearts'} {'diamond'}
```

The probability that the player gets 2 cards of spades, 1 card of diamonds and 2 cards of clubs is

$$\frac{C(13, 2)C(13, 1)C(13, 2)C(13, 0)}{C(52, 5)},$$

which can be computed in Matlab, using the chain rule from Theorem 1.7, as follows:

```
>>hygepdf(2,52,13,5)*hygepdf(1,39,13,3)*hygepdf(2,26,13,2)
ans = 0.0304
```

or by using

```
>>nchoosek(13,2)*nchoosek(13,1)*nchoosek(13,2)/nchoosek(52,5)
ans = 0.0304
```

▲

2.4.7   *Poisson Distribution*

A random variable X has **Poisson distribution** with parameter $\lambda > 0$, if its probability mass function is

$$P(X = k) = \frac{\lambda^k}{k!} e^{-\lambda} \quad \text{for } k \in \{0, 1, 2, \ldots\}.$$

Matlab functions: `poisspdf`, `poisscdf`, `poissinv`, `poissrnd`.
Notation: $X \sim Poiss(\lambda)$, $\lambda > 0$.

Example 2.20. The number of failures X in a certain electricity power station during a period of 120 days is modeled by using a Poisson distribution with parameter $\lambda = 0.02$. The probability that there are at least two failures during the 120 days is

$$1 - P(X \leq 1) = 1 - 1.02 \cdot \exp\{-0.02\} \approx 0.0002. \qquad \blacktriangle$$

2.4.8 *Negative Binomial Distribution*

Consider an infinite sequence of Bernoulli trials in which the outcome of any trial is either success with probability p or failure with probability $1-p$ and let X be the random variable denoting the total number of failures that occur before the nth success, where $p \in (0,1)$ and $n \in \mathbb{N}^*$. Then X has **negative binomial distribution** with parameters n and p. Its probability mass function is

$$P(X = k) = \mathrm{C}(n + k - 1, k)p^n(1 - p)^k \quad \text{for } k \in \mathbb{N}.$$

Matlab functions: `nbinpdf`, `nbincdf`, `nbininv`, `nbinrnd`.
Notation: $X \sim NBin(n, p)$, $n \in \mathbb{N}^*, p \in (0, 1)$.

Remark 2.13. Consider the following generalization: for $k \in \{0, 1, \dots\}$ and $a \in \mathbb{R}^*$ define

$$\mathrm{C}(a, k) = \begin{cases} 1, & \text{if } k = 0 \\ \dfrac{a(a - 1)\dots(a - k + 1)}{k!}, & \text{if } k \in \mathbb{N}^*. \end{cases}$$

We write the Taylor series expansion

$$(1 + t)^a = \sum_{k=0}^{\infty} \mathrm{C}(a, k)t^k \quad \text{for } |t| < 1.$$

If we take $a = -n$ and $t = -(1 - p)x$, then by some computations we have

$$\frac{p^n}{\left(1 - (1 - p)x\right)^n} = \sum_{k=0}^{\infty} \mathrm{C}(n + k - 1, k)p^n(1 - p)^k x^k.$$

Hence, for $k \in \mathbb{N}$ the number $\mathrm{C}(n + k - 1, k)p^n(1 - p)^k$ represents the coefficient of x^k in the series expansion of

$$\frac{p^n}{\left(1 - (1 - p)x\right)^n},$$

which is convergent if $|(1 - p)x| < 1$. $\qquad \blacktriangledown$

Example 2.21. We consider the random experiment of tossing a fair coin. The probability that it gives a head in one toss is 0.5. We compute the probability for getting seven times tails before we get the third head. Let X denote the number of tails before we get the third head. Then $X \sim NBin(3, 0.5)$ and thus

$$P(X = 7) = C(9, 7)(0.5)^{10} \approx 0.0352. \qquad \blacktriangle$$

2.4.9 Geometric Distribution

Consider an infinite sequence of Bernoulli trials in which the outcome of any trial is either success with probability p or failure with probability $1 - p$, where $p \in (0, 1)$, and let X be a random variable denoting the total number of failures that occur before the first success. Then X has **geometric distribution** with parameter p. Its probability mass function is

$$P(X = k) = p(1 - p)^k \quad \text{for } k \in \mathbb{N}.$$

Note that $X \sim NBin(1, p)$.
Matlab functions: `geopdf`, `geocdf`, `geoinv`, `geornd`.
Notation: $X \sim Geo(p)$, $p \in (0, 1)$.

Remark 2.14. For $k \in \mathbb{N}$ the number $p(1 - p)^k$ represents the coefficient of x^k in the geometric series

$$\frac{p}{1 - (1 - p)x} = \sum_{k=0}^{\infty} p(1 - p)^k x^k,$$

which is convergent if $|(1 - p)x| < 1$. $\qquad \blacktriangledown$

Example 2.22. A binary code is received one bit at a time such that the probability that the next bit is the last one is 0.6. Let X be the number of bits received before the last one is received. Then $X \sim Geo(0.6)$ and the probability that the received code has 5 bits is

$$P(X = 4) = 0.6 \cdot 0.4^4 \approx 0.0154. \qquad \blacktriangle$$

2.4.10 Solved Problems

Problem 2.4.1. Let $X \sim NBin(n, p)$, $n \in \mathbb{N}^*, p \in (0, 1)$. Give a recursion formula for the probabilities $P(X = k)$, $k \in \mathbb{N}$.

Solution 2.4.1: Let $k \in \mathbb{N}$. It holds $P(X = 0) = p^n$. We write successively

$$P(X = k + 1) = C(n + k, k + 1)p^n(1 - p)^{k+1}$$

$$= \frac{(n + k)!}{(n - 1)!(k + 1)!}p^n(1 - p)^{k+1}$$

$$= \frac{k + n}{k + 1} \cdot \frac{(n + k - 1)!}{(n - 1)!k!}p^n(1 - p)^{k+1}$$

$$= \frac{k + n}{k + 1}(1 - p)C(k + n - 1, k)p^n(1 - p)^k$$

$$= \frac{k + n}{k + 1}(1 - p)P(X = k).$$

Problem 2.4.2. Johnny buys a bag of gummy bears from a shop. He chooses randomly the bag. The shop is selling two types of bags — each one contains 100 pieces of gummy bears of various flavors/colors:
• type 1: 60% of the bags in the shop have the following proportions: 20% raspberry (red), 25% pineapple (white), 30% orange (orange) and 25% lemon (yellow);
• type 2: 40% of the bags in the shop have the following proportions: 20% pineapple (white), 25% orange (orange), 20% lemon (yellow) and 35% strawberry (green).

Johnny randomly picks 10 gummy bears from the bag. Determine the joint probability mass function of the number of gummy bears picked by Johnny, for each flavor/color, and then write a Matlab function that computes the probability for a set of possible values (given as input). Test the function by computing the probability that Johnny picks 3 pineapple, 4 orange and 3 lemon gummy bears.

Solution 2.4.2: Let X be the vector whose components are the random variables R, W, O, Y and G that give the number of gummy bears picked by Johnny with the corresponding colors. Let $r, w, o, l, g \in \{0, 1, \ldots, 10\}$. Let T_i be the event that Johnny chooses a bag of type $i \in \{1, 2\}$. Taking into account the law of total probability (see Theorem 1.8), we have:

$$P(R = r, W = w, O = o, L = l, G = g)$$
$$= P(R = r, W = w, O = o, L = l, G = g|T_1)P(T_1)$$
$$+ P(R = r, W = w, O = o, L = l, G = g|T_2)P(T_2).$$

Next, we consider the multivariate hypergeometric distribution (see Section 2.4.6). If $r = g = 0$, then

$$P(R = r, W = w, O = o, L = l, G = g)$$
$$= \frac{C(25, w)C(30, o)C(25, l)}{C(100, 10)} \cdot \frac{60}{100} + \frac{C(20, w)C(25, o)C(20, l)}{C(100, 10)} \cdot \frac{40}{100}.$$

If $r > 0$, then

$$P(R{=}r, W{=}w, O{=}o, L{=}l, G{=}g) = \frac{C(20, r)C(25, w)C(30, o)C(25, l)}{C(100, 10)} \cdot \frac{60}{100}.$$

If $g > 0$, then

$$P(R{=}r, W{=}w, O{=}o, L{=}l, G{=}g) = \frac{C(20, w)C(25, o)C(20, l)C(35, g)}{C(100, 10)} \cdot \frac{40}{100}.$$

```
function gummy_bears(r,w,o,l,g)
 if r==0&&g==0
 p=(nchoosek(25,w)*nchoosek(30,o)*nchoosek(25,l))/...
             nchoosek(100,10)*60/100+...
 (nchoosek(20,w)*nchoosek(25,o)*nchoosek(20,l))/...
             nchoosek(100,10)*40/100;
 elseif r>0
 p=(nchoosek(20,r)*nchoosek(25,w)*nchoosek(30,o)*...
             nchoosek(25,l))/nchoosek(100,10)*60/100;
 else
 p=(nchoosek(20,w)*nchoosek(25,o)*nchoosek(20,l)*...
             nchoosek(35,g))/nchoosek(100,10)*40/100;
 end
 fprintf(['Johnny picks %d red, %d white, %d orange, %d yellow,\n'...
          '%d green gummy bears with probability %5.4f.\n'],r,w,o,l,g,p);
end
```

```
>> gummy_bears(0,3,4,3,0)
Johnny picks 0 red, 3 white, 4 orange, 3 yellow,
0 green gummy bears with probability 0.0054.
```

2.5   Continuous Distributions

In this section, some classic continuous distributions are given by their (joint) probability density functions. Also, as in Section 2.4.1, we point out the corresponding Matlab functions (whenever they are available) for:
• (joint) probability density functions, which have the extension `pdf`
• (joint) cumulative distribution functions, which have the extension `cdf`
• quantile functions, which have the extension `inv`
• random number generators, which have the extension `rnd`.

2.5.1 *Uniform Distribution*

Let $a, b \in \mathbb{R}, a < b$. A random variable X has **uniform distribution** on the interval $[a, b]$, if the function

$$f_X(x) = \begin{cases} \dfrac{1}{b-a}, & \text{if } x \in [a, b] \\ 0, & \text{if } x \notin [a, b] \end{cases}$$

is a density function of X. Note that the distribution function of X is

$$F_X(x) = \begin{cases} 0, & \text{if } x < a \\ \dfrac{x-a}{b-a}, & \text{if } a \leq x < b \\ 1, & \text{if } b \leq x. \end{cases}$$

Matlab functions: `unifpdf`, `unifcdf`, `unifinv`, `unifrnd`.
Notation: $X \sim Unif[a, b]$, $a, b \in \mathbb{R}$, $a < b$.

2.5.2 *Multivariate Uniform Distribution*

Let $a_i, b_i \in \mathbb{R}$, $a_i < b_i$, $i \in \{1, \ldots, n\}$, $n \in \mathbb{N}^*$. A random vector $\mathbb{X} = (X_1, \ldots, X_n)$ has n-dimensional **multivariate uniform distribution** on

$$I = [a_1, b_1] \times \ldots \times [a_n, b_n],$$

if the function

$$f_{\mathbb{X}}(x_1, \ldots, x_n) = \begin{cases} \dfrac{1}{(b_1 - a_1) \ldots (b_n - a_n)}, & \text{if } (x_1, \ldots, x_n) \in I \\ 0, & \text{if } (x_1, \ldots, x_n) \notin I \end{cases}$$

is a joint density function of $\mathbb{X}$. The joint distribution function of $\mathbb{X}$ has for $(x_1, \ldots, x_n) \in \mathbb{R}^n$ the form

$$F_{\mathbb{X}}(x_1, \ldots, x_n) = \int_{-\infty}^{x_1} \cdots \int_{-\infty}^{x_n} f(t_1, \ldots, t_n) dt_1 \ldots dt_n$$

$$= \left(\frac{x_1 - a_1}{b_1 - a_1} \right)_* \cdots \left(\frac{x_n - a_n}{b_n - a_n} \right)_*,$$

where we use the notation

$$u_* = \begin{cases} 0, & \text{if } u < 0 \\ u, & \text{if } 0 \leq u < 1 \\ 1, & \text{if } 1 \leq u. \end{cases}$$

Notation: $(X_1, \ldots X_n) \sim MVUnif([a_1, b_1] \times \ldots \times [a_n, b_n])$, $a_i, b_i \in \mathbb{R}$, $a_i < b_i, i \in \{1, \ldots, n\}$.

2.5.3 Normal Distribution

Let $m \in \mathbb{R}$ and $\sigma > 0$. A random variable X has **normal distribution** with parameters m and σ^2, if the function

$$f_X(x) = \frac{1}{\sqrt{2\pi}\sigma} \exp\left\{ -\frac{1}{2}\left(\frac{x-m}{\sigma} \right)^2 \right\} \text{ for } x \in \mathbb{R}$$

is a density function of X. In the case $m = 0$ and $\sigma = 1$, we say X has **standard normal distribution**.
Matlab functions: `normpdf`, `normcdf`, `norminv`, `normrnd`.
Notation: $X \sim N(m, \sigma^2)$, $m \in \mathbb{R}, \sigma > 0$.

2.5.4 Multivariate Normal Distribution

Let $\mathrm{m} \in \mathbb{R}^n$ and let $A \in \mathcal{M}_{n \times n}(\mathbb{R})$ be a real-valued positive definite matrix. A random vector $\mathbb{X} = (X_1, \dots, X_n)$ has n-dimensional **multivariate normal distribution** with parameters m and A, if the function

$$f_{\mathbb{X}}(\mathrm{x}) = \frac{1}{(2\pi)^{n/2}\sqrt{\det(A)}} \exp\left\{ -\frac{1}{2}(\mathrm{x} - \mathrm{m})A^{-1}(\mathrm{x} - \mathrm{m})^T \right\} \text{ for } \mathrm{x} \in \mathbb{R}^n$$

is a joint density function of $\mathbb{X}$. This is the definition of the nondegenerate case of the multivariate normal distribution. The degenarate case of the multivariate normal distribution is discussed in Remark 3.7. In the case $\mathrm{m} = 0_n$ and $A = I_n$, we say $\mathbb{X}$ has n-dimensional **standard multivariate normal distribution**. The function

$$f_{\mathbb{X}}(\mathrm{x}) = \frac{1}{(2\pi)^{n/2}} \exp\left\{ -\frac{1}{2}\mathrm{x}\mathrm{x}^T \right\} = \frac{1}{(2\pi)^{n/2}} \exp\left\{ -\frac{1}{2}\sum_{i=1}^{n} x_i^2 \right\}$$

for $\mathrm{x} = (x_1, \dots, x_n) \in \mathbb{R}^n$ is a joint density function of $\mathbb{X}$.
Matlab functions: `mvnpdf`, `mvncdf`, `mvnrnd`.
Notation: $\mathbb{X} = (X_1, .., X_n) \sim MVN(\mathrm{m}, A)$, $\mathrm{m} \in \mathbb{R}^n$, $A \in \mathcal{M}_{n \times n}(\mathbb{R})$ positive definite matrix.

2.5.5 Gamma Distribution

Let $a, b > 0$. A random variable X has **Gamma distribution** with parameters a and b, if the function

$$f_X(x) = \begin{cases} 0, & \text{if } x \le 0 \\ \dfrac{1}{\Gamma(a)b^a} x^{a-1} \exp\left\{ -\dfrac{x}{b} \right\}, & \text{if } x > 0 \end{cases}$$

is a density function of X, where Γ is Euler's Gamma function (see Appendix A.3).

Matlab functions: `gampdf`, `gamcdf`, `gaminv`, `gamrnd`.

Notation: $X \sim Gamma(a, b)$, $a, b > 0$.

2.5.6 Exponential Distribution

Let $\lambda > 0$. A random variable X has **exponential distribution** with parameter λ, if the function

$$f_X(x) = \begin{cases} 0, & \text{if } x \leq 0 \\ \lambda e^{-\lambda x}, & \text{if } x > 0 \end{cases}$$

is a density function of X. The distribution function of X is

$$F_X(x) = \begin{cases} 0, & \text{if } x \leq 0 \\ 1 - e^{-\lambda x}, & \text{if } x > 0. \end{cases}$$

Note that $X \sim Gamma\left(1, \dfrac{1}{\lambda}\right)$.

Matlab functions: `exppdf`, `expcdf`, `expinv`, `exprnd`.

Notation: $X \sim Exp(\lambda)$, $\lambda > 0$.

2.5.7 χ^2 Distribution

Let $n \in \mathbb{N}^*$ and $\sigma > 0$. A random variable X has **χ^2 distribution** with n degrees of freedom and parameter σ, if the function

$$f_X(x) = \begin{cases} 0, & \text{if } x \leq 0 \\ \dfrac{1}{\Gamma(\frac{n}{2}) 2^{\frac{n}{2}} \sigma^n} x^{\frac{n}{2}-1} \exp\left\{-\dfrac{x}{2\sigma^2}\right\}, & \text{if } x > 0. \end{cases}$$

is a density function of X. Note that $X \sim Gamma\left(\dfrac{n}{2}, 2\sigma^2\right)$.

Matlab functions (for $\sigma = 1$): `chi2pdf`, `chi2cdf`, `chi2inv`, `chi2rnd`.

Notation: $X \sim \chi^2(n, \sigma)$, $n \in \mathbb{N}^*$, $\sigma > 0$.

2.5.8 Beta Distribution

Let $a, b > 0$. A random variable X has **Beta distribution** with parameters a and b, if the function

$$f_X(x) = \begin{cases} \dfrac{1}{B(a, b)} x^{a-1}(1 - x)^{b-1}, & \text{if } x \in [0, 1] \\ 0, & \text{if } x \notin [0, 1] \end{cases}$$

is a density function of X, where B is Euler's Beta function (see Appendix A.3).

Matlab functions: `betapdf, betacdf, betainv, betarnd`.

Notation: $X \sim Beta(a, b)$, $a, b > 0$.

2.5.9   Student Distribution

Let $n \in \mathbb{N}^*$. A random variable X has **Student distribution** (or **T distribution**) with n degrees of freedom, if the function

$$f_X(x) = \frac{\Gamma\left(\frac{n+1}{2}\right)}{\sqrt{n\pi}\,\Gamma\left(\frac{n}{2}\right)} \left(1 + \frac{x^2}{n}\right)^{-\frac{n+1}{2}} \quad \text{for } x \in \mathbb{R}$$

is a density function of X.

Matlab function: `tpdf, tcdf, tinv, trnd`.

Notation: $X \sim T(n)$, $n \in \mathbb{N}^*$.

2.5.10   Snedecor–Fischer Distribution

Let $m, n \in \mathbb{N}^*$. A random variable X has **Snedecor–Fischer distribution** (or **F distribution**) with m and n degrees of freedom, if the function

$$f_X(x) = \begin{cases} 0, & \text{if } x \le 0 \\ \left(\dfrac{m}{n}\right)^{\frac{m}{2}} \dfrac{x^{\frac{m}{2}-1}}{B(\frac{m}{2}, \frac{n}{2})(1 + \frac{m}{n}x)^{\frac{m+n}{2}}}, & \text{if } x > 0 \end{cases}$$

is a density function of X.

Matlab functions: `fpdf, fcdf, finv, frnd`.

Notation: $X \sim F(m, n)$, $m, n \in \mathbb{N}^*$.

2.5.11   Cauchy Distribution

Let $a \in \mathbb{R}$ and $b > 0$. A random variable X has **Cauchy distribution** with parameters a and b, if the function

$$f_X(x) = \frac{1}{\pi b \left(1 + \left(\dfrac{x-a}{b}\right)^2\right)} \quad \text{for } x \in \mathbb{R}$$

is a density function of X. The distribution function of X is

$$F_X(x) = \frac{1}{\pi} \arctan\left(\frac{x-a}{b}\right) + \frac{1}{2} \quad \text{for } x \in \mathbb{R}.$$

Note that, if X has Cauchy distribution with parameters $a = 0$ and $b = 1$, then $X \sim T(1)$.

Notation: $X \sim Cauchy(a, b)$, $a \in \mathbb{R}$, $b > 0$.

Example 2.23. (1) Let $I \subseteq \mathbb{R}$ be an open interval and let $g : I \to \mathbb{R}$ be a strictly monotone, differentiable function with continuous derivative and $0 \notin g'(I)$. Consider X to be a continuous random variable such that $P(X \in I) = 1$ and let $Y = g(X)$. Then, by Theorem 2.14, we get that the function

$$f_Y(y) = \begin{cases} \dfrac{f_X(g^{-1}(y))}{|g'(g^{-1}(y))|}, & \text{if } y \in g(I) \\ 0, & \text{if } y \notin g(I) \end{cases}$$

is a density function of Y.

(2) Let $X \sim Unif[a, b]$, where $0 < a < b$, and let f_X be a density function of X. Let $Y = \log\left(\dfrac{1}{X}\right)$. By Theorem 2.5-(4),

$$P\big(X \in (a, b)\big) = \int_a^b f_X(t) dt = \int_a^b \frac{1}{b - a} dt = 1.$$

In view of (1), we get that

$$f_Y(y) = \begin{cases} \dfrac{e^{-y}}{b - a}, & \text{if } y \in (-\log b, -\log a) \\ 0, & \text{otherwise} \end{cases}$$

is a density function of the random variable Y. ▲

Example 2.24.

(1) Let $\mathbb{X} = (X_1, ..., X_n) \sim MVN(0_n, I_n)$, $A \in \mathcal{M}_{n \times n}(\mathbb{R})$ be a positive definite matrix and $\mathrm{m} \in \mathbb{R}^n$ be a vector. Consider the random vector $\mathbb{Y} = \mathbb{X} A^{\frac{1}{2}} + \mathrm{m}$. Then $\mathbb{Y} \sim MVN(\mathrm{m}, A)$, where $A^{\frac{1}{2}}$ is the square root of the matrix A, see Proposition A.4.

Proof: Let $f_{\mathbb{X}}$ be a joint density function of $\mathbb{X}$. Using Proposition A.4 and Example 2.14 with $B = A^{\frac{1}{2}}$ and $b = \mathrm{m}$ we deduce that

$$f_{\mathbb{Y}}(\mathrm{y}) = \frac{1}{|\det(A^{\frac{1}{2}})|} f_{\mathbb{X}}((\mathrm{y} - \mathrm{m}) A^{-\frac{1}{2}})$$

$$= \frac{1}{(2\pi)^{n/2} \sqrt{\det(A)}} \exp\left\{-\frac{1}{2}(\mathrm{y} - \mathrm{m}) A^{-1} (\mathrm{y} - \mathrm{m})^T\right\}$$

for $\mathrm{y} \in \mathbb{R}^n$ is a joint density function of $\mathbb{Y}$. This is a joint density function of the $MVN(\mathrm{m}, A)$ distribution.

(2) Consider $\mathbb{X} \sim MVN(\mathrm{m}, A)$ to be an n-dimensional random vector, where $\mathrm{m} \in \mathbb{R}^n$ and $A \in \mathcal{M}_{n \times n}(\mathbb{R})$ is a positive definite matrix. Let $B \in \mathcal{M}_{n \times n}(\mathbb{R})$ be an invertible matrix and let $\mathrm{b} \in \mathbb{R}^n$. Then, by Example 2.14 it follows that for $\mathbb{Y} = \mathbb{X}B + \mathrm{b}$ we have $\mathbb{Y} \sim MVN(\mathrm{m}B + \mathrm{b}, B^T AB)$. Note that $B^T AB$ is a positive definite matrix, see Proposition A.3.

(3) If $X \sim N(m, \sigma^2)$, where $m \in \mathbb{R}$, $\sigma > 0$, then for $a, b \in \mathbb{R}$, $b \neq 0$ it follows by the above result (with $n = 1$) that $aX + b \sim N(am + b, a^2 \sigma^2)$. ▲

2.5.12   Solved Problems

Problem 2.5.1. Prove that $\mathbb{X} = (X_1, ..., X_n) \sim MVN(0_n, I_n)$ if and only if $X_1, ..., X_n$ are independent random variables with standard normal distribution.

Solution 2.5.1: If $\mathbb{X} = (X_1, ..., X_n) \sim MVN(0_n, I_n)$, then

$$f_{\mathbb{X}}(\mathrm{x}) = \frac{1}{(2\pi)^{n/2}} \exp\left\{ -\frac{1}{2} \sum_{i=1}^{n} x_i^2 \right\}, \ \mathrm{x} = (x_1, \ldots, x_n) \in \mathbb{R}^n$$

is a joint density function of $\mathbb{X}$ (see Section 2.5.4). By Theorem 2.10, we have that

$$f_{X_i}(x_i) = \frac{1}{(2\pi)^{1/2}} \exp\left\{ -\frac{1}{2} x_i^2 \right\}, \ x_i \in \mathbb{R}^n$$

is a density function of X_i for $i \in \{1, \ldots, n\}$. Hence,

$$f_{\mathbb{X}}(\mathrm{x}) = f_{X_1}(x_1) \cdot \ldots \cdot f_{X_n}(x_n), \ \mathrm{x} = (x_1, \ldots, x_n) \in \mathbb{R}^n.$$

By Theorem 2.13, it follows that $X_1, ..., X_n$ are independent random variables. Moreover, $X_i \sim N(0, 1)$ for each $i \in \{1, \ldots, n\}$, by the above computation of f_{X_i}.

If $X_1, ..., X_n \sim N(0, 1)$ are independent random variables, then, by Theorem 2.13, it follows that

$$f_{\mathbb{X}}(\mathrm{x}) = f_{X_1}(x_1) \cdot \ldots \cdot f_{X_n}(x_n) \ \text{ for a.e. } \mathrm{x} = (x_1, \ldots, x_n) \in \mathbb{R}^n,$$

where $f_{\mathbb{X}}$ is a joint density function of $\mathbb{X}$ and f_{X_i} is a density function of X_i for $i \in \{1, \ldots, n\}$. Hence,

$$f(\mathrm{x}) = \frac{1}{(2\pi)^{n/2}} \exp\left\{ -\frac{1}{2} \sum_{i=1}^{n} x_i^2 \right\} \ \text{ for a.e. } \mathrm{x} = (x_1, \ldots, x_n) \in \mathbb{R}^n.$$

It follows that $\mathbb{X} \sim MVN(0_n, I_n)$.

Problem 2.5.2. Mrs. Y and Mr. X take their dogs for a walk every day in a pocket park. Mrs. Y arrives in the park at 12 and T minutes and walks her dog for 15 minutes. Mr. X arrives in the park at 13 and S minutes and walks his dog for 10 minutes. Suppose that T and S are independent uniformly distributed random variables on the intervals $[30, 50]$, respectively $[-20, 10]$.

1) Estimate by Matlab simulations the probability that Mrs. Y and Mr. X meet in the park when they walk their dogs.

2) Compute the probability that Mrs. Y and Mr. X meet in the park when they walk their dogs.

Solution 2.5.2: 1)

```
function walking_dogs(N)
count=0;
for i=1:N
    Y=unifrnd(30,50);
    X=60+unifrnd(-20,10);
    if max(X+10,Y+15)-min(X,Y)<25
        count=count+1;
    end
end
fprintf(['Mrs. Y and Mr. X have met each other '...
        'in %3.1f%% of the simulations.\n'],count/N*100);
end
```

```
>> walking_dogs(10000)
Mrs. Y and Mr. X have met each other in 49.7% of the simulations.
```

2) Mrs. Y and Mr. X meet in the park when they walk their dogs if and only if $T < 60 + S < T + 15$ or $60 + S < T < 70 + S$. We compute the desired probability:

$$P(T < 60 + S < T + 15) + P(60 + S < T < 70 + S)$$

$$= \int_{30}^{40} \frac{1}{20} \left(\int_{-20}^{t-45} \frac{1}{30} ds \right) dt + \int_{40}^{50} \frac{1}{20} \left(\int_{t-60}^{t-45} \frac{1}{30} ds \right) dt$$

$$+ \int_{-20}^{-10} \frac{1}{30} \left(\int_{60+s}^{50} \frac{1}{20} dt \right) ds$$

$$= \frac{1}{600} \int_{30}^{40} t - 25 \, dt + \frac{1}{600} \int_{40}^{50} 15 \, dt + \frac{1}{600} \int_{-20}^{-10} -10 - s \, ds$$

$$= \frac{100}{600} + \frac{150}{600} + \frac{50}{600} = 0.5,$$

hence there are 50% chances for Mrs. Y and Mr. X to meet in the park when they walk their dogs.

Problem 2.5.3. Let $\lambda > 0$ and let $(X_k)_{k\in\mathbb{N}}$ be a sequence of independent random variables that follow the $Exp(\lambda)$ distribution. For every $k \in \mathbb{N}$ let

$$S_k = X_0 + X_1 + \ldots + X_k.$$

For every $z > 0$ let

$$M_z = \inf\{l \in \mathbb{N} : S_l > z\}$$

and

$$N_z = \sup\{l \in \mathbb{N} : S_l < z\}.$$

1) For $k \in \mathbb{N}$ find a density function of S_k.

2) Taking into account that $\inf \emptyset = \infty$ and $\sup \emptyset = -\infty$, prove that for $z > 0$ we have

$$P(M_z = \infty) = 0 \quad \text{and} \quad P(N_z = -\infty) = e^{-\lambda z}.$$

3) For $z > 0$ and $k \in \mathbb{N}$ find $P(M_z = k)$ and $P(N_z = k)$.

Solution 2.5.3: 1) We prove by induction that $S_k \sim Gamma\left(k+1, \dfrac{1}{\lambda}\right)$ for $k \in \mathbb{N}$. Clearly, $S_0 \sim Gamma\left(1, \dfrac{1}{\lambda}\right)$. Next, let $k \in \mathbb{N}$ and suppose that $S_k \sim Gamma\left(k+1, \dfrac{1}{\lambda}\right)$. By Example 2.15, we have a density function of S_{k+1} given for $z > 0$ by

$$f_{S_{k+1}}(z) = f_{S_k + X_{k+1}}(z) = \int_{\mathbb{R}} f_{S_k}(u) f_{X_{k+1}}(z-u)\,du$$

$$= \int_0^z \frac{\lambda^{k+1}}{k!} \cdot u^k e^{-\lambda z}\,du = \frac{\lambda^{k+1}}{k!} \cdot \lambda e^{-\lambda z} \cdot \frac{z^{k+1}}{k+1},$$

and for $z \leq 0$ by $f_{S_{k+1}}(z) = 0$, where f_{X_k} and f_{S_k} are density functions of X_k, respectively S_k. Hence, $S_{k+1} \sim Gamma\left(k+2, \dfrac{1}{\lambda}\right)$.

2) Let $z > 0$. We note that $\{l \in \mathbb{N} : S_l > z\} = \emptyset$ if and only if the event $\bigcap_{k=0}^{\infty} \{S_k \leq z\}$ occurs. Hence, in view of Theorem 1.6, we get

$$P(M_z = \infty) = P\left(\bigcap_{k=0}^{\infty} \{S_k \leq z\}\right)$$

$$= \lim_{k \to \infty} P(S_k \leq z) = \lim_{k \to \infty} \int_0^z \frac{\lambda^{k+1}}{k!} u^k e^{-\lambda z}\,du$$

$$\leq \lim_{k \to \infty} \frac{\lambda^{k+1}}{k!} \int_0^z u^k\,du = \lim_{k \to \infty} \frac{(\lambda z)^{k+1}}{(k+1)!} = 0.$$

Next, we note that $\{l \in \mathbb{N} : S_l < z\} = \emptyset$ if and only if the event $\{S_0 \geq z\}$ occurs. Hence,

$$P(N_z = -\infty) = P(S_0 \geq z) = e^{-\lambda z}.$$

3) For $k \in \mathbb{N}$ let f_{S_k} be a density function of S_k. Let $z > 0$. We have
$P(M_z = 0) = \int_z^\infty \lambda e^{-\lambda u} du = e^{-\lambda z}$. Next, let $k \in \mathbb{N}^*$. Then

$$P(M_z = k) = P(S_k > z, S_{k-1} \leq z) = P(S_k > z) - P(S_k > z, S_{k-1} > z)$$
$$= P(S_k > z) - P(S_{k-1} > z) = P(S_{k-1} \leq z) - P(S_k \leq z)$$
$$= \int_0^z f_{S_{k-1}}(u)du - \int_0^z f_{S_k}(u)du$$
$$= \int_0^z \frac{\lambda^k}{(k-1)!} \cdot u^{k-1}e^{-\lambda u}du - \frac{\lambda^{k+1}}{k!} \cdot u^k \frac{e^{-\lambda u}}{-\lambda}\bigg|_0^z - \int_0^z \frac{\lambda^k}{(k-1)!} \cdot u^{k-1}e^{-\lambda u}du$$
$$= \frac{(\lambda z)^k}{k!}e^{-\lambda z}.$$

Hence, we have $M_z \sim Poiss(\lambda z)$. Let $k \in \mathbb{N}$. Then

$$P(N_z = k) = P(S_{k+1} \geq z, S_k < z) = P(S_{k+1} \geq z) - P(S_{k+1} \geq z, S_k \geq z)$$
$$= P(S_{k+1} \geq z) - P(S_k \geq z) = P(S_k < z) - P(S_{k+1} < z)$$
$$= \frac{(\lambda z)^{k+1}}{(k+1)!}e^{-\lambda z}.$$

Problem 2.5.4. Let $w_1, \ldots, w_n \in \mathbb{N}^*$, $n \in \mathbb{N}^*$. Consider two experiments, regarding n balls marked with the consecutive integers from 1 to n:
A. Put in an urn w_i balls marked with the number i for every $i \in \{1, \ldots, n\}$. Extract a ball and let X_1 be the number marked on it. Next, discard all balls marked with the number X_1. Extract a new ball from the urn and let X_2 be the number marked on it. Again, discard all balls marked with the number X_2 and continue with the same procedure until the last ball marked with the number X_n is extracted.
B. Place an alarm-clock which is set to ring at a time that follows the $Exp(w_i)$ distribution in front of the ball marked with the number i for every $i \in \{1, \ldots, n\}$. Every time an alarm-clock rings pick the corresponding ball. Let Y_i be the number marked on the ith picked ball for $i \in \{1, \ldots, n\}$.
1) Simulate in Matlab the above experiments and compare the histograms for X_i and Y_i for some $i \in \{1, \ldots, n\}$ given as input.

2) Prove that X_i and Y_i have the same distribution for every $i \in \{1, \ldots, n\}$.

3) Prove that

$$\sum_{\sigma \in \mathcal{S}_n} \frac{1}{\sigma(1) \cdot (\sigma(1) + \sigma(2)) \cdot \ldots \cdot (\sigma(1) + \ldots + \sigma(n))} = \frac{1}{n!},$$

where $\mathcal{S}_n$ is the family of all permutations of $(1, \ldots, n)$.

Solution 2.5.4: 1)

```
function weighted_perm(N_sim,w,i)
n=length(w);
X=NaN(1,N_sim);Y=NaN(1,N_sim);
for j=1:N_sim
    %experiment A
    urn=repelem(1:n,w);
    for k=1:i
        X(j)=randsample(urn,1);
        urn(urn==X(j))=[ ];
    end
    %experiment B
    clocks=zeros(1,n);
    for k=1:n
        clocks(k)=exprnd(1/w(k));
    end
    ordered_clocks=sort(clocks);
    Y(j)=find(clocks==ordered_clocks(i));
end
subplot(1,2,1);
histogram(X,'normalization','probability','FaceColor','k');
grid on; legend('X_i');
subplot(1,2,2);
histogram(Y,'normalization','probability','FaceColor','k');
grid on; legend('Y_i');
fprintf('The data for experiment A:\n'); tabulate(X);
fprintf('The data for experiment B:\n'); tabulate(Y);
end
```

```
>> weighted_perm(100000,[30 50 20],2)
The data for experiment A:
  Value Count Percent
      1 37748 37.75%
      2 33666 33.67%
      3 28586 28.59%
The data for experiment B:
  Value Count Percent
      1 37845 37.84%
      2 33773 33.77%
      3 28382 28.38%
```

2) Let σ be a permutation of $(1, \ldots, n)$. We shall prove that

$$P\big(X_1 = \sigma(1), \ldots, X_n = \sigma(n)\big) = P\big(Y_1 = \sigma(1), \ldots, Y_n = \sigma(n)\big).$$

This relation implies that X_i and Y_i follow the same distribution for every $i \in \{1, \ldots, n\}$. In view of the chain rule from Theorem 1.7, we have

$$P\left(X_1 = \sigma(1), \ldots, X_n = \sigma(n)\right)$$
$$= P\left(X_1 = \sigma(1)\right) \cdot P\left(X_2 = \sigma(2) \big| X_1 = \sigma(1)\right) \cdots$$
$$\cdot P\left(X_n = \sigma(n) \big| X_1 = \sigma(1), \ldots, X_{n-1} = \sigma(n-1)\right)$$

and the analogous relation for $P\left(Y_1 = \sigma(1), \ldots, Y_n = \sigma(n)\right)$. Taking into account the steps of experiment **A** and those of experiment **B**, we note that it suffices to prove that

$$P(X_1 = \sigma(1)) = P(Y_1 = \sigma(1))$$

and then apply the same reasoning to prove that

$$P\left(X_i = \sigma(i) \big| X_1 = \sigma(1), \ldots, X_{i-1} = \sigma(i-1)\right)$$
$$= P\left(Y_i = \sigma(i) \big| Y_1 = \sigma(1), \ldots, Y_{i-1} = \sigma(i-1)\right)$$

for $i \in \{2, \ldots, n\}$.

Clearly, we have $P\big(X_1 = \sigma(1)\big) = \dfrac{w_{\sigma(1)}}{\sum_{i=1}^n w_i}$.

Let $M_{\sigma(1)} = \min\left\{Y_i : i \in \{1, \ldots, n\} \setminus \{\sigma(1)\}\right\}$. Since $Y_1, \ldots, Y_n$ are independent (because the alarm-clocks are independent), we deduce that

$$P(M_{\sigma(1)} \le z) = 1 - P(M_{\sigma(1)} > z) = 1 - \prod_{\substack{i=1 \\ i \ne \sigma(1)}}^n P(Y_i > z) = 1 - \exp\{-u_{\sigma(1)}z\}$$

for $z > 0$, where $u_{\sigma(1)} = \displaystyle\sum_{\substack{i=1 \\ i \ne \sigma(1)}}^n w_i$, and $P(M_{\sigma(1)} \le z) = 0$ for $z \le 0$. Thus $M_{\sigma(1)} \sim Exp(u_{\sigma(1)})$.

Let f_{Y_1}, $f_{M_{\sigma(1)}}$ and $f_{\frac{Y_1}{M_{\sigma(1)}}}$ be probability density functions of the random variables Y_1, $M_{\sigma(1)}$, respectively $\dfrac{Y_1}{M_{\sigma(1)}}$. Using Example 2.15, we get for a.e. $z > 0$

$$f_{\frac{Y_1}{M_{\sigma(1)}}}(z) = \int_{\mathbb{R}} |v| f_{Y_1}(vz) f_{M_{\sigma(1)}}(v) dv$$
$$= \int_0^\infty v w_{\sigma(1)} \exp\{-w_{\sigma(1)} vz\} u_{\sigma(1)} \exp\{-u_{\sigma(1)} v\} dv$$

$$= w_{\sigma(1)} u_{\sigma(1)} \int_0^\infty v \exp\{-(w_{\sigma(1)}z + u_{\sigma(1)})v\}dv$$

$$= \frac{w_{\sigma(1)} u_{\sigma(1)}}{(w_{\sigma(1)}z + u_{\sigma(1)})^2}$$

and for a.e. $z \leq 0$ we have $f_{\frac{Y_1}{M_{\sigma(1)}}}(z) = 0$. Hence,

$$P(Y_1 = \sigma(1)) = P(Y_1 \leq M_{\sigma(1)}) = \int_0^1 \frac{w_{\sigma(1)} u_{\sigma(1)}}{(w_{\sigma(1)}z + u_{\sigma(1)})^2}dz = \frac{w_{\sigma(1)}}{\sum_{i=1}^n w_i}.$$

Therefore, $P(X_1 = \sigma(1)) = P(Y_1 = \sigma(1))$.

3) From 2), we deduce that

$$1 = \sum_{\phi \in \mathcal{S}_n} P(X_1 = \phi(1), \ldots, X_n = \phi(n))$$

$$= \sum_{\phi \in \mathcal{S}_n} P(X_1 = \phi(1)) \cdot P(X_2 = \phi(2)|X_1 = \phi(1)) \cdot \ldots$$

$$\cdot P(X_n = \phi(n)|X_1 = \phi(1), \ldots, X_{n-1} = \phi(n-1))$$

$$= \sum_{\phi \in \mathcal{S}_n} \frac{w_{\phi(1)}}{w_{\phi(1)} + \ldots + w_{\phi(n)}} \cdot \frac{w_{\phi(2)}}{w_{\phi(2)} + \ldots + w_{\phi(n)}} \cdot \ldots \cdot \frac{w_{\phi(n)}}{w_{\phi(n)}}$$

$$= \sum_{\sigma \in \mathcal{S}_n} \frac{w_1 \cdot w_2 \cdot \ldots \cdot w_n}{w_{\sigma(1)} \cdot (w_{\sigma(1)} + w_{\sigma(2)}) \cdot \ldots \cdot (w_{\sigma(1)} + \ldots + w_{\sigma(n)})}.$$

Now, taking $w_i = i$, for $i \in \{1, \ldots, n\}$, we get the desired result.

2.6 Conditional Distribution Functions and Conditional Density Functions

Throughout this section, let $(\Omega, \mathcal{F}, P)$ be a probability space and let $B \in \mathcal{F}$ be such that $P(B) > 0$.

Definition 2.17. Let $\mathbb{X} = (X_1, \ldots, X_n)$ be a random vector. The **conditional distribution function** of $\mathbb{X}$ given B is $F_{\mathbb{X}}(\cdot|B) : \mathbb{R}^n \to \mathbb{R}$ defined by $F_{\mathbb{X}}(\mathbb{x}|B) = P(X_1 \leq x_1, \ldots, X_n \leq x_n|B)$ for $\mathbb{x} = (x_1, \ldots, x_n) \in \mathbb{R}^n$. ◆

Definition 2.18. Let $\mathbb{X} = (X_1, \ldots, X_n)$ be a random vector. If there exists a function $f_{\mathbb{X}}(\cdot|B) : \mathbb{R}^n \to \mathbb{R}$ such that

$$F_{\mathbb{X}}(\mathbb{x}|B) = \int_{-\infty}^{x_1} \cdots \int_{-\infty}^{x_n} f_{\mathbb{X}}(t_1, \ldots, t_n|B)dt_1 \ldots dt_n$$

for all $\mathbb{x} = (x_1, \ldots, x_n) \in \mathbb{R}^n$, then $f_{\mathbb{X}}(\cdot|B)$ is called **conditional density function** of $\mathbb{X}$ given B. ◆

Remark 2.15. Note that any two conditional density functions of a random vector $\mathbb{X}$ given the event B are a.s. equal. Also, note that a conditional density function of $\mathbb{X}$ given B satisfies the corresponding properties given in Theorem 2.8. ▼

Theorem 2.15. *Let $\mathbb{X}$ be a random vector. If $(B_i)_{i \in I} \subseteq \mathcal{F}$, $I \subseteq \mathbb{N}$, is a partition of Ω with $P(B_i) > 0$ for each $i \in I$, then the joint distribution function $F_{\mathbb{X}}$ of $\mathbb{X}$ can be written as a weighted sum of conditional distribution functions*

$$F_{\mathbb{X}}(\mathbb{x}) = \sum_{i \in I} P(B_i) F_{\mathbb{X}}(\mathbb{x}|B_i) \ \ for \ \mathbb{x} \in \mathbb{R}^n.$$

Example 2.25. Let $\mathbb{X}$ be a continuous random vector and let $f_{\mathbb{X}}$ be a joint density function.
(1) If $A \in \mathcal{B}^n$ and $P(\mathbb{X} \in A) > 0$, then the function

$$f_{\mathbb{X}}(\mathbb{x}|\mathbb{X} \in A) = \frac{f_{\mathbb{X}}(\mathbb{x})\mathbb{I}_A(\mathbb{x})}{P(\mathbb{X} \in A)} = \begin{cases} \dfrac{f_{\mathbb{X}}(\mathbb{x})}{P(\mathbb{X} \in A)}, & \text{if } \mathbb{x} \in A \\ 0, & \text{if } \mathbb{x} \notin A \end{cases}$$

is a conditional density function of $\mathbb{X}$ given $\{\mathbb{X} \in A\}$.
(2) If $(A_i)_{i \in I}$, $I \subseteq \mathbb{N}$, is a family of disjoint sets in $\mathcal{B}^n$ such that $P(\mathbb{X} \in A_i) > 0$, $i \in I$, and $\mathbb{X} \in \bigcup_{i \in I} A_i$ a.s., then $f_{\mathbb{X}}$ can be written as a weighted sum of conditional density functions $f_{\mathbb{X}}(\cdot|A_i)$ of $\mathbb{X}$ given A_i, $i \in I$, i.e., for a.e. $\mathbb{x} \in \mathbb{R}^n$

$$f_{\mathbb{X}}(\mathbb{x}) = \sum_{i \in I} P(\mathbb{X} \in A_i) f_{\mathbb{X}}(\mathbb{x}|\mathbb{X} \in A_i).$$

(3) Let $A, S \in \mathcal{B}^n$ be such that $A \subseteq S$, $P(\mathbb{X} \in A) > 0$ and $P(\mathbb{X} \in S) = 1$ and let $g : S \to g(S)$ be a bijective function with $g(S) \subseteq \mathbb{R}^n$ such that $g(A)$ is open and the inverse $g^{-1} = h : g(S) \to S$ has continuous partial derivatives on $g(A)$. Denote by J_h the Jacobian matrix of h on $g(A)$.
If $\mathbb{Y}$ is a random vector such that $\mathbb{Y} = g(\mathbb{X})$ a.s. and $B = g(A)$, then $f_{\mathbb{Y}}(\cdot|\mathbb{Y} \in B) : \mathbb{R}^n \to \mathbb{R}$ defined by

$$f_{\mathbb{Y}}(\mathbb{y}|\mathbb{Y} \in B) = \begin{cases} \dfrac{f_{\mathbb{X}}(h(\mathbb{y}))|\det(J_h(\mathbb{y}))|}{P(\mathbb{Y} \in B)}, & \text{if } \mathbb{y} \in B \\ 0, & \text{if } \mathbb{y} \notin B \end{cases}$$

is a conditional density function of $\mathbb{Y}$ given $\{\mathbb{Y} \in B\}$.
Proof: (1) Let $F_{\mathbb{X}}$ be the distribution function of $\mathbb{X}$ and let $F_{\mathbb{X}}(\cdot|\mathbb{X} \in A)$

be the conditional distribution function of $\mathbb{X}$ given $\{\mathbb{X} \in A\}$. By Theorem 2.8-(4), we have for every $\mathbb{x} = (x_1, \ldots, x_n) \in \mathbb{R}^n$

$$F_\mathbb{X}(\mathbb{x}|\mathbb{X} \in A) = \frac{1}{P(\mathbb{X} \in A)} \int_{-\infty}^{x_1} \cdots \int_{-\infty}^{x_n} f_\mathbb{X}(t_1, \ldots, t_n)\mathbb{I}_A(t_1, \ldots, t_n)dt_1 \ldots dt_n$$

and thus, in view of Definition 2.18 and Theorem 2.8-(1), we have that

$$f_\mathbb{X}(\mathbb{x}|\mathbb{X} \in A) = \frac{f_\mathbb{X}(\mathbb{x})\mathbb{I}_A(\mathbb{x})}{P(\mathbb{X} \in A)} \quad \text{for } \mathbb{x} \in \mathbb{R}^n$$

is a conditional density function of $\mathbb{X}$ given $\{\mathbb{X} \in A\}$.

(2) In view of (1), we have to prove that for a.e. $\mathbb{x} \in \mathbb{R}^n$

$$f_\mathbb{X}(\mathbb{x}) = \sum_{i \in I} f_\mathbb{X}(\mathbb{x})\mathbb{I}_{A_i}(\mathbb{x}).$$

Clearly, the above relation holds for each $\mathbb{x} \in \bigcup_{i \in I} A_i$.

Since $P\left(\mathbb{X} \notin \bigcup_{i \in I} A_i\right) = 0$, we have

$$\underbrace{\int \cdots \int}_{\mathbb{R}^n \backslash \bigcup_{i \in I} A_i} f_\mathbb{X}(t_1, \ldots, t_n)dt_1 \ldots dt_n = 0$$

and thus by the property $f_\mathbb{X}(\mathbb{x}) \geq 0$ for a.e. $\mathbb{x} \in \mathbb{R}^n$ we obtain

$$f_\mathbb{X}(\mathbb{x}) = 0 \text{ for a.e. } \mathbb{x} \in \mathbb{R}^n \setminus \bigcup_{i \in I} A_i.$$

Since, for every $i \in I$, $\mathbb{I}_{A_i}(\mathbb{x}) = 0$ for $\mathbb{x} \notin \bigcup_{i \in I} A_i$, the proof is complete.

(3) Let $\mathbb{y} = (y_1, \ldots, y_n) \in \mathbb{R}^n$ and denote $D_\mathbb{y} = (-\infty, y_1] \times \ldots \times (-\infty, y_n]$. In view of Example 2.25-(1) and [Rudin (1987), Theorem 7.26, p. 153] (change of variables theorem), we have

$$P(\mathbb{Y} \in D_\mathbb{y}|\mathbb{Y} \in B) = P(\mathbb{X} \in h(D_\mathbb{y})|\mathbb{X} \in A)$$

$$= \underbrace{\int \cdots \int}_{h(D_\mathbb{y})} f_\mathbb{X}(\mathbb{t}|\mathbb{X} \in A)dt_1 \ldots dt_n = \underbrace{\int \cdots \int}_{h(D_\mathbb{y}) \cap A} \frac{f_\mathbb{X}(\mathbb{t})}{P(\mathbb{X} \in A)}dt_1 \ldots dt_n$$

$$= \underbrace{\int \cdots \int}_{g(h(D_\mathbb{y}) \cap A)} \frac{f_\mathbb{X}(h(\mathbb{t}))|\det(J_h(\mathbb{t}))|}{P(\mathbb{Y} \in B)}dt_1 \ldots dt_n = \underbrace{\int \cdots \int}_{D_\mathbb{y}} f_\mathbb{Y}(\mathbb{t}|\mathbb{Y} \in B)dt_1 \ldots dt_n,$$

where we use the notation $\mathbb{t} = (t_1, \ldots, t_n)$. In view of Definition 2.18, $f_\mathbb{Y}(\cdot|\mathbb{Y} \in B)$ is a conditional density function of $\mathbb{Y}$ given $\{\mathbb{Y} \in B\}$. $\blacktriangle$

Example 2.26. Let $\mathbb{X}$ be a continuous random vector with joint density function $f_{\mathbb{X}}$. If $(B_i)_{i \in I} \subseteq \mathcal{F}$, $I \subseteq \mathbb{N}$, is a partition of Ω such that $P(B_i) > 0$ for each $i \in I$ and $f_{\mathbb{X}}(\cdot | B_i)$ is a conditional density function of $\mathbb{X}$ given B_i, then $f_{\mathbb{X}}$ can be written as a weighted sum of conditional density functions

$$f_{\mathbb{X}}(\mathbb{x}) = \sum_{i \in I} P(B_i) f_{\mathbb{X}}(\mathbb{x} | B_i) \ \text{ for a.e. } \mathbb{x} \in \mathbb{R}^n.$$

Proof: For every $i \in I$ let $F_{\mathbb{X}}(\cdot | B_i)$ be the conditional distribution function of $\mathbb{X}$ given B_i and let $F_{\mathbb{X}}$ be the joint distribution function of $\mathbb{X}$.

If I is finite, then we deduce from Definition 2.18 and Theorem 2.15 that $\mathbb{x} \in \mathbb{R}^n \mapsto \sum_{i \in I} P(B_i) f_{\mathbb{X}}(\mathbb{x} | B_i)$ is a joint density function of $\mathbb{X}$ and thus the desired relation holds.

In the following, we consider the case when I is infinite and we assume without loss of generality that $I = \mathbb{N}^*$.

We denote $D_{\mathbb{x}, \mathbb{h}} = \{\mathbb{x} + t\mathbb{h} : t \in [0,1]\}$ for each $\mathbb{x} \in \mathbb{R}^n$ and $\mathbb{h} \in \mathbb{R}_+^n$, where $\mathbb{R}_+ = (0, \infty)$. By Theorem 1.8 we have for $\mathbb{x} \in \mathbb{R}^n$, $\mathbb{h} = (h_1, \ldots, h_n) \in \mathbb{R}_+^n$ and $k \in \mathbb{N}^*$

$$\frac{1}{h_1 \cdot \ldots \cdot h_n} \sum_{i=1}^{k} P(B_i) P\big(\mathbb{X} \in D_{\mathbb{x}, \mathbb{h}} | B_i\big) \leq \frac{P(\mathbb{X} \in D_{\mathbb{x}, \mathbb{h}})}{h_1 \cdot \ldots \cdot h_n}$$

and thus, in view of Theorem 2.8-(1), by taking $\mathbb{h} \to 0_n$, we deduce that for every $k \in \mathbb{N}^*$ there exists $A_k \subset \mathbb{R}^n$ with $\lambda_{\mathbb{R}^n}(A_k) = 0$ such that

$$\sum_{i=1}^{k} P(B_i) f_{\mathbb{X}}(\mathbb{x} | B_i) \leq f_{\mathbb{X}}(\mathbb{x}) \text{ for } \mathbb{x} \in \mathbb{R}^n \setminus A_k. \tag{2.3}$$

For every $k \in \mathbb{N}^*$ let $f_k = \sum_{i=1}^{k} P(B_i) f_{\mathbb{X}}(\cdot | B_i) \mathbb{I}_{\mathbb{R}^n \setminus A_k}$ and let $f = \lim_{k \to \infty} f_k$. In view of (2.3), the limit exists and thus f is well defined on $\mathbb{R}^n$. Also, if $A = \bigcup_{k=1}^{\infty} A_k$, then $\lambda_{\mathbb{R}^n}(A) = 0$ and

$$f(\mathbb{x}) = \sum_{i \in I} P(B_i) f_{\mathbb{X}}(\mathbb{x} | B_i) \text{ for } \mathbb{x} \in \mathbb{R}^n \setminus A. \tag{2.4}$$

Using the Lebesgue Dominated Convergence Theorem (see [Nelson (2015), Theorem 4.3.12]), we have for $\mathbb{x} = (x_1, \ldots, x_n) \in \mathbb{R}^n$

$$\int_{-\infty}^{x_1} \cdots \int_{-\infty}^{x_n} f(t_1, \ldots, t_n) dt_1 \ldots dt_n$$

$$= \lim_{k \to \infty} \int_{-\infty}^{x_1} \cdots \int_{-\infty}^{x_n} f_k(t_1, \ldots, t_n) dt_1 \ldots dt_n$$

$$= \sum_{i=1}^{\infty} \int_{-\infty}^{x_1} \cdots \int_{-\infty}^{x_n} P(B_i) f_{\mathbb{X}}(t_1, \ldots, t_n | B_i) dt_1 \ldots dt_n$$

$$= \sum_{i=1}^{\infty} P(B_i) F_{\mathbb{X}}(\mathbb{x} | B_i) = F_{\mathbb{X}}(\mathbb{x}),$$

where we use Theorem 2.15 for the last equality. Hence, f is a joint density function of $\mathbb{X}$. Since $f_{\mathbb{X}}$ is also a joint density function of $\mathbb{X}$ and taking into account (2.4), the desired relation follows.   ▲

Definition 2.19. If X is a random variable and Y is a discrete random variable, then the **conditional distribution function** of X given $\{Y = y\}$ with $y \in \mathbb{R}$ such that $P(Y = y) > 0$ is defined by

$$F_{X|Y}(x|y) = P(X \le x | Y = y) = \frac{P(X \le x, Y = y)}{P(Y = y)} \text{ for each } x \in \mathbb{R}.$$

If there exists a function $f_{X|Y}(\cdot|y) : \mathbb{R} \to \mathbb{R}$ such that

$$F_{X|Y}(x|y) = \int_{-\infty}^{x} f_{X|Y}(u|y) du \text{ for all } x \in \mathbb{R},$$

then $f_{X|Y}(\cdot|y)$ is called **conditional density function** of X given $\{Y = y\}$.   ◆

The above definition is in fact equivalent with Definitions 2.17 and 2.18 in the special case $n = 1$ and $B = \{Y = y\}$.

Definition 2.20. Let (X, Y) be a continuous random vector with joint density function $f_{X,Y}$ and consider

$$f_X(x) = \int_{\mathbb{R}} f_{X,Y}(x, y) dy \quad \text{for } x \in \mathbb{R}$$

and

$$f_Y(y) = \int_{\mathbb{R}} f_{X,Y}(x, y) dx \quad \text{for } y \in \mathbb{R},$$

which are density functions of X, respectively Y. Consider the function $f_{X|Y}(\cdot|\cdot) : \mathbb{R}^2 \to \mathbb{R}$ defined for each $x, y \in \mathbb{R}$ by

$$f_{X|Y}(x|y) = \begin{cases} \dfrac{f_{X,Y}(x, y)}{f_Y(y)}, & \text{if } f_Y(y) > 0 \\[2mm] f_X(x), & \text{if } f_Y(y) = 0. \end{cases}$$

Then $f_{X|Y}(\cdot|y)$ is called **conditional density function** of the random variable X given $\{Y = y\}$ for $y \in \mathbb{R}$.

Consider the function $F_{X|Y}(\cdot|\cdot) : \mathbb{R}^2 \to \mathbb{R}$ defined by

$$F_{X|Y}(x|y) = \int\limits_{-\infty}^{x} f_{X|Y}(u|y)du \text{ for all } x, y \in \mathbb{R}.$$

Then $F_{X|Y}(\cdot|y)$ is called **conditional distribution function** of X given $\{Y = y\}$ for $y \in \mathbb{R}$.

For $A \in \mathcal{B}$ and $y \in \mathbb{R}$ define

$$P(X \in A|Y = y) = \int_A f_{X|Y}(u|y)du$$

and call it **conditional probability** of $\{X \in A\}$ given $\{Y = y\}$. ◆

Note that the above conditional density function, conditional distribution function and conditional probability depend on $f_{X,Y}$. For more details regarding the above definition, see [Chow and Teicher (1997), Section 7.2].

Example 2.27. Let (X, Y) be a continuous random vector with joint density function $f_{X,Y}$. Consider

$$f_X(x) = \int\limits_{\mathbb{R}} f_{X,Y}(x, y)dy \quad \text{for } x \in \mathbb{R}$$

and

$$f_Y(y) = \int\limits_{\mathbb{R}} f_{X,Y}(x, y)dx \quad \text{for } y \in \mathbb{R},$$

which are density functions of X, respectively Y.

(1) In what follows we prove that the distribution function of X verifies the relation

$$F_X(x) = \int\limits_{\mathbb{R}} f_Y(y)F_{X|Y}(x|y)dy \text{ for each } x \in \mathbb{R}.$$

Denote $D = \{y \in \mathbb{R} : f_Y(y) > 0\}$. Let $y \in \mathbb{R} \setminus D$. Then we have

$$0 = f_Y(y) = \int_{\mathbb{R}} f_{X,Y}(u, y)du,$$

which implies $f_{X,Y}(u, y) = 0$ for a.e. $u \in \mathbb{R}$. Therefore, for $y \in \mathbb{R} \setminus D$

$$\int_{-\infty}^{x} f_{X,Y}(u, y)du = 0 \quad \text{for each } x \in \mathbb{R}.$$

We write for each $x \in \mathbb{R}$

$$\int_{\mathbb{R}} f_Y(y) F_{X|Y}(x|y) dy = \int_{\mathbb{R}} f_Y(y) \Big(\int_{-\infty}^{x} f_{X|Y}(u|y) du \Big) dy$$

$$= \int_{D} \Big(\int_{-\infty}^{x} f_{X,Y}(u,y) du \Big) dy + \int_{\mathbb{R} \setminus D} f_Y(y) \Big(\int_{-\infty}^{x} f_X(u) du \Big) dy$$

$$= \int_{D} \Big(\int_{-\infty}^{x} f_{X,Y}(u,y) du \Big) dy + \int_{\mathbb{R} \setminus D} \Big(\int_{-\infty}^{x} f_{X,Y}(u,y) du \Big) dy$$

$$= \int_{\mathbb{R}} \Big(\int_{-\infty}^{x} f_{X,Y}(u,y) du \Big) dy = \int_{-\infty}^{x} \Big(\int_{\mathbb{R}} f_{X,Y}(u,y) dy \Big) du$$

$$= \int_{-\infty}^{x} f_X(u) du = F_X(x).$$

(2) By Definition 2.20, we have

$$f_{X|Y}(x|y) = \frac{f_{X,Y}(x,y)}{f_Y(y)} = \frac{f_{Y|X}(y|x) f_X(x)}{f_Y(y)}, \text{ if } f_X(x) > 0, f_Y(y) > 0.$$

Then Bayes' formula for a conditional density function:

$$f_{X|Y}(x|y) = \frac{f_{Y|X}(y|x) f_X(x)}{f_Y(y)}$$

holds for each $x \in \mathbb{R}$ with $f_X(x) > 0$ and $y \in \mathbb{R}$ with $f_Y(y) > 0$.   ▲

Example 2.28. Let $c \in \mathbb{R}$. Suppose that a joint density function of the random vector (X, Y) is

$$f(x,y) = \begin{cases} cx^2 y, & \text{if } 0 \le x \le y \le 1 \\ 0, & \text{otherwise.} \end{cases}$$

We determine:

(1) the value of the constant c;

(2) f_X and f_Y, density functions of X, respectively of Y;

(3) $f_{X|Y}\left(\cdot \Big| \frac{1}{2} \right)$;

(4) $P\left(X > \frac{1}{4} \Big| Y \le \frac{1}{2} \right)$.

Solution: (1) We determine c from the condition

$$\int_{\mathbb{R}} \int_{\mathbb{R}} f(x,y) dx dy = 1$$

given in Theorem 2.8-(3). We write

$$\int_{\mathbb{R}} \int_{\mathbb{R}} f(x,y)dxdy = \int_0^1 \left(\int_x^1 cx^2 ydy \right) dx = \frac{c}{15}.$$

So, the value of c must be 15.

(2) We take

$$f_X(x) = \int_{\mathbb{R}} f(x,y)dy \text{ for each } x \in \mathbb{R},$$

which is a density function of X by Theorem 2.9. From the definition of f it follows that $f_X(x) = 0$ for $x \notin [0,1]$ and

$$f_X(x) = \int_{\mathbb{R}} f(x,y)dy = \int_x^1 15x^2 ydy = \frac{15}{2}x^2(1-x^2) \text{ for } x \in [0,1].$$

Hence,

$$f_X(x) = \begin{cases} \dfrac{15}{2}x^2(1-x^2), & \text{if } x \in [0,1] \\ 0, & \text{otherwise.} \end{cases}$$

Analogously, we obtain a density function f_Y of Y such that $f_Y(y) = 0$ for $y \notin [0,1]$, while

$$f_Y(y) = \int_{\mathbb{R}} f(x,y)dx = \int_0^y 15x^2 ydx = 5y^4 \text{ for } y \in [0,1].$$

Hence,

$$f_Y(y) = \begin{cases} 5y^4, & \text{if } 0 \leq y \leq 1 \\ 0, & \text{otherwise.} \end{cases}$$

(3) The conditional density function of X given $\left\{ Y = \dfrac{1}{2} \right\}$ is defined by

$$f_{X|Y}\left(x \Big| \frac{1}{2} \right) = \frac{f_{X,Y}\left(x, \frac{1}{2} \right)}{f_Y\left(\frac{1}{2} \right)}, \quad \text{since } f_Y\left(\frac{1}{2} \right) = \frac{5}{16} > 0.$$

Hence,

$$f_{X|Y}\left(x \Big| \frac{1}{2} \right) = \begin{cases} 24x^2, & \text{if } 0 \leq x \leq \frac{1}{2} \\ 0, & \text{otherwise.} \end{cases}$$

(4) We obtain

$$P\left(X > \frac{1}{4}\middle| Y \leq \frac{1}{2}\right) = 1 - F_X\left(\frac{1}{4}\middle| Y \leq \frac{1}{2}\right) = 1 - \frac{P\left(X \leq \frac{1}{4}, Y \leq \frac{1}{2}\right)}{P\left(Y \leq \frac{1}{2}\right)},$$

where

$$P\left(X \leq \frac{1}{4}, Y \leq \frac{1}{2}\right) = \int_0^{\frac{1}{4}}\left(\int_x^{\frac{1}{2}} 15x^2 y dy\right) dx = \frac{17}{2048}$$

and

$$P\left(Y \leq \frac{1}{2}\right) = \int_0^{\frac{1}{2}} 5y^4 dy = \frac{1}{32}.$$

Therefore,

$$P\left(X > \frac{1}{4}\middle| Y \leq \frac{1}{2}\right) = \frac{17}{64}. \qquad \blacktriangle$$

Definition 2.21. Let X be a discrete random variable.
(1) Let Y be a discrete random variable and define

$$p_{X|Y}(x|y) = \begin{cases} \dfrac{P(X = x, Y = y)}{P(Y = y)}, & \text{if } P(Y = y) > 0 \\ P(X = x), & \text{otherwise.} \end{cases}$$

Then $p_{X|Y}(\cdot|y)$ is called **conditional probability mass function** of X given $\{Y = y\}$ for $y \in \mathbb{R}$.
(2) Let Y be a continuous random variable. Consider the density function of Y

$$f_Y(y) = \sum_{x \in X(\Omega)} f_{Y|X}(y|x)P(X = x), \ y \in \mathbb{R}$$

(see Example 2.26) and define

$$p_{X|Y}(x|y) = \begin{cases} \dfrac{f_{Y|X}(y|x)P(X = x)}{f_Y(y)}, & \text{if } f_Y(y) > 0, P(X = x) > 0 \\ P(X = x), & \text{otherwise,} \end{cases}$$

where $f_{Y|X}(\cdot|x)$ is a conditional density function of Y given $\{X = x\}$ for $x \in \mathbb{R}$ (see Definition 2.19). Then $p_{X|Y}(\cdot|y)$ is called **conditional probability mass function** of X given $\{Y = y\}$ for $y \in \mathbb{R}$.
Define

$$F_{X|Y}(x|y) = \sum_{\substack{u \in X(\Omega) \\ u \leq x}} p_{X|Y}(u|y) \text{ for all } x, y \in \mathbb{R}.$$

Then $F_{X|Y}(\cdot|y)$ is called **conditional distribution function** of X given $\{Y = y\}$ for $y \in \mathbb{R}$. $\qquad \blacklozenge$

Note that the conditional mass function and the conditional distribution function given in Definition 2.21-(2) depend on $f_{Y|X}$.

Example 2.29. A signal X is transmitted such that

$$P(X = 1) = p, P(X = -1) = 1 - p, \text{ where } p \in (0, 1).$$

The received signal is $Y = X + Z$, where $Z \sim N(0, 1)$ is independent of X. We compute $f_{Y|X}(\cdot|x)$, a conditional density function of Y given $\{X = x\}$ for $x \in \{-1, 1\}$, and $p_{X|Y}(\cdot|y)$, the conditional probability mass function of X given $\{Y = y\}$ for $y \in \mathbb{R}$.

Solution: For $x \in \{-1, 1\}$ and $y \in \mathbb{R}$ we have, by Definition 2.19 and the independence of X and Z, that

$$F_{Y|X}(y|x) = \frac{P(X + Z \le y, X = x)}{P(X = x)} = P(Z \le y - x) = F_Z(y - x).$$

Hence, for each $y \in \mathbb{R}$ it holds

$$F_{Y|X}(y|x) = \int_{-\infty}^{y} \frac{1}{\sqrt{2\pi}} \exp\left\{ -\frac{1}{2} (u - x)^2 \right\} du$$

and thus

$$f_{Y|X}(y|x) = \frac{1}{\sqrt{2\pi}} \exp\left\{ -\frac{1}{2} (y - x)^2 \right\}, \quad y \in \mathbb{R},$$

is a conditional density function of Y given $\{X = x\}$. For $y \in \mathbb{R}$ the function

$$f_Y(y) = f_{Y|X}(y| - 1)P(X = -1) + f_{Y|X}(y|1)P(X = 1)$$

is a density function of Y (see Example 2.26). Hence, for $y \in \mathbb{R}$

$$f_Y(y) = \frac{1}{\sqrt{2\pi}} \left((1 - p) \exp\left\{ -\frac{1}{2} (y + 1)^2 \right\} + p \exp\left\{ -\frac{1}{2} (y - 1)^2 \right\} \right).$$

We use that $f_Y(y) > 0$ for each $y \in \mathbb{R}$ and Definition 2.21 to obtain

$$p_{X|Y}(-1|y) = \frac{f_{Y|X}(y| - 1)P(X = -1)}{f_Y(y)} = \frac{(1 - p)e^{-y}}{(1 - p)e^{-y} + pe^y}, \quad y \in \mathbb{R},$$

and

$$p_{X|Y}(1|y) = \frac{f_{Y|X}(y|1)P(X = 1)}{f_Y(y)} = \frac{pe^y}{(1 - p)e^{-y} + pe^y}, \quad y \in \mathbb{R}. \quad \blacktriangle$$

2.6.1   Solved Problems

Problem 2.6.1. A gambler interested in a horse race of 1 mile (≈ 1.61 km) studies three horses: Aljazzi, Defoe and Stradivarius. Each horse finishes the race independently in $100 + T_i$ seconds, where $T_1 \sim Exp(0.1)$ for Aljazzi, $T_2 \sim Exp(0.2)$ for Defoe and $T_3 \sim Exp(0.3)$ for Stradivarius.

1) Simulate in Matlab a race between the three horses and print the ranking and the corresponding times.

2) Using simulations, plot the empirical cumulative distribution function of the time of the ith placed horse, using as input the ranking of the race, for $i \in \{1, 2, 3\}$.

3) Find the conditional distribution function of the time of Aljazzi, given that the ranking of the race is: 1st is Stradivarius, 2nd is Aljazzi and 3rd is Defoe. Plot it on the figure from 2) for the corresponding input.

4) Find a joint density function of $\mathbb{W} = (W_1, W_2, W_3)$, where W_i is the time of the ith placed horse for $i \in \{1, 2, 3\}$.

5) Using simulations, plot the empirical cumulative distribution function of the time of the ith placed horse for $i \in \{1, 2, 3\}$.

6) Find the distribution function of the time of the 2nd placed horse. Plot it on the figure from 5) for the corresponding input.

Solution 2.6.1: 1)

```
function horace
horses={'Aljazzi','Defoe','Stradivarius'};
T=[exprnd(1/0.1),exprnd(1/0.2),exprnd(1/0.3)];
[W,X]=sort(T);
fprintf('1st place: %s with %4.1f seconds\n',horses{X(1)},100+W(1));
fprintf('2nd place: %s with %4.1f seconds\n',horses{X(2)},100+W(2));
fprintf('3rd place: %s with %4.1f seconds\n',horses{X(3)},100+W(3));
end
```

```
>> horace
1st place: Aljazzi with 100.9 seconds
2nd place: Stradivarius with 110.7 seconds
3rd place: Defoe with 111.3 seconds
```

2)

```
function horace2(R,i,N_sim)
%R=the ranking; 1=Aljazzi, 2=Defoe, 3=Stradivarius;
Wi=[ ];
for k=1:N_sim
    T=[exprnd(1/0.1),exprnd(1/0.2),exprnd(1/0.3)];
    [W,X]=sort(T);
```

```
    if X==R
        Wi=[Wi,W(i)];
    end
end
ecdf(Wi);
end
```

3) In the following, we assume that two horses cannot finish the race at the same time, since the probability of this event is 0. For every $i \in \{1,2,3\}$ let

$$X_i = \begin{cases} 1, & \text{if Aljazzi is on the } i\text{th place} \\ 2, & \text{if Defoe is on the } i\text{th place} \\ 3, & \text{if Stradivarius is on the } i\text{th place.} \end{cases}$$

Let $\mathbb{T} = (T_1, T_2, T_3)$, $\mathbb{X} = (X_1, X_2, X_3)$ and let $\mathcal{S}_3$ be the family of all permutations of $(1, 2, 3)$.

In view of the solution of Problem 2.5.4, we have

$$P\big(\mathbb{X} = (\sigma(1), \sigma(2), \sigma(3))\big) = \frac{w_{\sigma(1)}}{w_{\sigma(1)} + w_{\sigma(2)} + w_{\sigma(3)}} \cdot \frac{w_{\sigma(2)}}{w_{\sigma(2)} + w_{\sigma(3)}} \cdot \frac{w_{\sigma(3)}}{w_{\sigma(3)}}$$
$$= \frac{w_{\sigma(1)} w_{\sigma(2)}}{0.6\big(w_{\sigma(2)} + w_{\sigma(3)}\big)},$$

for $\sigma \in \mathcal{S}_3$, where $w_i = \dfrac{i}{10}$ for $i \in \{1, 2, 3\}$.

For every $\sigma \in \mathcal{S}_3$, let $g_\sigma : \mathbb{R}^3 \to \mathbb{R}^3$ be given by

$$g_\sigma(t_1, t_2, t_3) = (t_{\sigma(1)}, t_{\sigma(2)}, t_{\sigma(3)}), \quad (t_1, t_2, t_3) \in \mathbb{R}^3.$$

Also, let $f_{\mathbb{T}}$ be a joint density function of $\mathbb{T}$ given by

$$f_{\mathbb{T}}(t_1, t_2, t_3) = \begin{cases} 0.006 \cdot \exp\{-0.1 \cdot t_1 - 0.2 \cdot t_2 - 0.3 \cdot t_3\}, & \text{if } t_1, t_2, t_3 > 0 \\ 0, & \text{otherwise.} \end{cases}$$

Let $\mathbb{W} = (W_1, W_2, W_3)$, where W_i is the time of the ith placed horse, for $i \in \{1, 2, 3\}$. For every $\sigma \in \mathcal{S}_3$, by Example 2.25-(3), we have a conditional density function of $\mathbb{W}$ given the ranking $\mathbb{X} = (\sigma(1), \sigma(2), \sigma(3))$:

$$f_{\mathbb{W}}\big(t_1, t_2, t_3 \mid \mathbb{X} = (\sigma(1), \sigma(2), \sigma(3))\big)$$
$$= \begin{cases} \dfrac{f_{\mathbb{T}}\big(g_\sigma^{-1}(t_1, t_2, t_3)\big)}{P\big(\mathbb{X} = (\sigma(1), \sigma(2), \sigma(3))\big)}, & \text{if } 0 < t_1 < t_2 < t_3 \\ 0, & \text{otherwise,} \end{cases}$$
$$= \begin{cases} \dfrac{f_{\mathbb{T}}\big(t_{\sigma^{-1}(1)}, t_{\sigma^{-1}(2)}, t_{\sigma^{-1}(3)}\big)}{P\big(\mathbb{X} = (\sigma(1), \sigma(2), \sigma(3))\big)}, & \text{if } 0 < t_1 < t_2 < t_3 \\ 0, & \text{otherwise.} \end{cases}$$

Now, in view of the hypothesis of 3), we take: $\sigma(1) = 3, \sigma(2) = 1, \sigma(3) = 2$. In particular, we have $P(\mathbb{X} = (\sigma(1), \sigma(2), \sigma(3))) = \dfrac{1}{6}$ and $\sigma^{-1}(1) = 2, \sigma^{-1}(2) = 3, \sigma^{-1}(3) = 1$. The conditional distribution function of the time of Aljazzi given the ranking in the hypothesis satisfies

$$
F_{W_2}(x|\mathbb{X} = (3,1,2))
$$

$$
= \int_{-\infty}^{\infty} \left(\int_{-\infty}^{x} \left(\int_{-\infty}^{\infty} f_{\mathbb{W}}(t_1,t_2,t_3|\mathbb{X} = (3,1,2)) dt_3 \right) dt_2 \right) dt_1
$$

$$
= \int_{0}^{x} \left(\int_{t_1}^{x} \left(\int_{t_2}^{\infty} 6 f_{\mathbb{T}}(t_2,t_3,t_1) dt_3 \right) dt_2 \right) dt_1
$$

$$
= 0.036 \int_{0}^{x} \left(\int_{t_1}^{x} \left(\int_{t_2}^{\infty} \exp\{-0.1 \cdot t_2 - 0.2 \cdot t_3 - 0.3 \cdot t_1\} dt_3 \right) dt_2 \right) dt_1
$$

$$
= 0.18 \int_{0}^{x} \left(\int_{t_1}^{x} \exp\{-0.3 \cdot t_2 - 0.3 \cdot t_1\} dt_2 \right) dt_1
$$

$$
= 0.6 \int_{0}^{x} \exp\{-0.6 \cdot t_1\} - \exp\{-0.3 \cdot x - 0.3 \cdot t_1\} dt_1, \quad x > 0.
$$

Hence, $F_{W_2}(x|\mathbb{X} = (3,1,2)) = \begin{cases} (1 - \exp\{-0.3x\})^2, & \text{if } x > 0 \\ 0, & \text{if } x \leq 0. \end{cases}$

%we call the function from 2)
```
>>horace2([3,1,2],2,10000);
```
%and then we call
```
>>hold on; fplot(@(x) (1-exp(-0.3*x)).^2,[0,max(Wi)]);
```

4) In view of Example 2.25-(2), we obtain the following joint density function of $\mathbb{W}$

$$
f_{\mathbb{W}}(t_1,t_2,t_3) = \sum_{\sigma \in \mathcal{S}_3} f_{\mathbb{W}}\big(t_1,t_2,t_3|\mathbb{X} = (\sigma(1),\sigma(2),\sigma(3))\big)
$$

$$
\cdot P\big(\mathbb{X} = (\sigma(1),\sigma(2),\sigma(3))\big), \text{ for } (t_1,t_2,t_3) \in \mathbb{R}^3.
$$

From the solution of 3), we deduce that

$$
f_{\mathbb{W}}(t_1,t_2,t_3) = \sum_{\sigma \in \mathcal{S}_3} f_{\mathbb{T}}\big(t_{\sigma^{-1}(1)}, t_{\sigma^{-1}(2)}, t_{\sigma^{-1}(3)}\big)
$$

$$
= \frac{6}{1000} \cdot \sum_{\sigma \in \mathcal{S}_3} \exp\left\{ -\frac{\sigma(1)}{10} \cdot t_1 - \frac{\sigma(2)}{10} \cdot t_2 - \frac{\sigma(3)}{10} \cdot t_3 \right\},
$$

if $0 < t_1 < t_2 < t_3$, and $f_{\mathbb{W}}(t_1,t_2,t_3) = 0$, otherwise.

5)

```
function horace5(i,N_sim)
Wi=zeros(1,N_sim);
for k=1:N_sim
    T=[exprnd(1/0.1),exprnd(1/0.2),exprnd(1/0.3)];
    W=sort(T);
    Wi(k)=W(i);
end
ecdf(Wi);
end
```

6) By arguments and computations similar to those in the solution of 3), we deduce that the distribution function of the time of the 2nd placed horse satisfies

$$F_{W_2}(x) = \frac{6}{1000} \sum_{\sigma \in S_3} \int_0^x \left(\int_{t_1}^x \left(\int_{t_2}^\infty \exp\left\{ -\frac{\sigma(1)}{10} \cdot t_1 \right\} \right.\right.$$
$$\left.\left. \cdot \exp\left\{ -\frac{\sigma(2)}{10} \cdot t_2 \right\} \cdot \exp\left\{ -\frac{\sigma(3)}{10} \cdot t_3 \right\} dt_3 \right) dt_2 \right) dt_1$$
$$= 2\exp\{-0.6x\} - \exp\{-0.5x\} - \exp\{-0.4x\} - \exp\{-0.3x\} + 1,$$

if $x > 0$, and $F_{W_2}(x) = 0$, otherwise.

```
%we call the function from 5)
>>horace5(2,10000);
%and then we call
>>F=@(x) 2*exp(-0.6*x)-exp(-0.5*x)-exp(-0.4*x)-exp(-0.3*x)+1;
>>hold on; fplot(F,[0,max(Wi)]);
```

2.7 Generation of Random Numbers

In this section we give some methods to simulate random values for given distributions based on uniformly distributed random values from $[0,1]$. We shall use the **rand** function in Matlab and avoid using the random generators for the distributions mentioned in the Sections 2.4 and 2.5. For simplicity we shall use the expression random numbers instead of pseudo-random numbers.

Example 2.30. Let $k, n \in \mathbb{N}^*$ and $p \in (0,1)$. We show some examples in Matlab for generating k values for the Bernoulli distribution $Ber(p)$ and for the binomial distribution $Bino(n,p)$, which are based on the property that if $X_1, \ldots, X_n \sim Ber(p)$ are independent random variables, then $X_1 + \ldots + X_n \sim Bino(n,p)$:

```
>>k=100; p=0.4; n=10;
>>X1=rand(1,k)<=p; % k values for Ber(p) distribution
>>X2=floor(rand(1,k)+p); % k values for Ber(p) distribution
>>Y1=sum(rand(n,k)<=p); % k values for Bino(n,p) distribution
>>Y2=sum(floor(rand(n,k)+p)); % k values for Bino(n,p) distribution
```

▲

2.7.1 Inversion Method

Let X be a random variable and denote by F its distribution function. Denote by Q the corresponding quantile function. Let $U \sim Unif[0,1]$ and consider $Y = Q(U)$. For each $y \in \mathbb{R}$ we have

$$F_Y(y) = P(Y \leq y) = P\big(Q(U) \leq y\big) = P\big(U \leq F(y)\big) = F(y),$$

see Theorem 2.6-(2), hence Y has the same distribution function as X. Note that $U(\omega) \in (0,1)$ for a.e. $\omega \in \Omega$, hence $Q(U(\omega))$ is well defined for a.e. $\omega \in \Omega$. Also, we mention that one can use the implemented quantile functions in Matlab for various classic distributions, which have the extension **inv**, see Sections 2.4 and 2.5.

Example 2.31. If $U \sim Unif[0,1]$, then $-\frac{1}{\lambda}\log(U) \sim Exp(\lambda)$, where $\lambda > 0$, because the quantile function for the $Exp(\lambda)$ distribution is

$$Q(y) = -\frac{1}{\lambda}\log(1-y), \ y \in (0,1),$$

and we have: $U \sim Unif[0,1] \Leftrightarrow 1 - U \sim Unif[0,1]$. Next, we generate k values for the $Exp(\lambda)$ distribution, for some $k \in \mathbb{N}^*$ and $\lambda > 0$, first by using the quantile function **expinv** from Matlab, and then by applying the above expression of Q:

```
>>lambda=2; k=100;
>>U=rand(1,k);
>>X1=expinv(U,1/lambda);% k values for Exp(lambda) distribution
>>X2=-1/lambda*log(U); % k values for Exp(lambda) distribution
```

▲

Example 2.32. (The inversion method for discrete distributions) Given the values $\{x_1, \ldots, x_n\} \subset \mathbb{R}$ and the probabilities $p_1, \ldots, p_n \in [0,1]$ such that $p_1 + \ldots + p_n = 1$, $n \in \mathbb{N}^*$, we generate random numbers Y_i, $i \in \{1, \ldots, k\}$, $k \in \mathbb{N}^*$, for the discrete random variable X whose probability mass function is

$$P(X = x_i) = p_i \text{ for each } i \in \{1, \ldots, n\},$$

by using uniformly distributed random numbers on $[0, 1]$.

The procedure to generate the random numbers is:

• Input: the numbers $x_1, x_2, \ldots, x_n$, their corresponding probabilities $p_1, p_2, \ldots, p_n$ and the number k. Set $p_0 = 0$.

• Generate k random numbers: $U_i \sim Unif[0, 1]$, $i \in \{1, \ldots, k\}$.

• For each $i \in \{1, \ldots, k\}$ if

$$p_0 + p_1 + \cdots + p_{j-1} < U_i \le p_0 + p_1 + \cdots + p_j$$

for some $j \in \{1, \ldots, n\}$, then we take $Y_i = x_j$.

Note that for each fixed $i \in \{1, \ldots, k\}$ the index j is the smallest value from $\{1, \ldots, n\}$ such that $U_i \le p_1 + \cdots + p_j$. Also, we mention that $Y_i = Q(U_i)$, $i \in \{1, \ldots, k\}$, where Q is the quantile function of X.

• Output: the random numbers $Y_i, i \in \{1, \ldots, k\}$.

We verify the above procedure: for $j \in \{1, \ldots, n\}$ and $i \in \{1, \ldots, k\}$ we have

$$P(``x_j \text{ is generated}") = P(p_0 + p_1 + \cdots + p_{j-1} < U_i \le p_0 + p_1 + \cdots + p_j) = p_j,$$

since $U_i \sim Unif[0, 1]$. In the following, we give the implementation in Matlab and test it.

```
function Y=discret(x,p,k)
% k random numbers are generated in the vector Y
% according to the values in the vector x and their probabilities in the vector p
p=[0 p]; q=cumsum(p);
Y=NaN(1,k); U=rand(1,k);
for i=1:k
    j=0;
    while q(j+1)<U(i)
        j=j+1;
    end
    Y(i)=x(j);
end
end
```

```
>> Y=discret([1,2,3],[0.25,0.25,0.5],1000);
>> tabulate(Y)
 Value Count Percent
     1 233 23.30%
     2 249 24.90%
     3 518 51.80%
```

Note that one can use in Matlab also the **randsample** function, which generates random numbers following a given discrete distribution. ▲

Example 2.33. We simulate a random value for the geometric distribution $Geo(p)$ by using $Unif[0, 1]$, where $p \in (0, 1)$.

Solution: We give three methods.

For the first one, we shall use the inversion method. Let $V \sim Unif[0,1]$. Recall that for $X \sim Geo(p)$ we have for $k \in \mathbb{N}$ the probability $P(X = k) = p(1-p)^k$, hence

$$\sum_{j=0}^{k} P(X = j) = \sum_{j=0}^{k} p(1-p)^j = 1 - (1-p)^{k+1}.$$

If we find $k \in \mathbb{N}$ such that $1 - (1-p)^k < V \le 1 - (1-p)^{k+1}$, then we take $X = k$. Observe that for $k \in \mathbb{N}$ we have

$$1 - (1-p)^k < V \le 1 - (1-p)^{k+1} \Leftrightarrow k < \frac{\log(1-V)}{\log(1-p)} \le k + 1.$$

By the above reasoning, we have $X = \left\lceil \dfrac{\log(1-V)}{\log(1-p)} \right\rceil - 1 \sim Geo(p)$.

We generate n values following the $Geo(p)$ distribution in Matlab with the above method, for some $n \in \mathbb{N}^*$ and $p \in (0,1)$:

```
>>n=100; p=0.6;
>>X=ceil(log(rand(1,n))/log(1-p))-1; % n values for Geo(p) distribution
```

For the second method, we use the quantile function, as described at beginning of this section. Note that, in view of the above reasoning, the quantile function Q of $X \sim Geo(p)$ is $Q(v) = \left\lceil \dfrac{\log(1-v)}{\log(1-p)} \right\rceil - 1$, $v \in (0,1)$. The corresponding Matlab code is:

```
>>n=100; p=0.6;
>>X=geoinv(rand(1,n),p); % n values for Geo(p) distribution
```

For the third method, we use the description of the distribution in Section 2.4.9. We generate independent random variables $U_1, U_2, \ldots \sim Unif[0,1]$. If we find the first index $k \in \mathbb{N}^*$ such that $U_k \le p$ (i.e., we have a success), then we take $X = k - 1$. Since X counts the number of failures that occur before the first success, $X \sim Geo(p)$. The corresponding Matlab function is:

```
function X=geo_sim(n,p)
X=zeros(1,n);
for i=1:n
  while rand>p
   X(i)=X(i)+1;
  end
end
end

>>X=geo_sim(100,0.6);
```

▲

Example 2.34. We generarate a random number according to the Poisson distribution with given parameter $\lambda > 0$, by three methods.

(1) We generate the random variables $T_1, T_2, \ldots \sim Exp(1)$ until we obtain the smallest index $k \in \mathbb{N}^*$ such that $T_1 + \ldots + T_k > \lambda$. Then we take $X = k - 1$. So, $X \sim Poiss(\lambda)$ (for the proof see Problem 2.5.3-3)).

(2) We generate the random variables $U_1, U_2, \ldots \sim Unif[0, 1]$ until we obtain the smallest index $k \in \mathbb{N}^*$ such that $U_1 \cdot U_2 \cdot \ldots \cdot U_k < e^{-\lambda}$ and we take $X = k - 1$. Then $X \sim Poiss(\lambda)$ (for the proof see (1) and Example 2.31).

(3) We apply the inversion method. We take $U \sim Unif[0, 1]$ and let $p_j = \dfrac{\lambda^j}{j!} e^{-\lambda}$, $j \in \mathbb{N}$. We find the smallest index $k \in \mathbb{N}^*$ such that $U \leq p_0 + p_1 + \ldots + p_k$ and take $X = k$. Then $X \sim Poiss(\lambda)$. ▲

2.7.2 Acceptance-Rejection Method

We present the acceptance-rejection method, also called accept-reject algorithm or rejection sampling, that generates random values for a given continuous distribution in $\mathbb{R}$. The method works similarly for joint density functions in higher dimensions.

We want to generate random values for the random variable X with density function f_X, by using another random variable Y with density function f_Y, which is "close" to f_X and for which we already have an efficient algorithm to generate random values. Let

$$D_X = \{x \in \mathbb{R} : f_X(x) > 0\} \subseteq D_Y = \{y \in \mathbb{R} : f_Y(y) > 0\}$$

and assume that there exists a constant $c > 1$ such that $\sup\limits_{y \in D_Y} \dfrac{f_X(y)}{f_Y(y)} = c$.

The accept-reject algorithm for generating random values for X is:

(1) Generate a random value for Y.

(2) Generate a random value for $U \sim Unif[0, 1]$ independent of Y.

(3) If $U \leq \dfrac{f_X(Y)}{c f_Y(Y)}$,

then, set $X = Y$ ("accept");

else, go back to step (1) ("reject").

Since $P\big(f_Y(Y) = 0\big) = \displaystyle\int_{f_Y^{-1}(\{0\})} f_Y(y) dy = 0$, we have $f_Y(Y) > 0$ a.s. and

thus $\dfrac{f_X(Y)}{c f_Y(Y)}$ is well defined a.s. The number of iterations that are required until a value of X is generated (i.e., a value of Y is accepted) is a random variable Z, where $Z - 1 \sim Geo(p)$ with parameter p equal to

$$P\left(U \leq \frac{f_X(Y)}{cf_Y(Y)}\right) = P\left((U,Y) \in \left\{(u,y) \in \mathbb{R}^2 : y \in D_Y, u \leq \frac{f_X(y)}{cf_Y(y)}\right\}\right)$$

$$= \int_{D_Y}\left(\int_0^{\frac{f_X(y)}{cf_Y(y)}} 1du\right) f_Y(y)dy = \frac{1}{c}\int_{D_Y} f_X(y)dy = \frac{1}{c}\int_{D_X} f_X(y)dy = \frac{1}{c}.$$

In practice, the values of c that are close to 1 are preferred, because the algorithm should terminate faster when the parameter p (i.e., the probability of accept) is closer to 1.

Writing successively for each $y \in \mathbb{R}$

$$F_Y\left(y\Big|U \leq \frac{f_X(Y)}{cf_Y(Y)}\right) = P\left(Y \leq y\Big|U \leq \frac{f_X(Y)}{cf_Y(Y)}\right) = \frac{P\left(Y \leq y, U \leq \frac{f_X(Y)}{cf_Y(Y)}\right)}{P\left(U \leq \frac{f_X(Y)}{cf_Y(Y)}\right)}$$

$$= c\underbrace{\int}_{D_Y \cap (-\infty, y]}\left(\int_0^{\frac{f_X(v)}{cf_Y(v)}} 1du\right) f_Y(v)dv = \int_{-\infty}^y f_X(v)dv = P(X \leq y) = F_X(y),$$

where we use the fact that $D_X \subseteq D_Y$, we deduce that the algorithm generates values for the random variable X.

Example 2.35. Let $f_X : \mathbb{R} \to \mathbb{R}$

$$f_X(x) = \begin{cases} x(1 + 2x^2), & \text{if } 0 \leq x \leq 1 \\ 0, & \text{otherwise} \end{cases}$$

be a density function of the random variable X. Generate random values of X by using the accept-reject algorithm and $Y \sim Unif[0,1]$.
Solution: The function $f_Y(y) = 1$, if $y \in [0,1]$, and $f_Y(y) = 0$, otherwise, is a density function of Y. Then $D_X = (0,1] \subset D_Y = [0,1]$,

$$\sup_{y \in D_Y} \frac{f_X(y)}{f_Y(y)} = f_X(1) = 3.$$

The accept-reject algorithm for generating random values for X is:
(1) Generate independent random values for $Y, U \sim Unif[0,1]$.
(2) If $U \leq \frac{1}{3}Y(1 + 2Y^2)$,
then, set $X = Y$ ("accept");
else, go back to step (1) ("reject").   ▲

Example 2.36. We will describe the accept-reject algorithm for a random vector which generates random values uniformly distributed within the unit circle $\mathcal{C}[(0,0);1]$ (the circle with center in $(0,0)$ and radius 1).

Solution: The random vector $\mathbb{X} = (X_1, X_2)$ follows the uniform distribution within the unit circle $\mathcal{C}[(0,0);1]$, if

$$f_{\mathbb{X}}(x_1, x_2) = \begin{cases} \dfrac{1}{\pi}, & \text{if } x_1^2 + x_2^2 \leq 1 \\ 0, & \text{otherwise} \end{cases}$$

is a joint density function of $\mathbb{X}$. Let $Y_1, Y_2 \sim Unif[-1,1]$ be independent random variables and consider a joint density function of $\mathbb{Y} = (Y_1, Y_2)$ given by

$$f_{\mathbb{Y}}(y_1, y_2) = \begin{cases} \dfrac{1}{4}, & \text{if } (y_1, y_2) \in [-1,1] \times [-1,1] \\ 0, & \text{otherwise.} \end{cases}$$

We have $D_{\mathbb{X}} = \mathcal{C}[(0,0);1] \subseteq D_{\mathbb{Y}} = [-1,1] \times [-1,1]$ and $\sup\limits_{\mathbb{y} \in D_{\mathbb{Y}}} \dfrac{f_{\mathbb{X}}(\mathbb{y})}{f_{\mathbb{Y}}(\mathbb{y})} = \dfrac{4}{\pi}$.

The accept-reject algorithm for generating random values for (X_1, X_2) is:

(1) Generate $\mathbb{Y} = (Y_1, Y_2)$, where $Y_1, Y_2 \sim Unif[-1,1]$ are independent.

(2) Generate a random value for $U \sim Unif[0,1]$ independent of $\mathbb{Y}$.

(3) If $U \leq \dfrac{f_{\mathbb{X}}(\mathbb{Y})}{\frac{4}{\pi} f_{\mathbb{Y}}(\mathbb{Y})}$,

then, set $(X_1, X_2) = (Y_1, Y_2)$ ("accept") ;

else, go back to step (1) ("reject").

Note that

$$\dfrac{f_{\mathbb{X}}(y_1, y_2)}{\frac{4}{\pi} f_{\mathbb{Y}}(y_1, y_2)} = \begin{cases} 1, & \text{if } y_1^2 + y_2^2 \leq 1 \\ 0, & \text{otherwise.} \end{cases}$$

But $U \in (0,1)$ a.s., hence the inequality in step (3) is in fact equivalent to the inequality $Y_1^2 + Y_2^2 \leq 1$. So, we can rewrite the above accept-reject algorithm for generating a random point (X_1, X_2) within the unit circle $\mathcal{C}[(0,0);1]$:

(1) Generate (Y_1, Y_2), where $Y_1, Y_2 \sim Unif[-1,1]$ are independent.

(2) If $Y_1^2 + Y_2^2 \leq 1$,

then, the point is within the unit circle (it is accepted);

set $(X_1, X_2) = (Y_1, Y_2)$;

else, reject the point and go back to step (1).

```
function accept_reject_circle(N_sim)
clf;axis equal;hold on;axis([-1 1 -1 1]);
for i=1:N_sim
  accept=0;
  while ~accept
  Y=unifrnd(-1,1,1,2);
  if Y(1)^2+Y(2)^2<=1
   accept=1; X=Y;
  end
  end
  plot(X(1),X(2),'ok',...
      'MarkerFaceColor','k','Markersize',4);
end
>>accept_reject_circle(2000)
```

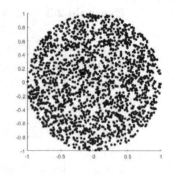

Fig. 2.2: Example 2.36 ▲

2.7.3 Box–Muller Algorithm

The Box–Muller algorithm is a random number sampling method for generating pairs of independent, standard normally distributed random numbers, given a generator of uniformly distributed random numbers. This algorithm first appeared in [Box and Muller (1958)] (see also [Ross (2014), Example 7b, p. 280], [Ghahramani (2005), Example 8.27, p. 358]).

Example 2.37. Let $U, V \sim Unif[0,1]$ be independent random variables. Define

$$X = \sqrt{-2\log(U)}\cos(2\pi V), \quad Y = \sqrt{-2\log(U)}\sin(2\pi V).$$

Then X and Y are independent $N(0,1)$ distributed random variables.
Proof: Define the random variables $R = \sqrt{-2\log(U)}$ and $\Phi = 2\pi V$. Then for $r \in \mathbb{R}$

$$P(R \le r) = P\Big(-2\log(U) \le r^2\Big) = 1 - P\Big(U < \exp\Big\{-\frac{r^2}{2}\Big\}\Big).$$

Since $U \sim Unif[0,1]$, we have

$$P(R \le r) = 1 - \exp\Big\{-\frac{r^2}{2}\Big\}, \; r \in \mathbb{R},$$

which implies that the following function is a density function of R

$$f_R(r) = \begin{cases} r\exp\Big\{-\dfrac{r^2}{2}\Big\}, & \text{if } r > 0 \\ 0, & \text{if } r \le 0. \end{cases}$$

Since $V \sim Unif[0,1]$, we have the following density function of Φ

$$f_\Phi(\varphi) = \begin{cases} \dfrac{1}{2\pi}, & \text{if } 0 \le \varphi \le 2\pi \\ 0, & \text{otherwise.} \end{cases}$$

Since U and V are independent, R and Φ are also independent (see Theorem 2.12) and the following function is a joint density function of the random vector (R, Φ)

$$f_{R,\Phi}(r, \varphi) = f_R(r)f_\Phi(\varphi) = \begin{cases} \dfrac{r}{2\pi}\exp\left\{-\dfrac{r^2}{2}\right\}, & \text{if } r > 0, \varphi \in (0, 2\pi) \\ 0, & \text{otherwise.} \end{cases}$$

Consider the polar coordinates transform

$$g\colon (0, \infty) \times \big((0, \pi) \cup (\pi, 2\pi)\big) \to \big(\mathbb{R} \times (0, \infty)\big) \cup \big(\mathbb{R} \times (-\infty, 0)\big)$$

given by

$$g(r, \varphi) = \big(r\cos(\varphi), r\sin(\varphi)\big), \quad r > 0, \varphi \in (0, \pi) \cup (\pi, 2\pi).$$

Its inverse $g^{-1} = h = (h_1, h_2)$ is the function

$$h\colon \big(\mathbb{R} \times (0, \infty)\big) \cup \big(\mathbb{R} \times (-\infty, 0)\big) \to (0, \infty) \times \big((0, \pi) \cup (\pi, 2\pi)\big)$$

defined by

$$h_1(x, y) = \sqrt{x^2 + y^2}, \quad h_2(x, y) = \begin{cases} \arccos\left(\dfrac{x}{\sqrt{x^2+y^2}}\right), & \text{if } y > 0 \\ 2\pi - \arccos\left(\dfrac{x}{\sqrt{x^2+y^2}}\right), & \text{if } y < 0. \end{cases}$$

We see that

$$(R, \Phi) \in (0, \infty) \times \big((0, \pi) \cup (\pi, 2\pi)\big) \quad \text{a.s.,}$$

$(X, Y) = g(R, \Phi)$ a.s. and thus $(R, \Phi) = g^{-1}(X, Y) = h(X, Y)$ a.s. Denote by J_h the Jacobian matrix of $h = g^{-1}$. Since $\big|\det(J_h(x, y))\big| = \dfrac{1}{\sqrt{x^2+y^2}}$ for $x \in \mathbb{R}$, $y \in \mathbb{R}^*$, by Theorem 2.14, we have a joint density function of (X, Y) given by

$$\begin{aligned} f_{X,Y}(x, y) &= f_{R,\Phi}(h(x, y))|\det(J_h(x, y))| \\ &= \frac{1}{2\pi}\exp\left(-\frac{x^2+y^2}{2}\right) \\ &= \underbrace{\frac{1}{\sqrt{2\pi}}\exp\left(-\frac{x^2}{2}\right)}_{=f_X(x)} \cdot \underbrace{\frac{1}{\sqrt{2\pi}}\exp\left(-\frac{y^2}{2}\right)}_{=f_Y(y)}, \end{aligned}$$

hence, X and Y are $N(0, 1)$ distributed and independent. ▲

The corresponding function in Matlab is:

```
function [X,Y]=normal(k)
U=rand(1,k); V=rand(1,k);
X=sqrt(-2*log(U)).*cos(2*pi*V); Y=sqrt(-2*log(U)).*sin(2*pi*V);
end
```

```
%we generate 100 pairs of values for two N(0,1) independent random variables
>>[X,Y]=normal(100);
```

2.7.4   *Random Shuffling*

In order to generate a random permutation (or shuffle) of a finite sequence, we consider two versions of the Fisher–Yates algorithm (see [Knuth (1998), pp. 145–146]) implemented in Matlab.

The backward version:

```
function X=bwdshuffle(X)
n=length(X);
for i = n:-1:2
    j=randi(i);
    X([j i])=X([i j]);
end
end
```

The forward version:

```
function X=fwdshuffle(X)
n=length(X);
for i = 1:n-1
    j=i-1+randi(n-i+1);
    X([j i])=X([i j]);
end
end
```

Each permutation of a finite sequence is generated with the same probability by each of the above shuffles. One can also use the Matlab function `randperm` to permute the indices of a finite sequence.

Example 2.38. Using the forward random shuffling algorithm in Matlab, we simulate the shuffling of a standard deck of 52 cards and some independent random hands (each hand is given by five cards dealt from the top of the shuffled deck). Then, we estimate the probability for exactly one pair in a random hand and compare it with the exact corresponding value.

Solution:

```
function p=poker_onepair(k)
%C=clubs, D=diamonds, H=hearts, S=spades
%J=jack, Q=queen, K=king, A=ace
cards={'2C','3C','4C','5C','6C','7C','8C','9C','10C','JC','QC','KC','AC'...
       '2D','3D','4D','5D','6D','7D','8D','9D','10D','JD','QD','KD','AD'...
       '2H','3H','4H','5H','6H','7H','8H','9H','10H','JH','QH','KH','AH'...
       '2S','3S','4S','5S','6S','7S','8S','9S','10S','JS','QS','KS','AS'};
count=0;
for i=1:k
  cards=fwdshuffle(cards);
  H=cards(1:5);
  H_num=H;
  for j=1:5
    H_num{j}=H{j}(1:end-1);
  end
  if length(unique(H_num))==4
      count=count+1;
  end
end
p=count/k;
end
```

```
>> poker_onepair(100000)
ans = 0.4223
```

Since exactly one pair in a poker hand means that exactly two cards have the same rank, the value of the exact probability is

$$\frac{C(13,1) \cdot C(4,2) \cdot C(12,3) \cdot 4^3}{C(52,5)} = \frac{352}{833} \approx 0.4226,$$

where, for the numerator: first we count the choices of the rank for the pair, next the choices of the suits for the pair, next the choices of the ranks for the remaining three cards and finally the choices of the suits for the remaining three cards.   ▲

2.7.5 *Solved Problems*

Problem 2.7.1. Using the acceptance-rejection method, write a function in Matlab that returns a random number that follows the $Beta(a,b)$, $a, b > 0$, distribution, by generating only random numbers that follow the $Unif[0,1]$ distribution.

Solution 2.7.1:

```
function X=my_beta(a,b)
fX=@(x) betapdf(x,a,b);
% which is equivalent with fX=@(x) x^(a-1)*(1-x)^(b-1)/beta(a,b);
[~,fval]=fminbnd(@(x)-fX(x),0,1);
c=-fval;
accept=0;
while ~accept
    Y=unifrnd(0,1);
    U=unifrnd(0,1);
    if U<=fX(Y)/c
        X=Y;
        accept=1;
    end
end
end
```

```
>>X=my_beta(1,2);
```

Problem 2.7.2. 1) If $U \sim Unif[0,1]$, then prove that $X = \lceil nU \rceil$ has $Unid(n)$ distribution, where $n \in \mathbb{N}^*$ and $\lceil \cdot \rceil$ is the ceiling function.

2) Let $Y \sim Unif[0,1]$. Find the probability mass function of $Z = \lfloor 10Y \rfloor$, where $\lfloor \cdot \rfloor$ is the floor function.

Solution 2.7.2: 1) Since $U \in [0,1]$ a.s., we have $\lceil nU \rceil \in \{1, \ldots, n\}$ a.s. and

$$P(X = k) = P\left(\frac{k-1}{n} < U \le \frac{k}{n}\right) = \frac{1}{n},$$

for each $k \in \{1, \ldots, n\}$. If we would like to simulate values for X without using the `ceil` function, we may find the smallest integer k such that $U \le \frac{k}{n}$ and then take $X = k$.

```
function X=my_unid(n)
U=rand; X=0;
while X/n<U
    X=X+1;
end
end
```

```
>>X=my_unid(10);
```

2) Since $Y \in [0,1]$ a.s., we have $\lfloor 10Y \rfloor \in \{0, \ldots, 9\}$ a.s. and

$$P(Z = k) = P\left(\frac{k}{10} \le Y < \frac{k+1}{10}\right) = \frac{1}{10}$$

for each $k \in \{0, \ldots, 9\}$. Hence, $Z + 1 \sim Unid(10)$. If we would like to simulate values for Z without using the **floor** function, we may find the smallest integer k such that $Y < \dfrac{k+1}{10}$ and then take $Z = k$.

```
function Z=my_unidplus1
Z=10; Y=rand;
while Y<Z/10
    Z=Z-1;
end
end
```

```
>>Z=my_unidplus1;
```

Problem 2.7.3. A darts player simulates throwing darts at a virtual dartboard centered in the origin of the cartesian coordinate system such that the coordinates of the hit are two independent random variables that follow the $N(0,1)$ distribution.

1) Using the Box–Muller algorithm, simulate graphically in Matlab independent throws. Estimate the probability of hitting the bullseye, which is the interior of the circle centered in the origin with radius 0.5, and the probability of hitting the enclosed region in the figure below, which is the intersection of the angle between 240° and 300° and the annulus bounded by the circles with radii of lengths 2 and 3.

2) Find the exact values of the probabilities from 1).

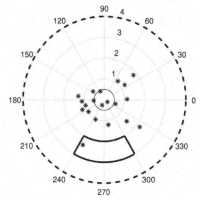

Fig. 2.3: Problem 2.7.3

Solution 2.7.3: 1)

```
function darts(k)
clf; t = 0:.01:2*pi; polar(t,4*ones(size(t)),'--k');
```

```
hold on; t = 4*pi/3:.01:5*pi/3;
polar(t,3*ones(size(t)),'-b'); polar(t,2*ones(size(t)),'-b');
plot([-3/2 -2/2],[-3*sqrt(3)/2 -2*sqrt(3)/2],'-b');
plot([3/2 2/2],[-3*sqrt(3)/2 -2*sqrt(3)/2],'-b');
set(findobj('type','line'),'linewidth',2);
rectangle('Position',[-0.5,-0.5,1,1],'Curvature',[1,1],'FaceColor','r');
R=sqrt(-2*log(rand(1,k)));T=2*pi*rand(1,k);
X=R.*cos(T); Y=R.*sin(T); plot(X,Y,'*k');
bullseye_estimated_probability=mean(R<0.5)
region_estimated_probability=mean(R<3&R>2&T>4*pi/3&T<5*pi/3)
end
```

```
>> darts(10000)
bullseye_estimated_probability = 0.1187
region_estimated_probability = 0.0195
```

2) Let $\left(R\cos(T), R\sin(T)\right)$ be the coordinates of the hit, where $R = \sqrt{-2\log(U)}$ and $T = 2\pi V$ with $U, V \sim Unif[0,1]$ independent random variables. For every $x \geq 0$ we have

$$P(R \leq x) = P\left(\sqrt{-2\log(U)} \leq x\right)$$

$$= P\left(U \geq \exp\left\{-\frac{x^2}{2}\right\}\right) = 1 - \exp\left\{-\frac{x^2}{2}\right\}.$$

The probability to hit the bullseye is

$$P\left(R \leq \frac{1}{2}\right) = 1 - \exp\left\{-\frac{1}{8}\right\} \approx 0.1175.$$

The probability to hit the enclosed region is

$$P\left(\{2 \leq R \leq 3\} \cap \left\{\frac{4\pi}{3} \leq T \leq \frac{5\pi}{3}\right\}\right)$$

$$= \left(\exp\left\{-\frac{2^2}{2}\right\} - \exp\left\{-\frac{3^2}{2}\right\}\right) \cdot \frac{\frac{5\pi}{3} - \frac{4\pi}{3}}{2\pi} \approx 0.0207.$$

Problem 2.7.4. A group of $n \in \mathbb{N}^*$ employees of a company decides to exchange Christmas gifts. Each employee brings one gift. The gifts are randomly shuffled and then distributed such that each employee gets exactly one gift.

1) Simulate in Matlab the gift exchange and estimate the probability that there is no employee that gets her/his gift back, for some n given as input.

2) Find the probability p_n that there is no employee that gets her/his gift back.

3) Prove that $\lim_{n\to\infty} p_n = \dfrac{1}{e}$.

Solution 2.7.4: 1) We shall use a shuffle algorithm.

```
function p=gifts_exchange(n,N_sim)
count=0;
for i=1:N_sim
    if (1:n)~=bwdshuffle(1:n)
        count=count+1;
    end
end
p=count/N_sim;
end
```

```
>> gifts_exchange(50,10000)
ans = 0.3676
```

2) For $i \in \{1, \ldots, n\}$ let A_i be the event that the ith employee gets her/his gift back. By Theorem 1.5, we have

$$P\left(\bigcup_{i=1}^{n} A_i\right) = \sum_{i=1}^{n} P(A_i) - \sum_{1 \le i < j \le n} P(A_i \cap A_j) + \ldots + (-1)^{n-1} P\left(\bigcap_{i=1}^{n} A_i\right)$$

$$= \sum_{i=1}^{n} \frac{(n-1)!}{n!} - \sum_{1 \le i < j \le n} \frac{(n-2)!}{n!} + \ldots + (-1)^{n-1} \frac{0!}{n!}$$

$$= \sum_{k=1}^{n} (-1)^{k-1} \cdot C(n,k) \cdot \frac{(n-k)!}{n!} = \sum_{k=1}^{n} \frac{(-1)^{k-1}}{k!}.$$

Hence, $p_n = 1 - P\left(\bigcup_{i=1}^{n} A_i\right) = \sum_{k=0}^{n} \frac{(-1)^k}{k!}.$

3) In view of the Taylor expansion of the exponential function, we have

$$e^{-1} = \sum_{k=0}^{\infty} \frac{(-1)^k}{k!} \quad \text{and thus} \quad \lim_{n \to \infty} p_n = \frac{1}{e} \approx 0.3679.$$

2.8 Bayesian Networks

Let $(\Omega, \mathcal{F}, P)$ be a probability space. All random variables and random vectors considered throughout this section are discrete and it is assumed that all conditional probabilities exist (i.e., the conditioning is with respect to events that have nonzero probabilities).

Definition 2.22. The events $A, B \in \mathcal{F}$ are **conditionally independent** given the event $C \in \mathcal{F}$, if $P(A \cap B | C) = P(A|C)P(B|C)$. ◆

Example 2.39. Let $A, B, C \in \mathcal{F}$. The following equivalences hold:

$$P(A \cap B | C) = P(A|C)P(B|C) \Leftrightarrow P(A|B \cap C) = P(A|C)$$

$$\Leftrightarrow P(B|A \cap C) = P(B|C).$$

Proof: First, we prove

$$P(A \cap B|C) = P(A|C)P(B|C) \implies P(A|B \cap C) = P(A|C).$$

We write

$$P(A|B \cap C) = \frac{P(A \cap B \cap C)}{P(B \cap C)} = \frac{P(A \cap B|C)P(C)}{P(B|C)P(C)}$$

$$= \frac{P(A|C)P(B|C)}{P(B|C)} = P(A|C).$$

Next, we prove

$$P(A|B \cap C) = P(A|C) \implies P(A \cap B|C) = P(A|C)P(B|C).$$

We have

$$P(A \cap B|C) = \frac{P(A \cap B \cap C)}{P(C)} = \frac{P(A|B \cap C)P(B \cap C)}{P(C)}$$

$$= \frac{P(A|C)P(B \cap C)}{P(C)} = P(A|C)P(B|C).$$

The second equivalence is obvious, since $B \cap A = A \cap B$. ▲

Example 2.40. We give an example of an event D and three events A, B, C such that each one is conditionally independent of the intersection of the other two given D (e.g., A is conditionally independent of $B \cap C$ given D, etc.), but A, B and C are not conditionally independent given D. Moreover, we choose the events such that D is neither the certain event Ω, nor $A \cup B \cup C$.

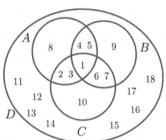

Fig. 2.4: Example 2.40

Solution: Let $\Omega = \{1, \ldots, 19\}$, $\mathcal{F} = \mathcal{P}(\Omega)$ and $P(S) = \dfrac{|S|}{19}$, for $S \subseteq \Omega$. Let $A = \{1, 2, 3, 4, 5, 8\}$, $B = \{1, 4, 5, 6, 7, 9\}$, $C = \{1, 2, 3, 6, 7, 10\}$ and $D = \{1, \ldots, 18\}$ (see Figure 2.4). We have

$$P(A \cap B \cap C|D) = P(A|D)P(B \cap C|D)$$

$$= P(B|D)P(A \cap C|D) = P(C|D)P(A \cap B|D) = \frac{1}{18},$$

but

$$P(A \cap B \cap C|D) = \frac{1}{18} \neq P(A|D)P(B|D)P(C|D) = \frac{1}{27}. \qquad \blacktriangle$$

Definition 2.23. Let $X, Y_1, \ldots, Y_m, Z_1, \ldots, Z_n$ be discrete random variables with ranges $\mathcal{X}, \mathcal{Y}_1, \ldots, \mathcal{Y}_m, \mathcal{Z}_1, \ldots, \mathcal{Z}_n$, respectively. The random variable X is **conditionally independent** of $Y_1, \ldots, Y_m$ given the random variables $Z_1, \ldots, Z_n$, if for $x \in \mathcal{X}, y_i \in \mathcal{Y}_i, i = \overline{1,m}, z_j \in \mathcal{Z}_j, j = \overline{1,n}$, it holds

$$P\left(X = x, \bigcap_{i=1}^{m}\{Y_i = y_i\} \,\middle|\, \bigcap_{j=1}^{n}\{Z_j = z_j\}\right)$$

$$= P\left(X = x \,\middle|\, \bigcap_{j=1}^{n}\{Z_j = z_j\}\right) P\left(\bigcap_{i=1}^{m}\{Y_i = y_i\} \,\middle|\, \bigcap_{j=1}^{n}\{Z_j = z_j\}\right). \qquad \blacklozenge$$

Notation: In order to simplify the writing of some formulae we use the following notations. Let $\mathbb{U}$ be a discrete random vector with range $\mathcal{U} \subset \mathbb{R}^m$ and let $\mathbb{V}$ be a discrete random vector with range $\mathcal{V} \subset \mathbb{R}^n$, $m, n \in \mathbb{N}^*$. We denote $P[\mathbb{U}] : \mathcal{U} \to [0,1]$,

$$P[\mathbb{U}](\mathbf{u}) = P(\mathbb{U} = \mathbf{u}) \text{ for all } \mathbf{u} \in \mathcal{U}$$

and $P[\mathbb{U}|\mathbb{V}] : \mathcal{U} \times \mathcal{V} \to [0,1]$,

$$P[\mathbb{U}|\mathbb{V}](\mathbf{u}, \mathbf{v}) = P(\mathbb{U} = \mathbf{u}|\mathbb{V} = \mathbf{v}) \text{ for all } \mathbf{u} \in \mathcal{U}, \mathbf{v} \in \mathcal{V}.$$

Note that $P[\mathbb{U}]$ is the joint probability mass function of $\mathbb{U}$, see Definition 2.6.

Remark 2.16. Let U_1 and U_2 be two discrete random variables.
(1) U_1 and U_2 have the same distribution if and only if $P[U_1] = P[U_2]$.
(2) If U_1 and U_2 are independent, then we have $P[U_1|U_2] = P[U_1]$ and $P[U_2|U_1] = P[U_2]$. $\qquad \blacktriangledown$

Remark 2.17. (1) Using the above notation, Definition 2.23 states:
The random variable X is conditionally independent of the random variables $Y_1, \ldots, Y_m$ given the random variables $Z_1, \ldots, Z_n$ if and only if

$$P[X, Y_1, \ldots, Y_m | Z_1, \ldots, Z_n] = P[X | Z_1, \ldots, Z_n] P[Y_1, \ldots, Y_m | Z_1, \ldots, Z_n].$$

(2) Let X be a random variable which is conditionally independent of the random variables $Y_1, \ldots, Y_m$ given the random variables $Z_1, \ldots, Z_n$. If $i_1, \ldots, i_k \in \{1, \ldots, m\}$ are distinct indices, then, by Example 2.39, we have

$$P[X | Y_{i_1}, \ldots, Y_{i_k}, Z_1, \ldots, Z_n] = P[X | Z_1, \ldots, Z_n],$$

$$P[Y_{i_1}, \ldots, Y_{i_k} | X, Z_1, \ldots, Z_n] = P[Y_{i_1}, \ldots, Y_{i_k} | Z_1, \ldots, Z_n]. \qquad \blacktriangledown$$

Example 2.41. The random variables X and Y are conditionally indepen-dent given the random variable $Z \Leftrightarrow P[X,Y|Z] = P[X|Z]P[Y|Z]$
$\Leftrightarrow P[X|Y,Z] = P[X|Z] \Leftrightarrow P[Y|X,Z] = P[Y|Z]$.　　　　▲

In what follows we will present a few aspects about discrete Bayesian networks. For the introductory notions we mainly follow Chapter 1 of the book [Neapolitan (2003)].

A Bayesian network is a probabilistic graphical model that represents a set of random variables and their probabilistic dependencies by using a directed acyclic graph.

A **directed graph** is a pair (V, E), where V is a finite nonempty set, whose elements are called **nodes** (or **vertices**), and E is a set of ordered pairs of distinct elements of V. Elements of E are called **edges** and if $(X, Y) \in E$, we say there is an edge from X to Y. Suppose we have a set of nodes $\{X_1, X_2, ..., X_k\}$, where $k \in \mathbb{N}, k \geq 2$, such that $(X_{i-1}, X_i) \in E$ for $i \in \{2, \ldots, k\}$. Then we call the set of edges connecting the k nodes a **path** from X_1 to X_k. A **directed cycle** is a path from a node to itself. A directed graph is called a **directed acyclic graph**, if it contains no directed cycles. Given a directed acyclic graph (V, E) and nodes X and Y in V such that $X \neq Y$: Y is called a **parent** of X, if there is an edge from Y to X; Y is called a **descendent** of X and X is called an **ancestor** of Y, if there is a path from X to Y; Y is called a **nondescendent** of X, if Y is not a descendent of X.

Denote the set of parents of X by $pa(X)$, the set of descendents by $de(X)$ and the set of nondescendents of X by $nd(X)$. Observe that $pa(X) \subseteq nd(X)$. A node X is a **root node**, if its parents set is empty $pa(X) = \emptyset$. We identify the nodes $X_1, \ldots, X_n$ of V $(n = \#V)$ with random variables defined on the same probability space $(\Omega, \mathcal{F}, P)$ and we assume that $P[X_j|pa(X_j)]$, $j \in \{1, \ldots, n\}$, are known; we make the convention $P[X_j|pa(X_j)] = P[X_j]$, if X_j is a root node. With this identification, we denote the directed acyclic graph by $((V, E),P)$. If we view $((V, E),P)$ as a causal structure, $pa(X_j)$ are the direct causes of X_j and $de(X_j)$ are the effects of X_j, $j \in \{1, \ldots, n\}$.

We say that a directed acyclic graph $((V, E),P)$ satisfies the so-called **Markov condition**, if each random variable $X \in V$ is conditionally in-dependent of the set of all its nondescendents $nd(X)$ given the set of all its parents $pa(X)$. If X is a root node (i.e., $pa(X) = \emptyset$), the Markov con-dition means that X is independent of $nd(X)$. If a directed acyclic graph $((V, E),P)$ satisfies the Markov condition, we call it a **Bayesian network**.

Theorem 2.16. *Let* $((V,E),P)$ *be a Bayesian network with* $V = \{X_1,\ldots,X_n\}$, $n \in \mathbb{N}^*$. *The joint probability mass function of* $(X_1,\ldots,X_n)$ *is given by*

$$P[X_1,\ldots,X_n] = P[X_1|pa(X_1)] \cdot \ldots \cdot P[X_n|pa(X_n)].$$

Example 2.42. A Bayesian network is given in Figure 2.5, where $X_1,\ldots,X_5$ are binary random variables. We list properties of this Bayesian network:

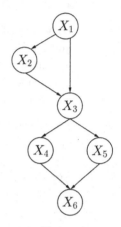

• the sets of parents, descendents and nondescendents are:

$pa(X_1)=\emptyset$, $pa(X_2)=\{X_1\}$, $pa(X_3)=\{X_1,X_2\}$,
$pa(X_4)=pa(X_5)=\{X_3\}$, $pa(X_6)=\{X_4,X_5\}$,
$de(X_1)=\{X_2,X_3,X_4,X_5,X_6\}$,
$de(X_2)=\{X_3,X_4,X_5,X_6\}$, $de(X_3)=\{X_4,X_5,X_6\}$,
$de(X_4)=de(X_5)=\{X_6\}$, $de(X_6)=\emptyset$,
$nd(X_1)=\emptyset$, $nd(X_2)=\{X_1\}$, $nd(X_3)=\{X_1,X_2\}$,
$nd(X_4)=\{X_1,X_2,X_3,X_5\}$,
$nd(X_5)=\{X_1,X_2,X_3,X_4\}$,
$nd(X_6)=\{X_1,X_2,X_3,X_4,X_5\}$;

Fig. 2.5: Example 2.42

• the probabilities associated to the nodes, which fully define the Bayesian network, are given by:

$P[X_1], P[X_2|X_1], P[X_3|X_1,X_2], P[X_4|X_3], P[X_5|X_3], P[X_6|X_4,X_5]$;

• conditional independent random variables according to the property of Bayesian networks are (see Remark 2.17):

  • X_4 is conditionally independent of X_1, X_2, X_5 given X_3

$$\Rightarrow P[X_4|X_1,X_2,X_3,X_5] = P[X_4|X_3],$$

  • X_5 is conditionally independent of X_1, X_2, X_4 given X_3

$$\Rightarrow P[X_5|X_1,X_2,X_3,X_4] = P[X_5|X_3],$$

  • X_6 is conditionally independent of X_1, X_2, X_3 given X_4, X_5

$$\Rightarrow P[X_6|X_1,X_2,X_3,X_4,X_5] = P[X_6|X_4,X_5];$$

• performing computations in a Bayesian network:
if $P(X_1{=}1) = 0.5$, $P(X_2{=}1|X_1{=}1) = 0.6$, $P(X_3{=}1|X_1{=}1,X_2{=}1) = 0.5$, $P(X_4{=}1|X_3{=}1) = 0.4$ and $P(X_4{=}1|X_3{=}0) = 0.3$, then

$P(X_4{=}1, X_2{=}1, X_1{=}1)$

$= P(X_4{=}1, X_3{=}1, X_2{=}1, X_1{=}1) + P(X_4{=}1, X_3{=}0, X_2{=}1, X_1{=}1)$

$$= P(X_1{=}1)P(X_2{=}1|X_1{=}1)P(X_3{=}1|X_1{=}1, X_2{=}1)P(X_4{=}1|X_3{=}1)$$
$$+ P(X_1{=}1)P(X_2{=}1|X_1{=}1)P(X_3{=}0|X_1{=}1, X_2{=}1)P(X_4{=}1|X_3{=}0)$$
$$= 0.105. \qquad \blacktriangle$$

2.8.1 Solved Problems

Problem 2.8.1. Let $n \in \mathbb{N}$, $n \geq 2$, let Z be a discrete random variable and let $X_1, \ldots, X_n$ be discrete random variables such that each one is conditionally independent of the other $n - 1$ given Z (e.g., X_1 is conditionally independent of $X_2, \ldots, X_n$ given Z). Prove that $X_1, \ldots, X_n$ are conditionally independent given Z.

Solution 2.8.1: For $n = 2$ we clearly have that X_1 and X_2 are conditionally independent given Z. Next, we proceed by induction. Suppose that the result is true for a fixed $n \in \mathbb{N}$, $n \geq 2$, and we want to prove it for $n+1$. Fix arbitrary values $x_1, x_2, \ldots, x_{n+1}, z$ of $X_1, X_2, \ldots, X_{n+1}$ respectively Z, and let $\mathcal{X}_1$ be the range of X_1. Then we have

$$P(X_2 = x_2, \ldots, X_{n+1} = x_{n+1}|Z = z)$$
$$= \sum_{x \in \mathcal{X}_1} P(X_1 = x, X_2 = x_2, \ldots, X_{n+1} = x_{n+1}|Z = z)$$
$$= \sum_{x \in \mathcal{X}_1} P(X_2 = x_2|Z = z)P(X_1 = x, X_3 = x_3, \ldots, X_{n+1} = x_{n+1}|Z = z)$$
$$= P(X_2 = x_2|Z = z)P(X_3 = x_3, \ldots, X_{n+1} = x_{n+1}|Z = z),$$

where we use that X_2 is conditionally independent of $X_1, X_3, \ldots, X_{n+1}$ given Z.

Hence X_2 is conditionally independent of $X_3, \ldots, X_{n+1}$ given Z. Similarly, for every $j \in \{2, \ldots, n + 1\}$, we deduce that X_j is conditionally independent of the random variables from the family $\{X_i : i \neq j, i \in \{2, \ldots, n + 1\}\}$ given Z. From the induction assumption we have that $X_2, \ldots, X_{n+1}$ are conditionally independent given Z. Then

$$P(X_1 = x_1, X_2 = x_2, \ldots, X_{n+1} = x_{n+1}|Z = z)$$
$$= P(X_1 = x_1|Z = z)P(X_2 = x_2, \ldots, X_{n+1} = x_{n+1}|Z = z)$$
$$= P(X_1 = x_1|Z = z)P(X_2 = x_2|Z = z)\ldots P(X_{n+1} = x_{n+1}|Z = z).$$

Therefore, we conclude that $X_1, X_2, \ldots, X_{n+1}$ are conditionally independent given Z.

An example of Bayesian network, where X_1, X_2, X_3, X_4 are conditionally independent given Z is given in Figure 2.6.

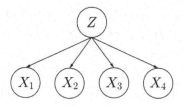

Fig. 2.6: Problem 2.8.1

Problem 2.8.2. Consider the following binary random variables which describe the evening of a person (1=yes,0=no):
- C: the person goes to a specific concert;
- T: there is an offer for cheap entrance tickets to the concert;
- R: the person goes to a restaurant;
- B: the person goes to a bar for a drink.

The above random variables depend on each other as in the Bayesian network shown in Figure 2.7 and the corresponding probabilities are:

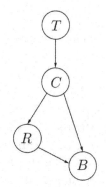

Fig. 2.7: Problem 2.8.2

$P(T{=}1) = 0.8, P(C{=}1|T{=}1) = 0.9, P(C{=}1|T{=}0) = 0.5,$
$P(R{=}1|C{=}1) = 0.3,\ P(R{=}1|C{=}0) = 0.5,$
$P(B{=}1|C{=}1) = 0.5,\ P(B{=}1|C{=}0) = 0.8,$
$P(B{=}1|C{=}1, R{=}1) = 0.2,\ P(B{=}1|C{=}1, R{=}0) = 0.7,$
$P(B{=}1|C{=}0, R{=}1) = 0.5,\ P(B{=}1|C{=}0, R{=}0) = 0.9.$

Compute the probabilities of the following events:
1) The person goes to the concert.
2) The person eats at a restaurant, knowing that the concert ticket was cheap.
3) The person does not go to a restaurant.
4) The concert ticket was expensive, knowing that the person went to a restaurant.

Solution 2.8.2: For 1), 2), 3) and 4) we use the properties of Bayesian networks and properties of conditional probabilities. We have

$$P(C{=}1) = P(C{=}1|T{=}1)P(T{=}1) + P(C{=}1|T{=}0)P(T{=}0) = 0.82\,,$$

$$P(R{=}1|T{=}1) = P(R{=}1, C{=}0|T{=}1) + P(R{=}1, C{=}1|T{=}1)$$
$$= P(R{=}1|C{=}0, T{=}1)P(C{=}0|T{=}1)$$

$$+ P(R{=}1|C{=}1, T{=}1)P(C{=}1|T{=}1)$$
$$= P(R{=}1|C{=}0)P(C{=}0|T{=}1) + P(R{=}1|C{=}1)P(C{=}1|T{=}1)$$
$$= 0.32 \,,$$

$$P(R{=}0) = P(R{=}0|C{=}1)P(C{=}1|T{=}1)P(T{=}1)$$
$$+ P(R{=}0|C{=}0)P(C{=}0|T{=}1)P(T{=}1)$$
$$+ P(R{=}0|C{=}1)P(C{=}1|T{=}0)P(T{=}0)$$
$$+ P(R{=}0|C{=}0)P(C{=}0|T{=}0)P(T = 0) = 0.664.$$

Then, $P(R{=}1) = 0.336$ and finally

$$P(T{=}0|R{=}1) = 1 - P(T{=}1|R{=}1) = 1 - \frac{P(R{=}1|T{=}1)P(T{=}1)}{P(R{=}1)} \approx 0.2381 \,.$$

Problem 2.8.3. Consider the Bayesian network given in Figure 2.8, where $B_1, \ldots, B_5$ are binary random variables. Let N be the random variable representing the decimal representation of the binary number $\overline{B_5 B_4 B_3 B_2 B_1}$. Assume that:

- $P(B_3{=}0|B_1{=}b_1, B_2{=}b_2) = \dfrac{1}{b_1 + b_2 + 2}$,
for each $b_1, b_2 \in \{0, 1\}$;

- $P(B_4{=}0|B_3{=}b_3){=}P(B_5{=}1|B_3{=}b_3){=}\dfrac{1}{b_3 + 2}$,
for each $b_3 \in \{0, 1\}$;

- $P(N{=}24_{(10)}) = P(N{=}17_{(10)}) = 0.025$.

Compute $P(N{=}12_{(10)})$ and $P(B_3 = 1)$.

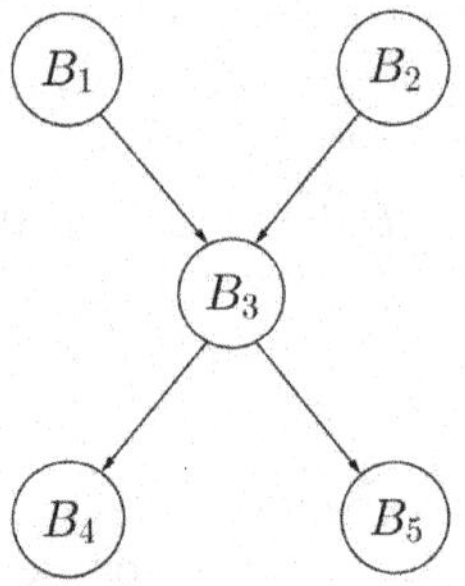

Fig. 2.8: Problem 2.8.3

Solution 2.8.3: We write

$$P(N{=}24_{(10)}) = P(B_1{=}0, B_2{=}0, B_3{=}0, B_4{=}1, B_5{=}1)$$
$$=P(B_1{=}0)P(B_2{=}0)P(B_3{=}0|B_1{=}0,B_2{=}0)P(B_4{=}1|B_3{=}0)P(B_5{=}0|B_3{=}0)$$
$$=P(B_1{=}0)P(B_2{=}0) \cdot 0.5 \cdot 0.5 \cdot 0.5 = 0.025 \,.$$

Hence $P(B_1{=}0)P(B_2{=}0) = 0.2$. Similarly, we compute

$$P(N{=}17_{(10)}) = P(B_1{=}1, B_2{=}0, B_3{=}0, B_4{=}0, B_5{=}1)$$
$$=P(B_1{=}1)P(B_2{=}0)P(B_3{=}0|B_1{=}1,B_2{=}0)P(B_4{=}0|B_3{=}0)P(B_5{=}1|B_3{=}0)$$
$$=P(B_1{=}1)P(B_2{=}0) \cdot 0.(3) \cdot 0.5 \cdot 0.5 = 0.025 \,.$$

Then, $(1 - P(B_1=0))P(B_2=0) = 0.3$ and we obtain

$$P(B_1=0) = 0.4, \ P(B_1=1) = 0.6, \ P(B_2=0) = P(B_2=1) = 0.5.$$

We have

$$P(N=12_{(10)}) = P(B_1=0, B_2=0, B_3=1, B_4=1, B_5=0)$$
$$=P(B_1=0)P(B_2=0)P(B_3=1|B_1=0,B_2=0)P(B_4=1|B_3=1)P(B_5=0|B_3=1)$$
$$\approx 0.0444.$$

Observe that

$$\begin{aligned} P(B_3 = 1) &= P(B_1=0)P(B_2=0)P(B_3=1|B_1=0,B_2=0) \\ &\quad + P(B_1=1)P(B_2=0)P(B_3=1|B_1=1,B_2=0) \\ &\quad + P(B_1=0)P(B_2=1)P(B_3=1|B_1=0,B_2=1) \\ &\quad + P(B_1=1)P(B_2=1)P(B_3=1|B_1=1,B_2=1) \\ &\approx 0.6583. \end{aligned}$$

2.9 Problems for Chapter 2

Problem 2.9.1. 1) Implement in Matlab the following random algorithm:
• generate four independent random numbers X_1, X_2, X_3 and X_4 such that $P(X_i = 1) = \dfrac{1}{3}$ and $P(X_i = 2) = \dfrac{2}{3}$, $i \in \{1,2,3,4\}$.
• convert to a decimal number Y the number $\overline{X_4 X_3 X_2 X_1}$ represented in base 3;
• return a random value Z that follows the uniform distribution on the interval $[Y, Y + 3]$.
2) Estimate $P(Z \leq 42)$, using the implementation of the above algorithm in Matlab.
3) Compute $P(Z \leq 42)$.

Solution 2.9.1: 1)

```
function Z=num2base3()
X=randsample([1 2],4,true,[1/3 2/3]);
Y=3.^(0:3)*X';
Z=unifrnd(Y,Y+3);
end
```

2)

```
function num2base3_sim(N)
count=0;
for i=1:N
    if num2base3<=42
        count=count+1;
    end
end
fprintf('The estimated value of P(Z<=42): %5.4f.\n',count/N);
end
```

```
>> num2base3_sim(100000)
The estimated value of P(Z<=42): 0.0166.
```

3) Since $P(Z \leq 42|Y \geq 42) = 0$ and Y is always larger than or equal to $40 = 3^3 + 3^2 + 3^1 + 3^0$, we have, in view of the law of total probability (see Theorem 1.8):

$$P(Z \leq 42) = P(Z \leq 42|Y = 40)P(Y = 40) + P(Z \leq 42|Y = 41)P(Y = 41)$$

$$= \frac{2}{3} \cdot \frac{1}{3} \cdot \frac{1}{3} \cdot \frac{1}{3} \cdot \frac{1}{3} + \frac{1}{3} \cdot \frac{1}{3} \cdot \frac{1}{3} \cdot \frac{1}{3} \cdot \frac{2}{3} = \frac{4}{243} \approx 0.0165.$$

Problem 2.9.2. Consider the following game:

• Roll a die and put in an urn as many white balls as the die is showing. Then complete the urn with black balls until you have 10 balls in total.

• Roll again a die and put in a second urn as many black balls as the die is showing. Then complete the urn with white balls until you have 10 balls in total.

• Next, flip a coin. If you get head, choose the first urn. If you get tail, choose the second urn.

• Now, extract 5 balls from the chosen urn, without replacement. Let X be the random variable that gives the number of extracted white balls.

1) Simulate $N \in \{1000, 5000, 10000\}$ times the above game in Matlab and estimate the probability to obtain k white balls at the end, where $k \in \{1, 2, 3, 4, 5\}$ is given as input.

2) Determine the probability mass function of X and write a function in Matlab that computes the corresponding probability of a value of X that is given as input.

Test the functions, from 1) and 2), for $k = 3$.

Solution 2.9.2: 1)

```
function die_coin(k,N)
X=zeros(1,N);
```

```
for i=1:N
    w=randi(6,1,2);
    if binornd(1,0.5)
      X(i)=hygernd(10,w(1),5);
    else
      X(i)=hygernd(10,10-w(2),5);
    end
end
fprintf('The probability to obtain %d white balls is %5.4f.\n',k,mean(X==k));
end
```

2) Let Y_1 and Y_2 be the numbers given by the first, respectively the second, rolled die and let Z be the result given by the coin, by identifying 1 and 0 with head, respectively tail. The law of total probability (see Theorem 1.8) implies

$$P(X = k)$$
$$= \frac{1}{12} \sum_{i=1}^{6} P(X = k|Y_1 = i, Z = 1) + \frac{1}{12} \sum_{i=1}^{6} P(X = k|Y_2 = i, Z = 0)$$
$$= \frac{1}{12} \sum_{i=k}^{6} \frac{C(i,k)C(10-i,5-k)}{C(10,5)} + \frac{1}{12} \sum_{i=5-k}^{6} \frac{C(10-i,k)C(i,5-k)}{C(10,5)},$$

for $k \in \{1, 2, 3, 4\}$, and

$$P(X = 5) = \frac{1}{12} \sum_{i=5}^{6} \frac{C(i,5)}{C(10,5)} + \frac{1}{12} \sum_{i=1}^{5} \frac{C(10-i,5)}{C(10,5)}.$$

```
function p=die_coin_prob(k)
p=0;
if k<5
    for i=k:6
        p=p+hygepdf(k,10,i,5);
    end
    for i=(5-k):6
        p=p+hygepdf(k,10,10-i,5);
    end
else
    for i=5:6
        p=p+hygepdf(5,10,i,5);
    end
    for i=1:5
        p=p+hygepdf(5,10,10-i,5);
    end
end
p=p/12;
end
```

```
>> die_coin(3,5000)
The probability to obtain 3 white balls is 0.2436.
> die_coin_prob(3)
ans = 0.2454
```

Problem 2.9.3. The random variable X has the distribution function $F : \mathbb{R} \to [0, 1]$ defined by

$$F(x) = \begin{cases} 0, & \text{if } x < 0 \\ \dfrac{x}{2}, & \text{if } 0 \leq x < 0.5 \\ x, & \text{if } 0.5 \leq x < 1 \\ 1, & \text{if } 1 \leq x. \end{cases}$$

Compute the corresponding quantile function and simulate in Matlab n random values of X for $n \in \mathbb{N}^*$ given as input. For $n = 10000$ plot the empirical cumulative distribution function of the simulated values.

Solution 2.9.3: The quantile function $Q : (0, 1) \to \mathbb{R}$ has the expression

$$Q(p) = \begin{cases} 2p, & \text{if } 0 < p \leq 0.25 \\ 0.5, & \text{if } 0.25 < p \leq 0.5 \\ p, & \text{if } 0.5 < p < 1. \end{cases}$$

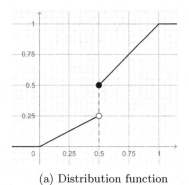

(a) Distribution function

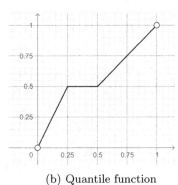

(b) Quantile function

Fig. 2.9: Problem 2.9.3

Note that the function F is strictly monotonically increasing on $[0, 1)$, but is discontinuous at 0.5. The function Q is continuous and monotone increasing on $(0, 1)$, but not strictly increasing.

```
%We use the inversion method:
function X=quantile_sim(n)
p=rand(1,n);
```

X=2*p.*(p<=0.25)+0.5.*and(0.25<p,p<=0.5)+p.*(0.5<p);
end

>> quantile_sim(10)
ans = 0.3152 0.9706 0.9572 0.5000 0.8003 0.2838 0.5000 0.9157 0.7922 0.9595
>> **ecdf**(quantile_sim(10000)); **axis equal**; **grid on**;

Problem 2.9.4. John goes home after hours by a train that is plying each day from town W, where John is working, to town H, John's hometown. The train departs at 18 o'clock sharp. John arrives at the railway station T minutes past 17, where $T = T_0 + 16X^2$, $T_0 \in \{0, \dots, 60\}$ is fixed and $X \sim Unif[-1, 1]$ (due to heavy traffic).
1) Find the probability that John misses the train, if $T_0 = 51$.
2) Find the maximum value of T_0 such that the probability that John catches the train is at least 95%.
3) Find the distribution of the number of consecutive working days when John catches the train before the working day he misses the train, for T_0 found at 2).
4) Using Matlab, find the minimum value $k \in \mathbb{N}$ such that the probability that there are at most k consecutive working days when John catches the train before the working day he misses the train is at least 90%, for T_0 found at 2).

Solution 2.9.4: John misses the train if and only if $T > 60$.
1) Thus we have to compute $P(T > 60)$. We have

$$P(T > 60) = P(16X^2 > 9) = P\left(X > \frac{3}{4}\right) + P\left(X < -\frac{3}{4}\right) = 0.25.$$

2) We write

$$P(T < 60) \geq 0.95 \iff P(16X^2 < 60 - T_0) \geq \frac{95}{100}$$

$$\iff P\left(|X| < \frac{\sqrt{60 - T_0}}{4}\right) \geq \frac{95}{100} \iff \frac{\sqrt{60 - T_0}}{4} \geq \frac{95}{100} \iff T_0 \leq 45.$$

Thus the maximum value of T_0 is 45.
3) We have the geometric distribution $D \sim Geo(p)$ with parameter $p = P(T > 60) = P(16X^2 > 15) = 1 - \frac{\sqrt{15}}{4}$, where D is the random variable that gives the number of consecutive working days when John catches the train before the working day he misses the train.
4) We have to find the minimum value of $k \in \mathbb{N}$ such that

$$P(D \leq k) \geq 0.9 \iff F(k) \geq 0.9,$$

where F denotes the distribution function of D. Hence, $k = Q(0.9)$, where Q is the quantile function of D, and we can use the **geoinv** function in Matlab with parameter $p = 1 - \frac{\sqrt{15}}{4}$:

```
>> k=geoinv(0.9,1-sqrt(15)/4)
k = 71
```

Problem 2.9.5. 1) A die is rolled nine times. Find the conditional probability that number 1 appears two times, number 2 appears one time and number 4 appears two times, given that number 3 appears one time and number 6 appears three times.

2) Simulate in Matlab the above experiment and estimate the desired probability.

Solution 1) Let X_i be the number of rolls that give the number i for $i \in \{1, \ldots, 6\}$. Also, let Y be the number of rolls that give a number in the set $\{1, 2, 4, 5\}$. Then $(X_1, \ldots, X_6) \sim Multinom\left(9, \frac{1}{6}, \ldots, \frac{1}{6}\right)$ and $(X_3, X_6, Y) \sim Multinom\left(9, \frac{1}{6}, \frac{1}{6}, \frac{2}{3}\right)$. Hence, the desired probability is

$$P(X_1 = 2, X_2 = 1, X_4 = 2 | X_3 = 1, X_6 = 3, Y = 5)$$

$$= \frac{P(X_1 = 2, X_2 = 1, X_3 = 1, X_4 = 2, X_5 = 0, X_6 = 3)}{P(X_3 = 1, X_6 = 3, Y = 5)}$$

$$= \frac{\dfrac{9!}{2!1!1!2!0!3!} \cdot \dfrac{1}{6^9}}{\dfrac{9!}{1!3!5!} \cdot \dfrac{1}{6^4} \cdot \dfrac{2^5}{3^5}} = \frac{15}{2^9}.$$

Compare the above probability with the result given by **mnpdf**:

```
>>mnpdf([2,1,1,2,0,3],[1/6 1/6 1/6 1/6 1/6 1/6])/mnpdf([1,3,5],[1/6 1/6 2/3])
ans = 0.0293
```

2) We present two methods in Matlab.
First method:

```
function die_sim(N_sims)
count1=0; count2=0;
for i=1:N_sims
 D=randi(6,1,9);
 if sum(D==3)==1&&sum(D==6)==3
  count1=count1+1;
  if sum(D==1)==2&&sum(D==2)==1&&sum(D==4)==2
   count2=count2+1;
  end
```

```
  end
 end
fprintf('The estimated probability is %4.3f.\n',count2/count1);
end
```

```
>> die_sim(10000)
The estimate probability is 0.027.
```

Second method:

```
function die_sim2(N_sims)
count1=0; count2=0;
for i=1:N_sims
 X=mnrnd(9,[1/6 1/6 1/6 1/6 1/6 1/6]);
 if X(3)==1&&X(6)==3
  count1=count1+1;
  if X(1)==2&&X(2)==1&&X(4)==2
   count2=count2+1;
  end
 end
end
fprintf('The estimate probability is %4.3f.\n',count2/count1);
end
```

```
>> die_sim2(10000)
The estimated probability is 0.029.
```

Problem 2.9.6. We shuffle a standard 52-card deck and then we divide it into two halves of 26 cards each. We draw a card from the first half-deck and we notice it is a heart. Next, we place the heart in the second half-deck. Then we shuffle this half-deck and we draw a card from it.

1) Simulate in Matlab the above steps and estimate the probability that the card drawn at the end is a heart.

2) Compute the probability that the card drawn at the end is a heart.

Solution 2.9.6 1)

```
function hearts(N)
count1=0; count2=0;
cards=1:52;%we assume that from 1 to 13 we have the hearts
for i=1:N
    deck1=randsample(cards,26);
    c1=randsample(deck1,1);
    if ismember(c1,1:13)
        count1=count1+1;
        c2=randsample([c1,setdiff(cards,deck1)],1);
        if ismember(c2,1:13)
            count2=count2+1;
        end
```

  end
end
fprintf('The estimated probability is %4.3f.\n',count2/count1);
end

\>\> hearts(100000)
The estimated probability is 0.263.

2) Let C_1 be the event that the card drawn from the first half-deck is a heart, C_2 be the event that the card drawn from the second half-deck is a heart and H_i be the event that there are i hearts in the first half-deck for $i \in \{1, \ldots, 13\}$. Let $i \in \{1, \ldots, 13\}$. Using the Bayes' formula from Theorem 1.9, we have

$$P(H_i|C_1) = \frac{P(C_1|H_i)P(H_i)}{\sum\limits_{j=1}^{13} P(C_1|H_j)P(H_j)}.$$

Taking into account the hypergeometric distribution, we deduce that

$$P(H_i|C_1) = \frac{\dfrac{i}{26} \cdot \dfrac{C(13,i)C(39,26-i)}{C(52,26)}}{\sum\limits_{j=1}^{13} \dfrac{j}{26} \cdot \dfrac{C(13,j)C(39,26-j)}{C(52,26)}}.$$

Next, we write

$$\sum_{j=1}^{13} \frac{jC(13,j)C(39,26-j)}{C(52,26)} = \sum_{j=1}^{13} \frac{13 \cdot C(12,j-1)C\big(39,25-(j-1)\big)}{\frac{52}{26} \cdot C(51,25)} = \frac{13}{2},$$
$$(2.5)$$

in view of the probability mass function of the $Hyge(25,12,39)$ distribution. Taking into account (2.5) and the probability mass function of the $Hyge(24,11,38)$ distribution, we also do the following computations:

$$\sum_{j=1}^{13} \frac{j^2 C(13,j)C(39,26-j)}{C(52,26)} = \sum_{l=0}^{12} \frac{(l+1)13C(12,l)C(39,25-l)}{\frac{52}{26} \cdot C(51,25)}$$

$$= \frac{13}{2}\left(1 + \sum_{l=1}^{12} \frac{lC(12,l)C(39,25-l)}{C(51,25)}\right)$$

$$= \frac{13}{2}\left(1 + \sum_{l=1}^{12} \frac{12C(11,l-1)C\big(39,24-(l-1)\big)}{\frac{51}{25} \cdot C(50,24)}\right) \qquad (2.6)$$

$$= \frac{13}{2}\left(1 + \frac{12 \cdot 25}{51}\right) = \frac{13}{2}\left(1 + \frac{100}{17}\right) = \frac{13^2 \cdot 9}{34}.$$

In view of the above, we compute the desired probability

$$P(C_2|C_1) = \frac{P(C_2 \cap C_1)}{P(C_1)} = \sum_{i=1}^{13} \frac{P(C_2 \cap C_1 \cap H_i)}{P(C_1)}$$

$$= \sum_{i=1}^{13} P(C_2|C_1 \cap H_i)P(H_i|C_1) = \sum_{i=1}^{13} \frac{14-i}{27} \cdot \frac{\dfrac{iC(13,i)C(39,26-i)}{C(52,26)}}{\dfrac{13}{2}}$$

$$= \frac{14}{27} - \frac{2}{27 \cdot 13} \sum_{i=1}^{13} \frac{i^2 C(13,i)C(39,26-i)}{C(52,26)} = \frac{14}{27} - \frac{2}{27 \cdot 13} \cdot \frac{13^2 \cdot 9}{34} \approx 0.2636.$$

Problem 2.9.7. Let X be a discrete random variable having the probability mass function: $P(X = -0.25) = 0.5$, $P(X = 0.25) = P(X = 0.5) = 0.25$.
1) Find the distribution function F and the quantile function Q of X.
2) Using the quantile function Q, simulate in Matlab n random values for X for $n \in \mathbb{N}^*$ given as input. For $n = 10000$ plot the empirical distribution function of the simulated values.

Solution 2.9.7: The distribution function $F : \mathbb{R} \to [0,1]$ has the expression

$$F(x) = \begin{cases} 0, & \text{if } x < -0.25 \\ 0.5, & \text{if } -0.25 \le x < 0.25 \\ 0.75, & \text{if } 0.25 \le x < 0.5 \\ 1, & \text{if } 0.5 \le x \end{cases}$$

while the quantile function $Q : (0,1) \to \mathbb{R}$ has the expression

$$Q(p) = \begin{cases} -0.25, & \text{if } 0 < p \le 0.5 \\ 0.25, & \text{if } 0.5 < p \le 0.75 \\ 0.5, & \text{if } 0.75 < p < 1. \end{cases}$$

%We use the inversion method:
```
function X=quantile_simul(n)
p=rand(1,n);
X=-0.25.*(p<=0.5)+0.25.*and(0.5<p,p<=0.75)+0.5.*(0.75<p);
end
```

```
>> quantile_simul(10)
ans = 0.5000 -0.2500 0.2500 0.5000 0.5000 0.5000 -0.2500 0.5000 -0.2500 -0.2500
>> ecdf(quantile_simul(10000)); axis equal; grid on;
```

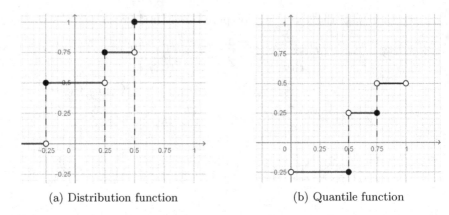

(a) Distribution function   (b) Quantile function

Fig. 2.10: Problem 2.9.7

Problem 2.9.8. Let $q : \mathbb{R} \to \mathbb{R}$ be a polynomial function given by

$$q(x) = x^6 - 14x^4 + 49x^2 - 36, \quad x \in \mathbb{R}.$$

Let X be a random variable such that:
1) $X \sim Unif[-2, 5]$;
2) $X \sim \chi^2(3, 1)$;
3) $X \sim Cauchy(0, 1)$.
Using the functions: **unifrnd**, **chi2rnd**, respectively **trnd**, estimate in Matlab the probability of the event $\{q(X) < 0\}$. Using the functions: **unifcdf**, **chi2cdf**, respectively **tcdf**, compute the value of the probability that $\{q(X) < 0\}$.

Solution 2.9.8:

```
function p=negpoly(N)
X=zeros(1,3);
p=zeros(1,3);
q=@(x) polyval([1,0,-14,0,49,0,-36],x);
for i=1:N
    X(1)=unifrnd(-2,5);
    X(2)=chi2rnd(3,1);
    X(3)=trnd(1);
    p(q(X)<0)=p(q(X)<0)+1;
end
p=p/N;
end


>> negpoly(100000)
ans = 0.4265 0.3806 0.5929
```

Using the Matlab Symbolic Math Toolbox:

```
>> syms x; q=x^6-14*x^4+49*x^2-36;
>> solve(q==0,x)'
ans = (sym) [-3 -2 -1 1 2 3]
```

we note that

$$q(x) = (x^2 - 1)(x^2 - 2^2)(x^2 - 3^2), \quad x \in \mathbb{R}.$$

Hence, $\{q(X) < 0\}$ if and only if $X \in (-3, -2) \cup (-1, 1) \cup (2, 3)$. Therefore, the value of the probabilities are:

```
>> sum(unifcdf([1,-2,3],-2,5)-unifcdf([-1,-3,2],-2,5))
ans = 0.4286
>> sum(chi2cdf([1,-2,3],3)-chi2cdf([-1,-3,2],3))
ans = 0.3795
>> sum(tcdf([1,-2,3],1)-tcdf([-1,-3,2],1))
ans = 0.5903
```

Problem 2.9.9. Two friends, Mary and Alice, do their shopping on Friday and arrive at a certain store independently of each other. Mary remains in the store for 15 minutes and Alice for 10 minutes. If the times of arrival of the two friends are uniformly distributed between 2 p.m. and 3 p.m., compute the probability p that Mary and Alice will be in the store at the same time. Estimate the probability p by Matlab simulations.

Solution 2.9.9: Let $X \sim Unif[2, 3]$ be the arrival time of Mary and $Y \sim Unif[2, 3]$ the arrival time of Alice in the store. Since the two friends arrive independently at the store, X and Y are independent. We note that

$$p = P\left(\left\{X \le Y \le X + \frac{1}{4}\right\} \cup \left\{Y \le X \le Y + \frac{1}{6}\right\}\right),$$

where we use the fact that 10 minutes are $\frac{1}{6}$ of 1 hour and 15 minutes are $\frac{1}{4}$ of 1 hour. Using the geometric probability, see Definition 1.7, we obtain

$$p = \frac{\lambda_{\mathbb{R}^2}\left(\left\{(x, y) \in [2, 3] \times [2, 3] : x \le y \le x + \frac{1}{4} \text{ or } y \le x \le y + \frac{1}{6}\right\}\right)}{\lambda_{\mathbb{R}^2}([2, 3] \times [2, 3])}$$

$$= 1 - \frac{1}{2}\left(\frac{5^2}{6^2} + \frac{3^2}{4^2}\right) \approx 0.3715.$$

```
function p=meeting_sim(n)
X=unifrnd(2,3,1,n); Y=unifrnd(2,3,1,n);
count=0;
for j=1:n
 if (X(j)<Y(j)&&Y(j)<=X(j)+1/4)||(Y(j)<=X(j)&&X(j)<=Y(j)+1/6)
```

```
   count=count+1;
 end
end
p=count/n; % estimated probability
end
```

```
>> meeting_sim(5000)
ans = 0.3716
```

Problem 2.9.10. We shuffle a standard 52-card deck and then we divide it into two halves of 26 cards each. We draw a card from the first half-deck and we notice it is a spade. Next, we draw a card from the second half-deck.
1) Simulate in Matlab the above steps and estimate the probability that the card drawn at the end is a spade.
2) Compute the probability that the card drawn at the end is a spade.

Solution 2.9.10: 1)

```
function spades(N)
count1=0; count2=0;
cards=1:52;
for i=1:N
    deck1=randsample(cards,26);
    c1=randsample(deck1,1);
    if ismember(c1,1:13)
        count1=count1+1;
        c2=randsample(setdiff(cards,deck1),1);
        if ismember(c2,1:13)
            count2=count2+1;
        end
    end
end
fprintf('The estimated probability is %4.3f.\n',count2/count1);
end
```

```
>> spades(100000)
The estimated probability is 0.236.
```

2) Let C be the event "the card drawn from the first half-deck is a spade", D be the event "the card drawn from the second half-deck is a spade" and S_i be the event "there are i spades in the first half-deck", for $i \in \{1, \ldots, 13\}$. Let $i \in \{1, \ldots, 13\}$. Using the Bayes' formula from Theorem 1.9, we have

$$P(S_i|C) = \frac{P(C|S_i)P(S_i)}{\displaystyle\sum_{j=1}^{13} P(C|S_j)P(S_j)}.$$

Taking into account the hypergeometric distribution, we deduce that

$$P(S_i|C) = \frac{\dfrac{i}{26} \cdot \dfrac{C(13,i)C(39,26-i)}{C(52,26)}}{\displaystyle\sum_{j=1}^{13} \dfrac{j}{26} \cdot \dfrac{C(13,j)C(39,26-j)}{C(52,26)}}.$$

From (2.5) and (2.6) given in the solution of Problem 2.9.6, we get

$$P(D|C) = \frac{P(D \cap C)}{P(C)} = \sum_{i=1}^{13} P(D|C \cap S_i)P(S_i|C)$$

$$= \sum_{i=1}^{13} \frac{13-i}{26} \cdot \frac{i \cdot \dfrac{C(13,i)C(39,26-i)}{C(52,26)}}{\dfrac{13}{2}} = \frac{1}{2} - \frac{1}{13^2} \cdot \frac{13^2 \cdot 9}{34} = \frac{4}{17} \approx 0.2353.$$

Problem 2.9.11. Let $X \sim N(0,1)$. Find density functions for the random variables $Y = |X|$ and $Z = X^2$.

Solution 2.9.11: The distribution function of Y is

$$F_Y(y) = P(Y \le y) = P(|X| \le y) = \begin{cases} 0, & \text{if } y \le 0 \\ P(-y \le X \le y), & \text{if } y > 0. \end{cases}$$

For $y > 0$ we have $F_Y(y) = P(-y \le X \le y) = F_X(y) - F_X(-y)$, where F_X is the distribution function of X. By Theorem 2.5 it follows for a.e. $y > 0$ that

$$f_Y(y) = F_Y'(y) = F_X'(y) + F_X'(-y) = f_X(y) + f_X(-y),$$

where f_Y is a density function of Y, while f_X is a density function of X. Therefore, for a.e. $y \in \mathbb{R}$

$$f_Y(y) = \begin{cases} 0, & \text{if } y \le 0 \\ \dfrac{2}{\sqrt{2\pi}} \exp\left\{-\dfrac{y^2}{2}\right\}, & \text{if } y > 0. \end{cases}$$

The distribution function of Z is

$$F_Z(z) = P(Z \le z) = P(X^2 \le z) = \begin{cases} 0, & \text{if } z \le 0 \\ P(|X| \le \sqrt{z}), & \text{if } z > 0. \end{cases}$$

For $z > 0$

$$F_Z(z) = P(-\sqrt{z} \le X \le \sqrt{z}) = F_X(\sqrt{z}) - F_X(-\sqrt{z}).$$

We have for a.e. $z > 0$

$$f_Z(z) = F_X'(\sqrt{z})\frac{1}{2\sqrt{z}} + F_X'(-\sqrt{z})\frac{1}{2\sqrt{z}} = f_X(\sqrt{z})\frac{1}{2\sqrt{z}} + f_X(-\sqrt{z})\frac{1}{2\sqrt{z}},$$

where f_Z is a density function of Z. Finally, for a.e. $z \in \mathbb{R}$

$$f_Z(z) = \begin{cases} 0, & \text{if } z \leq 0 \\ \dfrac{1}{\sqrt{2\pi z}}\exp\left\{-\dfrac{z}{2}\right\}, & \text{if } z > 0. \end{cases}$$

Problem 2.9.12. 1) A carpenter has nine measuring tapes in his workshop, three of each of the following lengths: 1.5, 2, respectively 3, meters. He randomly picks $k \in \{1, \ldots, 5\}$ tapes to measure a wooden plank of length L meters, where $L \sim Gamma(4, \frac{1}{2})$ is independent of the chosen tapes. Compute the probability p that the carpenter chooses at least one tape whose length is larger than the length of the plank, using Matlab.
2) Estimate the probability from 1), using Matlab, for $k = 4$.

Solution 2.9.12: 1) Let T_j be the length of the jth chosen tape, $j \in \{1, \ldots, k\}$. Then $p = P\left(L \leq \max_{j=\overline{1,k}} T_j\right)$. Moreover, note that if $t_1, \ldots, t_k$ are some lengths of the chosen tapes, then

$$P\left(L \leq \max_{j=\overline{1,k}} T_j \Big| T_1 = t_1, \ldots, T_k = t_k\right)$$
$$= P\left(L \leq \max_{i=\overline{1,k}} t_j \Big| T_1 = t_1, \ldots, T_k = t_k\right) = P\left(L \leq \max_{i=\overline{1,k}} t_j\right),$$

since L is independent of the chosen tapes. Furthermore, we observe that there are $C(9, k)$ possible ways to choose the k tapes. Hence, using the law of total probability (see Theorem 1.8) we have:

```
function p=carpenter(k)
T=repmat([1.5 2 3],1,3); chosenT=nchoosek(T,k);
p=0;
for i=1:nchoosek(9,k)
    p=p+gamcdf(max(chosenT(i,:)),4,1/2)/nchoosek(9,k);
end
end

>> carpenter(4)
ans = 0.8152
```

2)

```
function p=carpenter_sim(k,N)
T=repmat([1.5 2 3],1,3);
```

```
count=0;
for i=1:N
chosenT=randsample(T,k);
if gamrnd(4,1/2)<max(chosenT)
    count=count+1;
end
end
p=count/N;
end
```

```
>> carpenter_sim(4,10000)
ans = 0.8149
```

Problem 2.9.13. 1) Let X_1, $\ldots$, X_n be independent and identically distributed random variables, having the distribution function F, where $n \in \mathbb{N}^*$. Denote $U = \max\{X_1, \ldots, X_n\}$ and $V = \min\{X_1, \ldots, X_n\}$.
1) Find the distribution functions F_U and F_V of the random variables U and V, respectively.
2) Determine the expressions of F_U and F_V in the case when X_1, $\ldots$, $X_n \sim Unif[a, b]$, where $a, b \in \mathbb{R}$, $a < b$.

Solution 2.9.13: 1) For $u, v \in \mathbb{R}$ we write successively

$$F_U(u) = P(U \leq u) = P\big(\max\{X_1, \ldots, X_n\} \leq u\big) = P\Big(\bigcap_{i=1}^{n}\{X_i \leq u\}\Big)$$

$$= \prod_{i=1}^{n} P(X_i \leq u) = \prod_{i=1}^{n} F_{X_i}(u) = F^n(u),$$

$$F_V(v) = P(V \leq v) = P\big(\min\{X_1, \ldots, X_n\} \leq v\big)$$

$$= 1 - P\big(\min\{X_1, \ldots, X_n\} > v\big) = 1 - P\Big(\bigcap_{i=1}^{n}\{X_i > v\}\Big)$$

$$= 1 - \prod_{i=1}^{n} P(X_i > v) = 1 - \prod_{i=1}^{n}(1 - F_{X_i}(v)) = 1 - (1 - F(v))^n.$$

2) By the above computations we have

$$F_U(u) = \begin{cases} 0, & \text{if } u < a \\ \left(\dfrac{u - a}{b - a}\right)^n, & \text{if } a \leq u < b \\ 1, & \text{if } b \leq u \end{cases}$$

and

$$F_V(v) = \begin{cases} 0, & \text{if } v < a \\ 1 - \left(\dfrac{b-v}{b-a}\right)^n, & \text{if } a \leq v < b \\ 1, & \text{if } b \leq v. \end{cases}$$

Problem 2.9.14. The random variable X has Snedecor–Fischer distribution $F(m,n)$, $m, n \in \mathbb{N}^*$. Prove that $Y = \dfrac{1}{X} \sim F(n,m)$.

Solution 2.9.14: Let f_X, f_Y be density functions of X, respectively Y. Since $P(X = 0) = 0$, $\dfrac{1}{X}$ is well defined a.s.

We have

$$P(X > 0) = 1 - P(X \leq 0) = 1 - \int\limits_{-\infty}^{0} f_X(x)dx = 1.$$

So, $f_Y(y) = 0$ for a.e. $y \leq 0$. For $y > 0$ we write

$$P\left(\frac{1}{X} < y\right) = 1 - P\left(X \leq \frac{1}{y}\right).$$

We take the derivative and obtain for a.e. $y > 0$

$$f_Y(y) = \frac{1}{y^2} f_X\left(\frac{1}{y}\right).$$

By the definition of the density function f_X, as well as the symmetry property of the Beta function (see Section 2.5.10 and Proposition A.2), we get that f_Y is a density function for Snedecor–Fischer distribution $F(n,m)$. Hence, $Y \sim F(n,m)$.

Problem 2.9.15. The random variable X has Student distribution with n degrees of freedom, $n \in \mathbb{N}^*$. Show that the random variable $Y = X^2$ has Snedecor–Fischer distribution $F(1,n)$.

Solution 2.9.15: Let f_X, f_Y be density functions of X, respectively Y. Also, let F_X, F_Y be the distribution functions of X, respectively Y. Obviously, $F_Y(y) = P(X^2 \leq y) = 0$ for $y \leq 0$. For $y > 0$ we have

$$F_Y(y) = P(X^2 \leq y) = P(-\sqrt{y} \leq X \leq \sqrt{y}) = F_X(\sqrt{y}) - F_X(-\sqrt{y}).$$

Then for a.e. $y > 0$ it holds

$$f_Y(y) = F_Y'(y) = \frac{1}{2\sqrt{y}}\big(f_X(\sqrt{y}) + f_X(-\sqrt{y})\big).$$

Taking into account the Student distribution with n degrees of freedom (see Section 2.5.9), as well as the properties of the Gamma and Beta functions (see Proposition A.2-(5) and Proposition A.1-(3)), we obtain that f_Y is a density function of the Snedecor–Fischer distribution $F(1, n)$.

Problem 2.9.16. Let $X_1, \ldots, X_n \sim Exp(1)$, $n \in \mathbb{N}^*$, be independent random variables and let $Y_i = X_1 + \ldots + X_i$, for $i \in \{1, \ldots, n\}$.
1) Estimate in Matlab the probability
$$P\Big(Y_1 \in (0,1), Y_2 \in (1,2), \ldots, Y_n \in (n-1, n)\Big),$$
for some $n \in \{1, 2, 3, 4, 5\}$ given as input.
2) Find the probability
$$P\Big(Y_1 \in (0,1), Y_2 \in (1,2), \ldots, Y_n \in (n-1, n)\Big).$$

Solution 2.9.16:

```
function p=pb_cumsum(N_sim,n)
count=0;
for k=1:N_sim
    Y=cumsum(exprnd(1,1,n));
    if 0:n-1<Y & Y<1:n
        count=count+1;
    end
end
p=count/N_sim;
end
```

`>> pb_cumsum(10000, 4)`
ans = 0.0316

2) Let $f_{\mathbb{Y}}$ be a joint density function of the random vector $\mathbb{Y} = (Y_1, \ldots, Y_n)$, $f_{\mathbb{X}}$ be a joint density function of the random vector $\mathbb{X} = (X_1, \ldots, X_n)$ and

$$B = \begin{pmatrix} 1 & 1 & 1 & \ldots & 1 & 1 \\ 0 & 1 & 1 & \ldots & 1 & 1 \\ \vdots & \vdots & \vdots & \ddots & \vdots & \vdots \\ 0 & 0 & 0 & \ldots & 1 & 1 \\ 0 & 0 & 0 & \ldots & 0 & 1 \end{pmatrix}.$$

Then $\mathbb{Y} = \mathbb{X}B$, $\det B = 1$ and

$$B^{-1} = \begin{pmatrix} 1 & -1 & 0 & \ldots & 0 & 0 \\ 0 & 1 & -1 & \ldots & 0 & 0 \\ 0 & 0 & 1 & \ldots & 0 & 0 \\ \vdots & \vdots & \vdots & \ddots & \vdots & \vdots \\ 0 & 0 & 0 & \ldots & 1 & -1 \\ 0 & 0 & 0 & \ldots & 0 & 1 \end{pmatrix}.$$

By Example 2.14, $f_{\mathbb{Y}}(\mathbb{y}) = f_{\mathbb{X}}(\mathbb{y}B^{-1})$ for a.e. $\mathbb{y} \in \mathbb{R}^n$. So, $f_{\mathbb{Y}}(y_1, \ldots, y_n) = \exp\{-(y_1 + y_2 - y_1 + \ldots + y_n - y_{n-1})\} = e^{-y_n}$ for a.e. $y_1 > 0$, $y_2 - y_1 > 0, \ldots, y_n - y_{n-1} > 0$. Hence,

$$P\left(\bigcap_{i=1}^{n}\{Y_i \in (i-1, i)\}\right) = \int_0^1 \int_1^2 \cdots \int_{n-1}^n e^{-y_n} dy_1 dy_2 \ldots dy_n = e^{-(n-1)} - e^{-n}.$$

```
>> exp(-3)-exp(-4)
ans = 0.0315
```

Problem 2.9.17. Suppose that the numbers of misprints per page in a booklet are independent and they are modeled by a Poisson distribution with parameter $\lambda = 3$.
1) Determine the probability that a page contains at least 2 misprints.
2) Suppose that the booklet has 40 pages. What is the probability that there will be at least 10 pages which contain at least 2 misprints per page?

Solution 2.9.17: Let X be the random variable that shows the number of misprints on a page. Then

$$P(X = k) = \frac{3^k}{k!}e^{-3}, \ k \in \mathbb{N}.$$

1) The probability that a page contains at least 2 misprints is

$$p = \sum_{k=2}^{\infty} P(X = k) = 1 - 4e^{-3}.$$

2) Let Y denote the random variable that shows the number of pages in the booklet that contain at least 2 misprints per page. So, Y has a binomial distribution with parameters $n = 40$ and $p = 1 - 4e^{-3}$ and thus

$$P(Y = j) = C(40, j)p^j(1 - p)^{40-j} \ \text{for } j \in \{0, \ldots, 40\}.$$

Then the probability that there will be at least 10 pages having at least 2 misprints per page is

$$\sum_{j=10}^{40} P(Y = j) = \sum_{j=10}^{40} C(40, j)(1 - 4e^{-3})^j(4e^{-3})^{40-j}.$$

Problem 2.9.18. Let X be a random variable having exponential distribution with parameter $\lambda > 0$. Prove that it has "the forgetfulness property":

$$P(X > t + s | X > t) = P(X > s) \ \text{for each } s, t \geq 0.$$

This property is usually characterizing an "unreliable bus driver": after waiting for t minutes for the bus, the probability that we will wait for another s minutes is the same as if we would not have waited for the first t minutes.

Solution 2.9.18: Note that the conditional probability has meaning, since for each $t \geq 0$ we have (see Section 2.5.6)

$$P(X > t) = 1 - P(X \leq t) = 1 - (1 - e^{-\lambda t}) = e^{-\lambda t} > 0.$$

For $s, t \geq 0$ we compute

$$P(X > t + s | X > t) = \frac{P(X > s + t)}{P(X > t)} = e^{-\lambda s} = P(X > s).$$

Problem 2.9.19. Let X be a random variable such that

$$P(X > 0) = 1 \text{ and } P(X > t) > 0 \text{ for all } t > 0$$

having "the forgetfulness property":

$$P(X > t + s | X > t) = P(X > s) \text{ for each } s, t \geq 0.$$

Prove that X has exponential distribution.

Solution 2.9.19: We define the function $h : [0, \infty) \to \mathbb{R}$ by

$$h(t) = P(X > t) \text{ for all } t \geq 0.$$

Observe that h is a decreasing function, $h(t) > 0$ for all $t \geq 0$, $h(0) = 1$ and $\lim_{t \to \infty} h(t) = 0$. The "the forgetfulness property" implies

$$h(t + s) = h(t)h(s) \text{ for all } s, t \geq 0.$$

From this relation we conclude that

$$h(1) = \left[h\left(\frac{1}{n}\right) \right]^{\frac{1}{n}} \text{ for each } n \in \mathbb{N}^*.$$

Then for every positive rational number $r = \frac{p}{q}$ $(p, q \in \mathbb{N}, q \neq 0)$ we have

$$h\left(\frac{p}{q}\right) = \left[h\left(\frac{1}{q}\right) \right]^p = h^{\frac{p}{q}}(1).$$

We choose $\lambda \geq 0$ such that $h(1) = e^{-\lambda}$, then we get

$$h(r) = e^{-\lambda r} \text{ for each } r \in \mathbb{Q}, r \geq 0.$$

For arbitrary $t \in [0, \infty) \setminus \mathbb{Q}$ we choose $r, s \in [0, \infty) \cap \mathbb{Q}$ such that $r < t < s$ and by the monotonicity of h we have

$$e^{-\lambda r} = h(r) \geq h(t) \geq h(s) = e^{-\lambda s}.$$

We can take s and r arbitrarily close to t and thus, by the above inequality, it follows that

$$h(t) = e^{-\lambda t} \text{ for each } t \geq 0.$$

Since $\lim\limits_{t \to \infty} h(t) = 0$, we have $\lambda > 0$. The distribution function of X is

$$F(t) = 1 - P(X > t) = 1 - h(t) = 1 - e^{-\lambda t} \text{ for each } t \geq 0.$$

For $t < 0$ we have, by the monotonicity and positivity of the distribution function,

$$0 \leq F(t) \leq F(0) = 1 - P(X > 0) = 0.$$

So, $F(t) = 0$ for $t < 0$. Thus, X has exponential distribution with parameter $\lambda = -\log\big(P(X > 1)\big)$.

Problem 2.9.20. Suppose that n resistors which function independently of each other, are soldered in series on a circuit board, $n \in \mathbb{N}^*$. The circuit board functions properly, if all n resistors function properly.
a) The probability that the ith resistor will function properly is $p_i \in (0,1)$, $i \in \{1,\ldots,n\}$. Determine the reliability of the circuit board, i.e., the probability that the system will function properly.
b) Suppose now that the number of periods for which the ith resistor functions properly is a random variable R_i having geometric distribution with parameter p_i, $i \in \{1,\ldots,n\}$. Each resistor will function properly for a certain number of periods and then will fail. Determine the distribution of the number of periods for which the circuit board will function properly.

Solution 2.9.20:
a) The reliability of the circuit board is $p_1 \cdot p_2 \cdot \ldots \cdot p_n$.
b) Let X denote the random variable that shows the number of periods for which the circuit board functions properly, i.e., $X = k$ shows that for $k \in \mathbb{N}$ periods the board functions properly and then fails.
   We have $P(X = k) = P(X \leq k) - P(X \leq k-1)$ for $k \in \mathbb{N}^*$. Moreover,

$$P(X \leq k) = P(\min\{R_1,\ldots,R_n\} \leq k) = 1 - P\left(\bigcap_{i=1}^{n}\{R_i > k\}\right)$$

$$= 1 - \prod_{i=1}^{n}\big(1 - P(R_i \leq k)\big)$$

for $k \in \mathbb{N}$. In particular, we have

$$P(X = 0) = 1 - \prod_{i=1}^{n}\big(1 - P(R_i = 0)\big) = 1 - \prod_{i=1}^{n}(1 - p_i).$$

Since R_i has a geometric distribution,

$$P(R_i \leq k) = \sum_{j=0}^{k} P(R_i = j) = 1 - (1 - p_i)^{k+1}$$

for $i \in \{1, \ldots, n\}$ and $k \in \mathbb{N}$. Therefore, for $k \in \mathbb{N}^*$

$$P(X = k) = \prod_{i=1}^{n} \left(1 - P(R_i \leq k - 1)\right) - \prod_{i=1}^{n} \left(1 - P(R_i \leq k)\right)$$

$$= \prod_{i=1}^{n}(1 - p_i)^k - \prod_{i=1}^{n}(1 - p_i)^{k+1} = \left(1 - \prod_{i=1}^{n}(1 - p_i)\right) \prod_{i=1}^{n}(1 - p_i)^k.$$

Problem 2.9.21. If the random variable $X \sim N(0, 1)$ and the random variable $Y \sim \chi^2(n, 1)$ are independent, then

$$Z = \frac{X}{\sqrt{\dfrac{Y}{n}}}$$

has Student distribution $T(n)$ for $n \in \mathbb{N}^*$.

Solution 2.9.21: If f_Y is a density function of Y, then $f_Y(y) = 0$ for a.e. $y \leq 0$ and thus $Y > 0$ a.s. So, $Z = \dfrac{X}{\sqrt{\dfrac{Y}{n}}}$ is well defined a.s. The distribution function of the random variable $\sqrt{\dfrac{Y}{n}}$ satisfies

$$F_{\sqrt{\frac{Y}{n}}}(x) = P\left(\sqrt{\frac{Y}{n}} \leq x\right) = \begin{cases} 0, & \text{if } x \leq 0 \\ P(Y \leq nx^2), & \text{if } x > 0. \end{cases}$$

Then a density function of $\sqrt{\dfrac{Y}{n}}$ is

$$f_{\sqrt{\frac{Y}{n}}}(x) = \begin{cases} 0, & \text{if } x \leq 0 \\ 2\left(\dfrac{n}{2}\right)^{\frac{n}{2}} \dfrac{1}{\Gamma\left(\frac{n}{2}\right)} x^{n-1} \exp\left\{-\dfrac{nx^2}{2}\right\}, & \text{if } x > 0. \end{cases}$$

In order to determine a density function of Z, we use Example 2.15 as well as the fact that X and $\sqrt{\dfrac{Y}{n}}$ are independent random variables (see Theorem 2.12). We obtain that a density function of Z is

$$f_Z(x) = \int_{\mathbb{R}} |v| f_X(xv) f_{\sqrt{\frac{Y}{n}}}(v) dv = \frac{\Gamma\left(\frac{n+1}{2}\right)}{\sqrt{n\pi}\,\Gamma\left(\frac{n}{2}\right)} \left(1 + \frac{x^2}{n}\right)^{-\frac{n+1}{2}}, \quad x \in \mathbb{R},$$

where f_X is a density function of X. Therefore, $Z \sim T(n)$.

Problem 2.9.22. If the random variable $X \sim \chi^2(m, 1)$ and the random variable $Y \sim \chi^2(n, 1)$ are independent, then the random variable

$$Z = \frac{nX}{mY}$$

has Snedecor–Fischer distribution $F(m, n)$ for $m, n \in \mathbb{N}^*$.

Solution 2.9.22: Let f_X, f_Y be density functions of X, respectively Y. Since $P(Y = 0) = 0$, $Z = \dfrac{nX}{mY}$ is well defined a.s. Let $f_{\frac{1}{m}X}$, $f_{\frac{1}{n}Y}$ be density functions of $\frac{1}{m}X$, respectively $\frac{1}{n}Y$. Since $\frac{1}{m}X$ and $\frac{1}{n}Y$ are independent random variables, we have, by Example 2.15, that

$$f_Z(x) = \int_{\mathbb{R}} |v| f_{\frac{1}{m}X}(xv) f_{\frac{1}{n}Y}(v)dv = mn \int_0^\infty v f_X(mxv) f_Y(nv)dv$$

for a.e. $x \in \mathbb{R}$ is a density function of Z. Then

$$f_Z(x) = \begin{cases} 0, & \text{if } x \leq 0 \\ \left(\dfrac{m}{n}\right)^{\frac{m}{2}} \dfrac{x^{\frac{m}{2}-1}}{B(\frac{m}{2}, \frac{n}{2})(1 + \frac{m}{n}x)^{\frac{m+n}{2}}}, & \text{if } x > 0. \end{cases}$$

Hence, $Z \sim F(m, n)$.

Problem 2.9.23. Peter is standing in line at a ticket booth. Suppose that the times it takes for the customers to be served are independent and exponentially distributed with parameter $\lambda > 0$. Determine the probability that the client in front of Peter takes at least twice as much time to be served as Peter does. If the person in front of Peter needed more than 4 minutes to be served, what is the probability that the service for Peter takes more than 5 minutes?

Solution 2.9.23: Let X and Y denote the random variables that show the times of service for Peter and for the client in front of him, respectively. These two random variables are independent. The following are density functions of X and Y

$$f_X(x) = f_Y(x) = \begin{cases} \lambda e^{-\lambda x}, & \text{if } x > 0 \\ 0, & \text{otherwise.} \end{cases}$$

Let $D = \{(x, y) \in \mathbb{R}^2 : y \geq 2x\}$. The probability that the client in front of Peter takes at least twice as much time is

$$P(Y \geq 2X) = \iint_D f_X(x) f_Y(y) dx dy = \lambda^2 \int_0^\infty e^{-\lambda y} \left(\int_0^{\frac{y}{2}} e^{-\lambda x} dx \right) dy = \frac{1}{3}.$$

If the client in front of Peter took more than 4 minutes to be served, the probability that the service for Peter takes more than 5 minutes is

$$P(X > 5 | Y > 4) = P(X > 5) = 1 - P(X \le 5) = 1 - \int_0^5 \lambda e^{-\lambda x} dx = e^{-5\lambda},$$

where we use the fact that X and Y are independent.

Problem 2.9.24. The random variables X and Y are independent and have the same $N(0, 1)$ distribution. Define the random variables $U = X + Y$ and $V = X - Y$. Are U and V independent?

Solution 2.9.24: Let f_U, f_V be density functions of U, respectively V, and let $f_{U,V}$ be a joint density function of (U, V). Then U and V are independent if and only if $f_U(u)f_V(v) = f_{U,V}(u, v)$ for a.e. $u, v \in \mathbb{R}$.

Since X and Y are independent and have $N(0, 1)$ distribution, it follows that U and V have $N(0, 2)$ distribution (for this we can use Example 2.15 to compute density functions of U and V). Then we have

$$f_U(u)f_V(v) = \frac{1}{4\pi} e^{-\frac{u^2 + v^2}{4}} \quad \text{for a.e. } u, v \in \mathbb{R}.$$

Observe that the right hand side of the above relation is a joint density function of $(U, V) = (X, Y) \cdot \begin{pmatrix} 1 & 1 \\ 1 & -1 \end{pmatrix}$, in view of Example 2.14. Therefore, U and V are independent.

Problem 2.9.25. If the random variables X and Y are independent with an $N(0, 1)$ distribution, what distribution does $\dfrac{X + Y}{|X - Y|}$ have?

Solution 2.9.25: Since $P(X = Y) = 0$, we have $X \ne Y$ a.s. and thus $\dfrac{X + Y}{|X - Y|}$ is well defined a.s. Next, we note that one can show that $X + Y$ and $X - Y$ have an $N(0, 2)$ distribution and that they are independent (see Problem 2.9.24). Moreover, we see that $\dfrac{X + Y}{\sqrt{2}}$ and $\dfrac{X - Y}{\sqrt{2}}$ have $N(0, 1)$ distribution. By Problem 2.9.11, it follows that $Z = \dfrac{(X - Y)^2}{2} \sim \chi^2(1, 1)$ and, by Problem 2.9.21, it follows that

$$\frac{X + Y}{|X - Y|} = \frac{\dfrac{X + Y}{\sqrt{2}}}{\sqrt{Z}} \sim T(1).$$

Problem 2.9.26. Let $k \in \mathbb{N}^*$ and $\lambda > 0$. The function

$$f_{k,\lambda}(x) = \begin{cases} 0, & x \leq 0 \\ \dfrac{\lambda^k x^{k-1} e^{-\lambda x}}{(k-1)!}, & x > 0 \end{cases}$$

is a density function of the Erlang distribution with parameters k and λ.

1) Let $X_1, \ldots, X_k \sim Exp(\lambda)$ be independent random variables. Prove that $Z_k = X_1 + \cdots + X_k$ has Erlang distribution with parameters k and λ. Using the function **exprnd**, write a function in Matlab that generates a vector of random numbers following the Erlang distribution.

2) Let $U_1, \ldots, U_k \sim Unif[0,1]$ be independent random variables. Prove that $Z_k = -\dfrac{1}{\lambda} \log(U_1 \cdot \ldots \cdot U_k)$ follows the Erlang distribution with parameters k and λ. Using the function **rand**, write a function in Matlab that generates a vector of random numbers following the Erlang distribution.

3) Generate 1000 random numbers with each function and compare the corresponding empirical cumulative distribution functions and histograms.

Solution 2.9.26 1) We prove by induction that $f_{k,\lambda}$ is a density function of Z_k for $k \in \mathbb{N}^*$. We clearly have that $f_{1,\lambda}$ is a density function of Z_1, because $Z_1 \sim Exp(\lambda)$. Suppose that $f_{k,\lambda}$ is a density function of Z_k for a fixed $k \in \mathbb{N}^*$. By Example 2.15, we have that Z_{k+1} has a probability density function $f_{Z_{k+1}}$ given by

$$f_{Z_{k+1}}(z) = \int_{\mathbb{R}} f_{k,\lambda}(x) f_{1,\lambda}(z-x) dx = \frac{\lambda^{k+1} e^{-\lambda z}}{(k-1)!} \int_0^z x^{k-1} dx = f_{k+1,\lambda}(z),$$

if $z > 0$, and $f_{Z_{k+1}}(z) = 0$, if $z \leq 0$.

```
function Z=erlang1(lambda,k,n)
X=exprnd(1/lambda,k,n); Z=sum(X);
end
```

2) By Example 2.31, we have $-\frac{1}{\lambda} \log U_i \sim Exp(\lambda)$ for all $i \in \mathbb{N}^*$. Hence, $Z_k = -\dfrac{1}{\lambda} \log(U_1 \cdot \ldots \cdot U_k)$ has the Erlang distribution with parameters k and λ, in view of 1).

```
function Z=erlang2(lambda,k,n)
U=rand(k,n); Z=-log(prod(U))./lambda;
end
```

3)

```
>> Z1=erlang1(1/2,5,1000);
>> Z2=erlang2(1/2,5,1000);
>> subplot(1,2,1); hold on;
>> ecdf(Z1);ecdf(Z2);
>> subplot(1,2,2); hold on;
>> histogram(Z1);histogram(Z2);
```

Problem 2.9.27. 1) Write a programm in Matlab that generates random security codes with 4 decimal digits $C_1C_2C_3C_4$ (which can be used for a CAPTCHA verification on a website) according to a Bayesian network with the following probabilities:

- $P(C_1 = 1) = P(C_1 = 5) = 0.5$;
- $P(C_2 = 3|C_1 = 1) = P(C_3 = 5|C_1 = 1) = 0.2$;

$P(C_2 = 3|C_1 = 5) = P(C_3 = 5|C_1 = 5) = 0.6$;

$P(C_2 = 5|C_1 = 1) = P(C_3 = 7|C_1 = 1) = 0.8$;

$P(C_2 = 5|C_1 = 5) = P(C_3 = 7|C_1 = 5) = 0.4$;

- $P(C_4 = 1|C_2 = 3, C_3 = 5) = P(C_4 = 1|C_2 = 5, C_3 = 5) = 0.3$;

$P(C_4 = 1|C_2 = 3, C_3 = 7) = P(C_4 = 1|C_2 = 5, C_3 = 7) = 0.2$;

$P(C_4 = 5|C_2 = 3, C_3 = 5) = P(C_4 = 5|C_2 = 5, C_3 = 5) = 0.3$;

$P(C_4 = 5|C_2 = 3, C_3 = 7) = P(C_4 = 5|C_2 = 5, C_3 = 7) = 0.5$;

$P(C_4 = 9|C_2 = 3, C_3 = 5) = P(C_4 = 9|C_2 = 5, C_3 = 5) = 0.4$;

$P(C_4 = 9|C_2 = 3, C_3 = 7) = P(C_4 = 9|C_2 = 5, C_3 = 7) = 0.3$.

2) Represent graphically in Matlab the above Bayesian network.

3) Compute the probability that all digits of the code are equal.

4) Compute the conditional probability that all digits of the code are equal, given that the first three digits are equal.

Solution 2.9.27 1)

```
function captcha(k) %prints k CAPTCHA codes
C=ones(1,4);
for i=1:k
  if rand<0.5 C(1)=1; else C(1)=5; end
  if C(1)==1
    if rand<0.2 C(2)=3; else C(2)=5; end
    if rand<0.2 C(3)=5; else C(3)=7; end
  else
    if rand<0.6 C(2)=3; else C(2)=5; end
    if rand<0.6 C(3)=5; else C(3)=7; end
  end
  if (C(2)==3&&C(3)==5)||(C(2)==5&&C(3)==5)
    r=rand;
    if r<0.3 C(4)=1; elseif r<0.6 C(4)=5; else C(4)=9; end
  else
```

```
    r=rand;
    if r<0.2 C(4)=1; elseif r<0.7 C(4)=5; else C(4)=9; end
    end
    fprintf('%d',C);fprintf('\n');
end
end
```

>> captcha(3)
5371
1579
5379

2)

>>BN=**digraph**([0 1 1 0;0 0 0 1;0 0 0 1;0 0 0 0],{'C_1','C_2','C_3','C_4'});
>>**plot**(BN,'-k','LineWidth',1.2,'MarkerSize',8);

3)

$$P(C_1 = C_2 = C_3 = C_4) = P(C_1 = 5, C_2 = 5, C_3 = 5, C_4 = 5)$$
$$= P(C_1 = 5)P(C_2 = 5|C_1 = 5)P(C_3 = 5|C_1 = 5)P(C_4 = 5|C_2 = 5, C_3 = 5)$$
$$= 0.5 \cdot 0.4 \cdot 0.6 \cdot 0.3 = 0.036.$$

4)

$$P(C_1 = C_2 = C_3 = C_4|C_1 = C_2 = C_3) = P(C_4 = 5|C_1 = 5, C_2 = 5, C_3 = 5)$$
$$= P(C_4 = 5|C_2 = 5, C_3 = 5) = 0.3.$$

Problem 2.9.28. Let $(X_k)_{k\in\mathbb{N}}$ be a sequence of independent random variables that follow the $Unif[0,1]$ distribution. For every $k \in \mathbb{N}$ let

$$S_k = X_0 + X_1 + \ldots + X_k.$$

For every $z \in (0,1]$ let

$$M_z = \inf\{l \in \mathbb{N} : S_l > z\}$$

and

$$N_z = \sup\{l \in \mathbb{N} : S_l < z\}.$$

1) Let $k \in \mathbb{N}$ and let f_k be a density function of S_k. Prove that

$$f_k(z) = \frac{z^k}{k!} \quad \text{for a.e. } z \in [0,1].$$

2) Taking into account that $\inf \emptyset = \infty$ and $\sup \emptyset = -\infty$, prove that for $z \in (0,1]$ we have

$$P(M_z = \infty) = 0 \quad \text{and} \quad P(N_z = -\infty) = 1 - z.$$

3) For $z \in (0,1]$ and $k \in \mathbb{N}$ find $P(M_z = k)$ and $P(N_z = k)$.

Solution 2.9.28: 1) We prove the desired relation by induction. Clearly, $S_0 \sim Unif[0,1]$ and thus $f_0(z) = 1$ for a.e. $z \in [0,1]$. Suppose that $f_k(z) = \dfrac{z^k}{k!}$ for a.e. $z \in [0,1]$, where $k \in \mathbb{N}$ is fixed. By Example 2.15, we have for a.e. $z \in [0,1]$

$$f_{k+1}(z) = \int_{\mathbb{R}} f_k(u) f_{X_{k+1}}(z-u)du = \int_0^z \frac{u^k}{k!}du = \frac{z^{k+1}}{(k+1)!},$$

where $f_{X_{k+1}}$ is a density function of X_{k+1}.

2) Let $z \in (0,1]$. We note that $\{l \in \mathbb{N} : S_l > z\} = \emptyset$ if and only if $\bigcap_{k=0}^{\infty} \{S_k \leq z\}$ occurs. Hence, in view of Theorem 1.6, we get

$$P(M_z = \infty) = P\left(\bigcap_{k=0}^{\infty} \{S_k \leq z\}\right)$$

$$= \lim_{k \to \infty} P(S_k \leq z) = \lim_{k \to \infty} \int_0^z \frac{u^k}{k!}du = \lim_{k \to \infty} \frac{z^{k+1}}{(k+1)!} = 0.$$

Next, we note that $\{l \in \mathbb{N} : S_l < z\} = \emptyset$ if and only if $\{S_0 \geq z\}$ occurs. Hence, $P(N_z = -\infty) = P(S_0 \geq z) = 1 - z$.

3) Let $z \in (0,1]$. We have $P(M_z = 0) = 1 - z$. Next, let $k \in \mathbb{N}^*$. Then

$$P(M_z = k) = P(S_k > z, S_{k-1} \leq z) = P(S_k > z) - P(S_k > z, S_{k-1} > z)$$

$$= P(S_k > z) - P(S_{k-1} > z) = P(S_{k-1} \leq z) - P(S_k \leq z)$$

$$= \int_0^z f_{k-1}(u)du - \int_0^z f_k(u)du$$

$$= \int_0^z \frac{u^{k-1}}{(k-1)!}du - \int_0^z \frac{u^k}{k!}du = \frac{z^k}{k!} - \frac{z^{k+1}}{(k+1)!}.$$

Let $k \in \mathbb{N}$. Then

$$P(N_z = k) = P(S_{k+1} \geq z, S_k < z) = P(S_{k+1} \geq z) - P(S_{k+1} \geq z, S_k \geq z)$$

$$= P(S_{k+1} \geq z) - P(S_k \geq z) = P(S_k < z) - P(S_{k+1} < z)$$

$$= \frac{z^{k+1}}{(k+1)!} - \frac{z^{k+2}}{(k+2)!}.$$

Problem 2.9.29. An entrepreneur wants to start a business that brings him each year an independent profit of X million euros such that $X \sim Unif[0,10]$. Moreover, he wants to retire after he accumulates 10 million euros.

1) Estimate the probability the entrepreneur stays in business at most 5 years, using simulations in Matlab.

2) Find the probability the entrepreneur stays in business at most 5 years.

Solution 2.9.29: 1)

```
function p=entrepreneur(N)
count5=0;
for i=1:N
    profit=0; years=0;
    while profit<10
        profit=profit+unifrnd(0,10); years=years+1;
    end
    if years<=5
        count5=count5+1;
    end
end
p=count5/N;
end
```

```
>> entrepreneur(10000)
ans = 0.9918
```

2) Let $Z_l \sim Unif[0, 10]$ be the profit in the lth year, for $l \in \mathbb{N}^*$, and

$$Y = \inf\{l \in \mathbb{N}^* : Z_1 + \ldots + Z_l > 10\}.$$

Then Y is a random variable which indicates the number of years that the entrepreneur stays in business. Note that

$$Y = \inf\{l \in \mathbb{N}^* : X_0 + \ldots + X_{l-1} > 1\},$$

where $X_j = \dfrac{1}{10}Z_{j+1} \sim Unif[0,1]$ for $j \in \mathbb{N}$. In view of Problem 2.9.28, we have $P(Y = k + 1) = \dfrac{1}{k!} - \dfrac{1}{(k+1)!}$, $k \in \mathbb{N}$, and thus

$$P(Y \le 5) = \sum_{k=0}^{4} \left(\frac{1}{k!} - \frac{1}{(k+1)!} \right) = 1 - \frac{1}{5!} \approx 0.9917.$$

Problem 2.9.30. Consider the Bayesian network given in Figure 2.11, where $X_1, ..., X_5$ are binary random variables. It is known that:
$P(X_1=0)=0.4$, $P(X_2=0|X_1=0)=0.2$,
$P(X_2=0|X_1=1) = 0.5$, $P(X_3=0|X_1=0) = 0.3$,
$P(X_3=0|X_1=1) = 0.4$, $P(X_4=0|X_2=0) = 0.2$,
$P(X_4=0|X_2=1) = 0.5$, $P(X_5=0|X_2=0, X_3=0) = 0.5$,
$P(X_5=0|X_2=0, X_3=1)=0.2$,
$P(X_5=0|X_2=1, X_3=0) = 0.7$,
$P(X_5=0|X_2=1, X_3=1) = 0.4$.

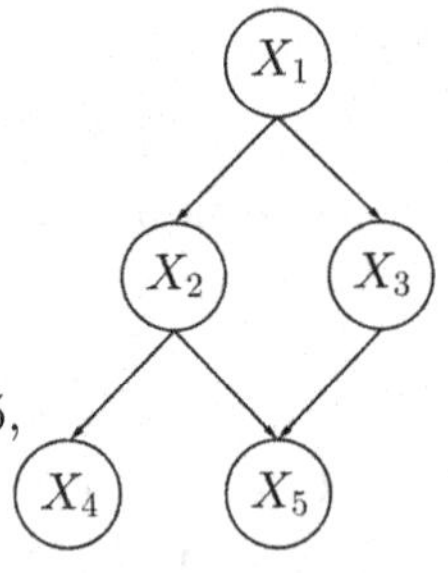

Fig. 2.11: Problem 2.9.30

1) Compute

$$P(X_3=1|X_2=1), \ P(X_1=0, X_3=1), \ P\left(\bigcap_{i=1}^{5}\{X_i=1\}\right).$$

2) Write the probability mass function of X_3.

Solution 2.9.30: We determine the probabilities needed for the computations:

$$P(X_1=1) = 1 - P(X_1=0)=0.6$$
$$P(X_2=1|X_1=0) = 1 - P(X_2=0|X_1=0) = 0.8$$
$$P(X_2=1|X_1=1) = 1 - P(X_2=0|X_1=1) = 0.5$$
$$P(X_3=1|X_1=0) = 1 - P(X_3=0|X_1=0) = 0.7$$
$$P(X_3=1|X_1=1) = 1 - P(X_3=0|X_1=1) = 0.6$$
$$P(X_4=1|X_2=1) = 1 - P(X_4=0|X_2=1) = 0.5$$
$$P(X_5=1|X_2=1, X_3=1) = 1 - P(X_5=0|X_2=1, X_3=1) = 0.6.$$

1) It holds

$$P(X_3=1|X_2=1) = \frac{P(X_3=1, X_2=1)}{P(X_2=1)}.$$

By the law of total probability (see Theorem 1.8) and the independence properties in a Bayesian network (X_2 is conditionally independent of X_3, given X_1) we have:

$$\begin{aligned} P(X_3=1, X_2=1) &= P(X_3=1, X_2=1|X_1=0)P(X_1=0) \\ &+ P(X_3=1, X_2=1|X_1=1)P(X_1=1) \\ &= P(X_3=1|X_1=0)P(X_2=1|X_1=0)P(X_1=0) \\ &+ P(X_3=1|X_1=1)P(X_2=1|X_1=1)P(X_1=1) = 0.404, \end{aligned}$$
$$P(X_2=1)=P(X_2=1|X_1=0)P(X_1=0) + P(X_2=1|X_1=1)P(X_1=1)=0.62.$$

Therefore, $P(X_3=1|X_2=1) \approx 0.6516$. It holds

$$P(X_1=0, X_3=1) = P(X_3=1|X_1=0)P(X_1=0) = 0.56.$$

By the chain rule for probabilities (see Theorem 1.7) and the independence properties in a Bayesian network (X_2 is conditionally independent of X_3, given X_1; X_4 is conditionally independent of X_1, X_3, given X_2; X_5 is conditionally independent of X_1, X_4, given X_2, X_3) we have:

$$P(X_1{=}1, X_2{=}1, X_3{=}1, X_4{=}1, X_5{=}1)$$
$$= P(X_1{=}1)P(X_2{=}1|X_1{=}1)P(X_3{=}1|X_1{=}1)P(X_4{=}1|X_2{=}1)$$
$$\cdot P(X_5{=}1|X_2{=}1, X_3{=}1) = 0.054\,.$$

2) We have

$$P(X_3{=}0) = P(X_3{=}0|X_1{=}0)P(X_1{=}0) + P(X_3{=}0|X_1{=}1)P(X_1{=}1) = 0.36$$

and $P(X_3{=}1){=}1 - P(X_3{=}0){=}0.64$.

Chapter 3

Numerical Characteristics of Random Variables and Vectors

3.1 Expectation

Throughout this chapter $(\Omega, \mathcal{F}, P)$ denotes a probability space.

In what follows we introduce the notion of **expectation** (**mean value, expected value**). Definition 3.1 is based on the integral with respect to the probability (measure) P (see Section A.1.1), while Definition 3.2 uses the Lebesgue–Stieltjes integral (see Section A.1.2).

Definition 3.1. Let X be a random variable, which is Lebesgue integrable with respect to the measure P. The **expectation** of the random variable X is

$$E(X) = \int_\Omega X(\omega)dP(\omega). \qquad \blacklozenge$$

Definition 3.2. The **expectation** of the random variable X is

$$E(X) = \int_\mathbb{R} x\,dF_X(x),$$

if the Lebesgue–Stieltjes integral exists and is finite. $\qquad \blacklozenge$

The above two definitions are equivalent. The reader (focused on applications) can skip these definitions and use directly Definition 3.3 for the expectation of a discrete random variable, respectively of a continuous random variable.

Definition 3.3. Let $X = \sum_{i \in I} x_i \mathbb{I}_{\{X=x_i\}}$ be a discrete random variable, then its **expectation** is

$$E(X) = \sum_{i \in I} x_i P(X = x_i),$$

if the series is absolutely convergent, i.e.,

$$\sum_{i \in I} |x_i| P(X = x_i) < \infty.$$

Let X be a continuous random variable and denote by f_X a density function of X. Then the **expectation** of X is

$$E(X) = \int_{\mathbb{R}} x f_X(x) dx,$$

if the integral is absolutely convergent, i.e.,

$$\int_{\mathbb{R}} |x| f_X(x) dx < \infty. \qquad \blacklozenge$$

Definition 3.4. We say that X is an **integrable random variable**, if $E(|X|)$ exists, i.e., $E(|X|) < \infty$. $\qquad \blacklozenge$

Definition 3.5. Let $\mathbb{X} = (X_1, \ldots, X_n)$ be a random vector. The n-dimensional vector

$$E(\mathbb{X}) = \big(E(X_1), \ldots, E(X_n)\big)$$

is called the **expectation** of the random vector $\mathbb{X}$, if each of the random variables X_i has expectation $E(X_i)$, $i \in \{1, \ldots, n\}$. $\qquad \blacklozenge$

Remark 3.1. (1) If $h : \mathbb{R} \to \mathbb{R}$ is a $\mathcal{B}/\mathcal{B}$ measurable function and $h(X)$ is an integrable random variable, then the expectation of the random variable $h(X)$ is

$$E\big(h(X)\big) = \int_{\mathbb{R}} h(x) dF_X(x).$$

If X is a discrete integrable random variable $X = \sum_{i \in I} x_i \mathbb{I}_{\{X = x_i\}}$, then

$$E\big(h(X)\big) = \sum_{i \in I} h(x_i) P(X = x_i),$$

while for a continuous integrable random variable X we have

$$E\big(h(X)\big) = \int_{\mathbb{R}} h(x) f_X(x) dx.$$

(2) Let $\mathbb{X} = (X_1, \ldots, X_n)$ be a random vector and let $h : \mathbb{R}^n \to \mathbb{R}$ be a $\mathcal{B}^n/\mathcal{B}$ measurable function such that $h(\mathbb{X})$ is integrable. Then the expectation of the random variable $h(\mathbb{X})$ is given by

$$E\big(h(\mathbb{X})\big) = \underbrace{\int \cdots \int}_{\mathbb{R}^n} h(x_1, \ldots, x_n) dF_{\mathbb{X}}(x_1, \ldots, x_n).$$

If $\mathbb{X}$ is a discrete random vector $\mathbb{X} = \sum_{i \in I} x^i \mathbb{I}_{\{\mathbb{X} = x^i\}}$, where $x^i = (x_1^i, \ldots, x_n^i)$, then

$$E\big(h(\mathbb{X})\big) = \sum_{i \in I} h(x_1^i, \ldots, x_n^i) P\big(\mathbb{X} = (x_1^i, \ldots, x_n^i)\big),$$

while for a continuous random vector $\mathbb{X}$, we have

$$E\big(h(\mathbb{X})\big) = \underbrace{\int \cdots \int}_{\mathbb{R}^n} h(x_1, \ldots, x_n) f_{\mathbb{X}}(x_1, \ldots, x_n) \, dx_1 \ldots dx_n. \qquad \blacktriangledown$$

Theorem 3.1. *If X and Y are integrable random variables, then the following properties hold:*

(1) $E(aX + b) = aE(X) + b$ *for all $a, b \in \mathbb{R}$.*
(2) $E(X + Y) = E(X) + E(Y)$.
(3) *If X and Y are independent, then*

$$E(X \cdot Y) = E(X)E(Y).$$

(4) *If $X(\omega) \leq Y(\omega)$ for a.e. $\omega \in \Omega$, then $E(X) \leq E(Y)$.*

Example 3.1. Peter has a bike combination lock with a 3-digit cipher code. He remembers that the first digit is 0 and the other two are even, but he forgot them. Let X denote the number of Peter's unsuccessful trials before opening the lock for the first time. We compute $E(X)$ for the following cases:

(1) unsuccessful codes are not eliminated from further selections (Peter is not a well organized person);
(2) unsuccessful codes are eliminated (Peter likes systematical search and notes the already tried codes).

Solution: In case **(1)**, the probability to open the lock in one trial is $\dfrac{1}{5^2}$. Since X is the number of unsuccessful trials before opening the lock for the first time, $X \sim Geo(p)$ with $p = \dfrac{1}{25}$. We have

$$E(X) = \sum_{k=0}^{\infty} kP(X = k) = p(1 - p) \sum_{k=1}^{\infty} k(1 - p)^{k-1}$$

$$= p(1-p)\left(\frac{1}{1-x}\right)'\bigg|_{x=1-p} = \frac{1-p}{p},$$

where we use [Schinazi (2012), Power series and differentiation, p. 216] for

the geometric series $\dfrac{1}{1-x} = \displaystyle\sum_{k=0}^{\infty} x^k$, $x \in (-1,1)$. We obtain $E(X) = 24$.

In case **(2)**, where unsuccessful codes are eliminated, it can take at least 0 and at most 24 unsuccessful attempts before opening the lock

$$P(X=0) = \frac{1}{25}, P(X=1) = \frac{24}{25}\cdot\frac{1}{24}, P(X=2) = \frac{24}{25}\cdot\frac{23}{24}\cdot\frac{1}{23}, \cdots,$$

$$P(X=24) = \frac{24}{25}\cdot\frac{23}{24}\cdot\ldots\cdot\frac{1}{2}\cdot 1.$$

We obtain

$$E(X) = \frac{0+1+\cdots+24}{25} = 12. \qquad \blacktriangle$$

Example 3.2. Compute the expectation of the random variable X having $Exp(\lambda)$, $\lambda > 0$, distribution (see Section 2.5.6).

Solution: Let f_X be a density function of X. Then $f_X(x) = \lambda e^{-\lambda x}$ for a.e. $x > 0$ and $f_X(x) = 0$ for a.e. $x \leq 0$. By using integration by parts formula we have

$$E(X) = \int_{\mathbb{R}} x f_X(x)dx = \int_0^{\infty} x\lambda e^{-\lambda x}dx$$

$$= x(-e^{-\lambda x})\bigg|_0^{\infty} + \int_0^{\infty} e^{-\lambda x}dx = -\frac{1}{\lambda}e^{-\lambda x}\bigg|_0^{\infty} = \frac{1}{\lambda}. \qquad \blacktriangle$$

3.1.1 *Solved Problems*

Problem 3.1.1. In a tombola bin there are 500 tickets from which 50 are winning.

1) 100 tickets were sold. Compute the expected number of winning tickets left in the tombola bin.

2) Peter is the 13th customer at the tombola bin and buys 10 tickets. Compute the expected number of winning tickets bought by Peter.

Solution 3.1.1: 1) Denote by T_{100} the number of sold winning tickets. Observe that in fact $T_{100} \sim Hyge(100, 50, 450)$. For $i \in \{1, 2, \ldots, 100\}$ denote

$$X_i = \begin{cases} 1, & \text{if the } i\text{th sold ticket is winning} \\ 0, & \text{otherwise.} \end{cases}$$

Moreover, for each $i \in \{1, 2, \ldots, 100\}$ we have

$$P(X_i = 1) = \frac{50 \cdot P(500 - 1, i - 1)}{P(500, i)} = 0.1,$$

because, for the number of possible outcomes, there are $P(500, i)$ possible ways to choose the first i sold tickets (i.e., i-permutations of 500 tickets without repetitions, see Section A.2), while, for the number of favorable outcomes, there are 50 possible choices for the ith sold winning ticket and $P(500 - 1, i - 1)$ possible ways to choose the first $i - 1$ sold tickets (since we take out the winning ticket chosen for the ith sold ticket). Then

$$E(T_{100}) = E(X_1 + \cdots + X_{100}) = 10$$

and the expected number of winning tickets left in the tombola bin is equal to $50 - E(T_{100}) = 40$.

2) The number 13 does not influence the answer, since the probability to get a winning ticket in the ith extraction is independent of i. The expected number of winning tickets, in view of the solution for 1), is $10 \cdot 0.1 = 1$.

Problem 3.1.2. John has to make a survey research in which 36 persons should take part. He decides to ask each adult passer-by in the street to participate in the survey. The probability that a passer agrees to answer the list of questions is 0.6.

1) Compute the probability that 50 persons are asked, before 36 are found to participate in the survey.

2) Which is the expected number of persons that reject to participate in the survey, before 36 persons agree to answer the questions in the survey?

3) Which is the expected number of persons that are asked, before 36 are found to participate in the survey?

Solution 3.1.2: Let X denote the number of persons which refused to participate in the survey, before 36 agreed.

1) We have $X \sim NBin(36, 0.6)$.

$$P(X = 14) = C(49, 14)(0.6)^{36}(0.4)^{14} \approx 0.0187.$$

2) Denote by X_1 the number of persons asked before the first person agrees, let X_2 be the number of additional persons asked before the second person agrees and so on until X_{36}. We have $X_i \sim Geo(0.6)$, $i \in \{1, \ldots, 36\}$. Let $Y = X_1 + X_2 + \cdots + X_{36}$ and, as computed in Example 3.1, we have $E(X_i) = \frac{1 - 0.6}{0.6} = \frac{2}{3}$, $i \in \{1, \ldots, 36\}$. Then $E(Y) = 36 \cdot \frac{2}{3} = 24$ is the expected number of persons that reject to participate in the survey, before

36 are found to participate in the survey.

3) The expected number of persons that are asked, before 36 are found to complete the survey is $35 + E(Y) = 59$.

Problem 3.1.3. Two points are independently and uniformly selected in the interval $[0, L]$, where $L > 0$ is given.

1) What is the expected value of the distance between the two points?

2) Estimate the expected value of the distance between the two points by using Matlab simulations.

Solution 3.1.3: 1) Let $X, Y \sim Unif[0, L]$ be the values of the two numbers. Then the distance between the two points is the random variable $Z = |X - Y|$. We consider the following joint density function of (X, Y)

$$
f_{X,Y}(x,y) = \begin{cases} \dfrac{1}{L^2}, & \text{if } (x,y) \in [0,L] \times [0,L] \\ 0, & \text{if } (x,y) \notin [0,L] \times [0,L] \end{cases}
$$

and we get

$$
\begin{aligned}
E(Z) &= \int_0^L \int_0^L |x - y| \, f_{X,Y}(x,y) \, dx \, dy \\
&= \frac{1}{L^2} \int_0^L \int_0^L |x - y| \, dx \, dy \\
&= \frac{1}{L^2} \int_0^L \left(\int_0^x (x - y) \, dy \right) dx + \frac{1}{L^2} \int_0^L \left(\int_x^L (y - x) \, dy \right) dx = \frac{L}{3}.
\end{aligned}
$$

2)

```
function distance(L,N)
% L length of the interval
% N random pairs of numbers are generated in [0,L]
X=unifrnd(0,L,1,N); Y=unifrnd(0,L,1,N);
E=mean(abs(X-Y));
fprintf('The estimated distance is %5.4f.\n',E)
end
```

\>\> distance(3,1000)
The estimated distance is 1.0056.

Problem 3.1.4. Let N be a discrete random variable having the distribution $P(N = 3) = \dfrac{1}{2}$, $P(N = 4) = P(N = 6) = \dfrac{1}{4}$ and let $L \sim Unif[0, 1]$. Suppose that these random variables are independent. Find the expected value of the area of the regular polygon of N sides having length L.

Solution 3.1.4: The area of the regular polygon of N sides having length L is $S = \dfrac{1}{4} \cdot N \cdot \cot\left(\dfrac{\pi}{N}\right) \cdot L^2$. Hence, by Theorem 3.1-(3),

$$E(S) = \frac{1}{4} \cdot E\left(N \cdot \cot\left(\frac{\pi}{N}\right)\right) \cdot E(L^2)$$

$$= \frac{1}{4} \cdot \left(\frac{3}{2\sqrt{3}} + \frac{4}{4} + \frac{6\sqrt{3}}{4}\right) \int_0^1 x^2 dx = \frac{1 + 2\sqrt{3}}{12}.$$

Problem 3.1.5. Suppose that an electrical circuit has n components ($n \in \mathbb{N}, n \geq 2$) connected in series, which function independently of each other. The lifetime of the ith component is given by a random variable having exponential distribution with parameter $\lambda_i > 0$ for $i \in \{1, \ldots, n\}$.
1) Find the expected value of the time until the system fails.
2) Let $I \subsetneq \{1, \ldots, n\}$ be nonempty. Consider the subsystem formed by the components corresponding to the indices in I. Find the probability that the system fails because of the failure of the subsystem.

Solution 3.1.5: Let X be the random variable that represents the time until the system fails and let X_i denote the lifetime of the ith component, $i \in \{1, \ldots, n\}$.
1) In this case, $X = \min\{X_1, \ldots, X_n\}$. For $x > 0$ we have

$$P(X \leq x) = P\left(\min\{X_1, \ldots, X_n\} \leq x\right) = 1 - P\left(\bigcap_{i=1}^{n}\{X_i > x\}\right)$$

$$= 1 - \prod_{i=1}^{n} P(X_i > x) = 1 - \exp\left\{-\sum_{i=1}^{n}\lambda_i x\right\}.$$

Then

$$f_X(x) = \begin{cases} \displaystyle\sum_{i=1}^{n}\lambda_i \exp\left\{-\sum_{i=1}^{n}\lambda_i x\right\}, & \text{if } x > 0 \\ 0, & \text{if } x \leq 0 \end{cases}$$

is a density function of X. We observe that X has an exponential distribution with parameter $\displaystyle\sum_{i=1}^{n}\lambda_i$ and thus by Example 3.2 its expectation is

$$E(X) = \frac{1}{\displaystyle\sum_{i=1}^{n}\lambda_i}.$$

2) Let $J = \{1, \ldots, n\} \setminus I$. The probability that the system fails because of the subsystem is

$$P\big(\min\{X_k : k \in I\} < \min\{X_k : \in J\}\big) = P(Y < Z),$$

where $Y = \min\{X_k : k \in I\}$ follows the exponential distribution with parameter $\sum_{k \in I} \lambda_k$ and $Z = \min\{X_k : k \in J\}$ follows the exponential distribution with parameter $\sum_{k \in J} \lambda_k$, in view of the proof of 1). Obviously, Y and Z are independent.

Let $D = \{(y, z) \in \mathbb{R}^2 : 0 \leq y \leq z\}$. Using Fubini's theorem, we deduce

$$P(Y < Z) = P(Y \leq Z) = \iint_D f_Y(y) f_Z(z) dy dz = \int_0^\infty f_Y(y) \left(\int_y^\infty f_Z(z) dz \right) dy$$

$$= \int_0^\infty f_Y(y)(1 - F_Z(y)) dy = \frac{\sum_{k \in I} \lambda_k}{\sum_{k=1}^n \lambda_k}.$$

3.2   Conditional Expectation with Respect to an Event

In the following, we consider $B \in \mathcal{F}$ with $P(B) > 0$.

Definition 3.6. The **conditional expectation** of the random variable X given the event B is defined by

$$E(X|B) = \frac{E(X \cdot \mathbb{I}_B)}{P(B)},$$

if $\mathbb{I}_B \cdot X$ is an integrable random variable.   ◆

An equivalent definition is the following.

Definition 3.7. Let $F_X(\cdot|B)$ be the conditional distribution function of X given event B (see Definition 2.17). The **conditional expectation** of a random variable X given event B is

$$E(X|B) = \int_{\mathbb{R}} x dF_X(x|B),$$

if the Lebesgue–Stieltjes integral exists and is finite.   ◆

Remark 3.2. **(1)** The **conditional expectation** of the random variable X given event B for
(1a) a discrete random variable $X = \sum_{i \in I} x_i \mathbb{I}_{\{X=x_i\}}$ is

$$E(X|B) = \sum_{i \in I} x_i P(X = x_i|B),$$

if the series is absolutely convergent;
(1b) a continuous random variable X with conditional density function $f_X(\cdot|B)$ is

$$E(X|B) = \int_{\mathbb{R}} x f_X(x|B)dx,$$

if the integral is absolutely convergent.
(2) The linearity properties of expectation stated in Theorem 3.1-(1)-(2) also hold for conditional expectation.
(3) Let X and Y be random variables and let $\mathcal{G}$ be the σ-field generated by Y and $\mathbb{I}_B$. If X is independent of $\mathcal{G}$, then

$$E(X \cdot Y|B) = E(X) \cdot E(Y|B),$$

in view of Definition 3.6, Remark 2.8-(1) and Theorem 3.1-(3).   ▼

Theorem 3.2. (1) *The expectation of a random variable X can be written as the weighted sum of conditional expectations*

$$E(X) = \sum_{i \in I} P(B_i)E(X|B_i),$$

where $(B_i)_{i \in I}$ is a partition of Ω with $P(B_i) > 0$ for $i \in I$.
(2) *The conditional expectation of a random variable X given event B can be written as*

$$E(X|B) = \frac{1}{P(B)} \sum_{i \in I} P(B_i)E(X|B_i),$$

where $(B_i)_{i \in I}$ is a partition of B with $P(B_i) > 0$ for $i \in I$.

Definition 3.8. Let X and Y be random variables such that $F_{X|Y}(\cdot|y)$, the conditional distribution function of X given $\{Y = y\}$, is defined (see Definitions 2.19, 2.20, 2.21). The **conditional expectation** $E(X|Y = y)$ of the random variable X given $\{Y = y\}$ is defined as

$$E(X|Y = y) = \int_{\mathbb{R}} x dF_{X|Y}(x|y),$$

if the Lebesgue–Stieltjes integral exists and is finite.   ♦

Remark 3.3. (1) Note that the **conditional expectation** of the random variable X given $\{Y = y\}$ is given by the expressions:

(1a) If X is a discrete random variable and $p_{X|Y}(\cdot|y)$ is a conditional probability mass function of X given $\{Y = y\}$ (see Definition 2.21), then

$$E(X|Y = y) = \sum_{x \in X(\Omega)} x p_{X|Y}(x|y),$$

if the series is absolutely convergent.

(1b) If $f_{X|Y}(\cdot|y)$ is a conditional density function of X given $\{Y = y\}$ (see Definitions 2.19 and 2.20), then

$$E(X|Y = y) = \int_{\mathbb{R}} x f_{X|Y}(x|y) dx,$$

if the integral is absolutely convergent.

(2) Let $h : \mathbb{R} \to \mathbb{R}$ be a $\mathcal{B}/\mathcal{B}$ measurable function and let X be a random variable, then the conditional expectation of a random variable $h(X)$ given $\{Y = y\}$ can be written as

$$E(h(X)|Y = y) = \int_{\mathbb{R}} h(x) dF_{X|Y}(x|y),$$

if the above integral exists and is finite.

(3) The linearity properties of expectation stated in Theorem 3.1-(1)-(2) also hold for the conditional expectation.   ▼

3.2.1   *Solved Problems*

Problem 3.2.1. We are rolling two dice. Let X be the random variable that represents the number of doubles until we get the first sum equal to 7 and let Y be the random variable that represents the number of rolls until we get the first sum equal to 7. Compute $E(X|Y = k)$ for $k \in \mathbb{N}^*$.

Solution 3.2.1: Obviously, $E(X|Y = 1) = 0$. By Remark 3.3-(1a), we get for $k \in \mathbb{N}, k \geq 2$,

$$E(X|Y = k) = \sum_{i \in X(\Omega)} i \cdot P(X = i|Y = k).$$

When we roll two dice, the conditional probability to obtain a double given that the sum of the two numbers is not 7 is $\dfrac{6}{30} = \dfrac{1}{5}$, while the conditional probability not to obtain a double given that the sum of the two numbers is not 7 is $\frac{4}{5}$. Hence, for $i \in \mathbb{N}$ and $k \in \mathbb{N}, k \geq 2$, we have

$$P(X = i | Y = k) = \begin{cases} C(k-1, i) \cdot \left(\frac{1}{5}\right)^i \cdot \left(\frac{4}{5}\right)^{k-1-i}, & \text{if } 0 \leq i \leq k-1 \\ 0, & \text{if } k \leq i. \end{cases}$$

Then for $k \in \mathbb{N}, k \geq 2$, we have

$$E(X|Y = k) = \left(\frac{4}{5}\right)^{k-1} \sum_{i=0}^{k-1} i \cdot C(k-1, i) \cdot \frac{1}{4^i}$$

$$= \left(\frac{4}{5}\right)^{k-1} \cdot \frac{k-1}{4} \sum_{i=0}^{k-2} C(k-2, i) \cdot \frac{1}{4^i} = \frac{k-1}{5}.$$

Problem 3.2.2. A number X is chosen according to $Bino\,(n, 0.5)$, where $n \in \mathbb{N}^*$. After the value $X = k$ was observed ($k \in \{0, 1, \ldots, n\}$), a number Y is chosen uniformly in the interval $[k, k+1]$. Find:
1) the conditional distribution function $F_{Y|X}(\cdot|k)$ of Y given $\{X = k\}$ for $k \in \{0, 1, \ldots, n\}$;
2) the conditional expectation of Y given $\{X = k\}$ for $k \in \{0, 1, \ldots, n\}$;
3) the expectation of Y.

Solution 3.2.2: 1) Let $k \in \{0, 1, \ldots, n\}$. Let $F_{Y|X}(\cdot|k)$ be the conditional distribution function of Y given $\{X = k\}$. Then

$$F_{Y|X}(y|k) = P(Y \leq y | X = k) = \begin{cases} 0, & \text{if } y < k \\ y - k, & \text{if } y \in [k, k+1) \\ 1, & \text{if } y \geq k+1. \end{cases}$$

2) The function

$$f_{Y|X}(y|k) = \begin{cases} 1, & \text{if } y \in [k, k+1] \\ 0, & \text{otherwise} \end{cases}$$

is clearly a conditional density function of Y given $\{X = k\}$. By Remark 3.3-(1b), we have for $k \in \{0, 1, \ldots, n\}$

$$E(Y|X = k) = \int_{\mathbb{R}} y f_{Y|X}(y|k) dy = k + 0.5.$$

3) By Theorem 3.2-(1), we get

$$E(Y) = \sum_{k=0}^{n} P(X = k) E(Y|X = k) = \sum_{k=0}^{n} C(n, k) 0.5^n (k + 0.5)$$

$$= 0.5^n n \sum_{k=1}^{n} C(n-1, k-1) + 0.5^{n+1} \sum_{k=0}^{n} C(n, k) = \frac{n+1}{2}.$$

3.3   Conditional Expectation Given a σ-Field

Definition 3.9. Let X be an integrable random variable on $(\Omega, \mathcal{F}, P)$ and let $\mathcal{G}$ be a σ-field contained in $\mathcal{F}$. Then the **conditional expectation of X given the σ-field** $\mathcal{G}$ is defined to be any random variable $Z = E(X|\mathcal{G})$ such that:
1) $E(X|\mathcal{G})$ is $\mathcal{G}/\mathcal{B}$ measurable.
2) For any $A \in \mathcal{G}$ it holds $E(Z \cdot \mathbb{I}_A) = E(X \cdot \mathbb{I}_A)$.   ♦

Remark 3.4. By the Radon–Nikodym Theorem from measure theory (see [Rudin (1987), Theorem 6.10, p. 121]) one can prove that $E(X|\mathcal{G})$ exists; moreover, $E(X|\mathcal{G})$ is unique in the sense that if Z and Z' are conditional expectations of X given $\mathcal{G}$, then $Z = Z'$ a.s. (see [Durrett (2010), Chapter 5, Section 5.1]).   ▼

We present some properties of the conditional expectation of a random variable given a σ-field (see [Breiman (1992), Chapter 4, Sections 2-3] and [Kolokoltsov (2011), Section 1.3]).

Theorem 3.3. *Let X and Y be integrable random variables on a probability space $(\Omega, \mathcal{F}, P)$ and $\mathcal{G}, \mathcal{H} \subseteq \mathcal{F}$ be σ-fields on Ω. The conditional expectation has the following properties:*
(1) *$E(X|\mathcal{G})$ is an integrable random variable;*
(2) *$E(E(X|\mathcal{G})) = E(X)$;*
(3) *if $\mathcal{G} = \{\emptyset, \Omega\}$, then $E(X|\mathcal{G}) = E(X)$ a.s.;*
(4) *if X is $\mathcal{G}$-measurable, then $E(X|\mathcal{G}) = X$ a.s.;*
(5) *$E(aX + bY|\mathcal{G}) = aE(X|\mathcal{G}) + bE(Y|\mathcal{G})$ a.s. for each $a, b \in \mathbb{R}$;*
(6) *if $X \cdot Y$ is integrable and X is $\mathcal{G}$-measurable, then*

$$E(X \cdot Y|\mathcal{G}) = X \cdot E(Y|\mathcal{G}) \ a.s.;$$

(7) *if X is independent of $\mathcal{G}$ (see Definition 2.14), then $E(X|\mathcal{G})=E(X)$ a.s.;*
(8) *if Y is $\mathcal{G}$-measurable, X is independent of $\mathcal{G}$ and $f : \mathbb{R}^2 \to \mathbb{R}$ is a bounded $\mathcal{B}^2/\mathcal{B}$ measurable function, then*

$$E(f(X, Y)|\mathcal{G})(\omega) = h(Y(\omega)) \ for \ a.e. \ \omega \in \Omega,$$

where $h(y) = E(f(X, y))$ for $y \in \mathbb{R}$, which is a $\mathcal{B}/\mathcal{B}$ measurable function;
(9) *if $\mathcal{H} \subseteq \mathcal{G}$, then $E\big(E(X|\mathcal{G})|\mathcal{H}\big) = E(X|\mathcal{H})$ and $E\big(E(X|\mathcal{H})|\mathcal{G}\big) = E(X|\mathcal{H})$;*
(10) *if $X \geq Y$ a.s., then $E(X|\mathcal{G}) \geq E(Y|\mathcal{G})$ a.s.;*
(11) *(Jensen's inequality) if $\Phi : \mathbb{R} \to \mathbb{R}$ is a convex function such that $\Phi(X)$ is integrable, then*

$$\Phi(E(X|\mathcal{G})) \leq E(\Phi(X)|\mathcal{G}) \ a.s.$$

Example 3.3. Consider X to be an integrable random variable on the probability space $(\Omega, \mathcal{F}, P)$.

(1) Let $B \in \mathcal{F}$ with $P(B) > 0$, $P(\bar{B}) > 0$ and choose $\mathcal{G} = \{\emptyset, B, \bar{B}, \Omega\}$. Then it holds

$$E(X|\mathcal{G}) = \frac{E(X \cdot \mathbb{I}_B)}{P(B)} \mathbb{I}_B + \frac{E(X \cdot \mathbb{I}_{\bar{B}})}{P(\bar{B})} \mathbb{I}_{\bar{B}} \quad \text{a.s.}$$

The discrete random variable from the right hand side of the above equality verifies Definition 3.9 and by the uniqueness property from Remark 3.4 it follows that it must coincide a.s. with $E(X|\mathcal{G})$.

(2) If $(A_i)_{i \in I}$ is a partition of Ω, where $\emptyset \neq I \subseteq \mathbb{N}$, with $P(A_i) > 0$ for each $i \in I$, then consider $\mathcal{A} = \{A_i : i \in I\}$ and $\mathcal{G} = \sigma(\mathcal{A})$ (see also Example 1.3-(2)). Then it holds

$$E(X|\mathcal{G}) = \sum_{i \in I} \frac{E(X \cdot \mathbb{I}_{A_i})}{P(A_i)} \mathbb{I}_{A_i} = \sum_{i \in I} E(X|A_i) \mathbb{I}_{A_i} \quad \text{a.s.}$$

The last equality follows by Definition 3.6. ▲

Example 3.4. Let X be a random variable on a probability space $(\Omega, \mathcal{F}, P)$ such that X^2 is integrable. Let $\mathcal{G}$ be a σ-field contained in $\mathcal{F}$. Denote the following space of random variables

$$\mathcal{L}^2(\mathcal{G}) = \{Y{:}\Omega \to \mathbb{R} : Y \text{ is a } \mathcal{G}/\mathcal{B} \text{ measurable such that } Y^2 \text{ is integrable}\}.$$

We prove that

$$E\big((X - E(X|\mathcal{G}))^2\big) = \min\Big\{E\big((X - Y)^2\big) : Y \in \mathcal{L}^2(\mathcal{G})\Big\}.$$

Proof: Denote $\hat{Y} = E(X|\mathcal{G})$. By Jensen's inequality (see Theorem 3.3-(11)) it holds a.s.

$$\hat{Y}^2 = \big(E(X|\mathcal{G})\big)^2 \leq E(X^2|\mathcal{G}),$$

which implies

$$E\big(\hat{Y}^2\big) \leq E\big(E(X^2|\mathcal{G})\big) = E(X^2) < \infty$$

by Theorem 3.3-(2). Hence, $\hat{Y} \in \mathcal{L}^2(\mathcal{G})$.

Consider $Y \in \mathcal{L}^2(\mathcal{G})$ arbitrary. Then, by Theorem 3.3-(6), we have a.s.

$$E(X \cdot Y|\mathcal{G}) = Y \cdot E(X|\mathcal{G}) = Y \cdot \hat{Y}.$$

By taking expectation, one has

$$E\big(E(X \cdot Y|\mathcal{G})\big) = E(Y \cdot \hat{Y}),$$

and using Theorem 3.3-(2) it follows

$$E(X \cdot Y) = E(Y \cdot \hat{Y}).$$

By taking $Y = \hat{Y}$, we also have

$$E(X \cdot \hat{Y}) = E(\hat{Y}^2).$$

Finally, we write by using the above computations

$$E((X - Y)^2) - E((X - \hat{Y})^2) = E((\hat{Y} - Y)^2) \geq 0,$$

while equality holds if and only if $\hat{Y} = Y$ a.s. We obtain

$$E((X - E(X|\mathcal{G}))^2) = \min \left\{ E((X - Y)^2) : Y \in \mathcal{L}^2(\mathcal{G}) \right\}.$$

Note that $\mathcal{G}$ represents a given information and $E(X|\mathcal{G})$ is
- the best approximation in mean square of X;
- the best prediction of X given the information in $\mathcal{G}$.   ▲

Definition 3.10. Let X be an integrable random variable and let Y be an arbitrary random variable. Then the **conditional expectation of X given the random variable Y** is defined to be the random variable $E(X|\mathcal{F}_Y)$ and is denoted by $E(X|Y)$, $\mathcal{F}_Y$ denotes the σ-field generated by the random variable Y, see Definition 2.4.   ♦

Example 3.5. Consider X to be an integrable random variable on the probability space $(\Omega, \mathcal{F}, P)$ and let Y be a discrete random variable with $Y(\Omega) = \{y_i : i \in I\}$ ($I \subseteq \mathbb{N}$ is the index set).
(1) The family $\mathcal{A} = \{Y^{-1}(\{y_i\}) : i \in I\}$ forms a partition of Ω and

$$\mathcal{F}_Y = \sigma(\mathcal{A}) = \{\emptyset\} \cup \left\{ \bigcup_{i \in J} Y^{-1}(\{y_i\}) : \emptyset \neq J \subseteq I \right\},$$

see Example 2.3. Therefore, by Example 3.3, we have a.s.

$$E(X|Y) = E(X|\mathcal{F}_Y) = \sum_{i \in I} \frac{E(X \cdot \mathbb{I}_{\{Y=y_i\}})}{P(Y = y_i)} \mathbb{I}_{\{Y=y_i\}} \tag{3.1}$$

$$= \sum_{i \in I} E(X|Y = y_i) \mathbb{I}_{\{Y=y_i\}}.$$

(2) The following relation holds

$$E(X|Y)(\omega) = h(Y(\omega)) \quad \text{for a.e. } \omega \in \Omega, \tag{3.2}$$

where the function $h : Y(\Omega) \to \mathbb{R}$ is defined by

$$h(y) = \frac{E(X \cdot \mathbb{I}_{\{Y=y\}})}{P(Y = y)}, \quad y \in Y(\Omega).$$

By Definition 3.6, we have $h(y) = E(X|Y = y)$ for $y \in Y(\Omega)$. By (3.1) we have a.s.

$$E(X|Y)(\omega) = \frac{E(X \cdot \mathbb{I}_{\{Y=y_i\}})}{P(Y = y_i)}, \quad \text{if } \omega \in \{Y = y_i\}, i \in I.$$

Therefore, (3.2) holds.   ▲

Example 3.6. Let (X, Y) be a continuous random vector on the probability space $(\Omega, \mathcal{F}, P)$ such that X is an integrable random variable. $f_{X,Y}$ denotes a joint density function and consider f_Y given by

$$f_Y(y) = \int_{\mathbb{R}} f_{X,Y}(x, y)dx \quad \text{for } y \in \mathbb{R},$$

which is a density function of Y, by Theorem 2.9. We prove that the following relation holds

$$E(X|Y)(\omega) = h(Y(\omega)) \text{ for a.e. } \omega \in \Omega, \tag{3.3}$$

where the function $h : \mathbb{R} \to \mathbb{R}$ is defined by

$$h(y) = E(X|Y = y) = \int_{\mathbb{R}} x f_{X|Y}(x|y)dx, \quad y \in \mathbb{R},$$

see Definition 2.20 and Remark 3.3-(1b).

Proof: Let $D = \{y \in \mathbb{R} : f_Y(y) > 0\}$. Observe that $P(Y \in \mathbb{R} \setminus D) = 0$, hence $Y(\omega) \in D$ for a.e. $\omega \in \Omega$. In order to prove (3.3), we check the properties from Definition 3.9. First, note that $h(Y)$ is $\mathcal{F}_Y/\mathcal{B}$ measurable (since h is $\mathcal{B}/\mathcal{B}$ measurable). Further, consider $A \in \mathcal{F}_Y$ arbitrary. Then there exists $B \in \mathcal{B}$ such that $A = Y^{-1}(B)$ and this implies $\mathbb{I}_A(\omega) = \mathbb{I}_B(Y(\omega))$ for each $\omega \in \Omega$. We write

$$E\big(h(Y) \cdot \mathbb{I}_A\big) = E\big(h(Y) \cdot \mathbb{I}_B(Y)\big) = \int_{\mathbb{R}} \mathbb{I}_B(y)h(y)f_Y(y)dy$$

$$= \int_{\mathbb{R}} \mathbb{I}_B(y)\bigg(\int_D x f_{X,Y}(x, y)dx\bigg)dy + \int_{\mathbb{R}} \mathbb{I}_B(y)\bigg(\int_{\mathbb{R}\setminus D} x f_X(x)f_Y(y)dx\bigg)dy$$

$$= \int_{\mathbb{R}}\int_D \mathbb{I}_B(y)x f_{X,Y}(x, y)dxdy + \int_{\mathbb{R}}\int_{\mathbb{R}\setminus D} \mathbb{I}_B(y)x f_{X,Y}(x, y)dxdy$$

$$= \int_{\mathbb{R}}\int_{\mathbb{R}} \mathbb{I}_B(y)x f_{X,Y}(x, y)dxdy = E(X \cdot \mathbb{I}_B(Y)) = E(X \cdot \mathbb{I}_A),$$

where we have used the fact that for $y \in \mathbb{R} \setminus D$ we have $f_{X,Y}(u, y) = 0$ for a.e. $u \in \mathbb{R}$ (see Example 2.27-(1)). The random variable $h(Y)$ verifies Definition 3.9 and, by the uniqueness property from Remark 3.4, it follows that it must coincide a.s. with $E(X|Y)$.   ▲

Example 3.7. Suppose that a point X is chosen based on the uniform distribution on the interval $[a, b]$. After the value $X = x$ was observed ($x \in [a, b]$), a point Y is chosen in accordance with the uniform distribution on the interval $[x, b]$. In what follows we compute a joint density function $f_{X,Y}$ for the random vector (X, Y), then $E(Y|X)$ and $E(Y)$.

Solution: Let $f_{X,Y}$ be a joint density function of (X, Y). We consider the triangle $\mathcal{T} = \{(x, y) \in \mathbb{R}^2 : x \in [a, b), y \in [x, b]\}$ and we observe that (X, Y) is a.s. chosen in $\mathcal{T}$. Hence,

$$f_{X,Y}(x, y) = 0 \text{ for a.e. } (x, y) \notin \mathcal{T}.$$

Taking into account the way Y is chosen, we note that

$$Y = (1 - U) \cdot X + U \cdot b,$$

where $U \sim Unif[0, 1]$ is such that X and U are independent. Thus, we have a joint density function of (X, U) given by

$$f_{X,U}(x, u) = \begin{cases} \dfrac{1}{b-a}, & \text{if } x \in [a, b], u \in [0, 1] \\ 0, & \text{otherwise.} \end{cases}$$

Let $(x, y) \in \mathcal{T}$ and $D_{x,y} = \left\{ (v, u) \in \mathbb{R}^2 : v \in [a, x], u \in \left[0, \dfrac{y-v}{b-v}\right] \right\}$. Then the joint distribution function of (X, U) satisfies

$$F_{X,Y}(x, y) = P\left(X \le x, (1 - U) \cdot X + U \cdot b \le y\right) = \underbrace{\int\int}_{D_{x,y}} f_{X,U}(v, u) dv du$$

$$= \int_a^x \left(\int_0^{\frac{y-v}{b-v}} \frac{1}{b-a} du \right) dv = \frac{b-y}{b-a} \cdot \log\left(\frac{b-x}{b-a}\right) + \frac{x-a}{b-a}.$$

So, a joint density function for (X, Y) is

$$f_{X,Y}(x, y) = \begin{cases} \dfrac{1}{(b-x)(b-a)}, & \text{if } x \in [a, b), y \in [x, b], \\ 0, & \text{otherwise.} \end{cases}$$

Note that

$$f_X(x) = \int_{\mathbb{R}} f_{X,Y}(x, y) dy = \begin{cases} \dfrac{1}{b-a}, & \text{if } x \in [a, b) \\ 0, & \text{if } x \notin [a, b) \end{cases}$$

is a density function of X. As in Example 3.6, we can prove that

$$E(Y|X)(\omega) = g(X(\omega)) \text{ for a.e. } \omega \in \Omega,$$

where the function $g : \mathbb{R} \to \mathbb{R}$ is defined by

$$g(x) = E(Y|X = x) = \int_{\mathbb{R}} y f_{Y|X}(y|x) dy \quad \text{for } x \in \mathbb{R}.$$

By computations we obtain $g(x) = \dfrac{x + b}{2}$ for $x \in [a, b)$ and $g(x) = E(Y)$ if $x \notin [a, b)$. But $X \in [a, b)$ a.s., hence

$$E(Y|X) = \frac{1}{2}X + \frac{b}{2} \quad \text{a.s.}$$

Therefore,

$$E(Y) = E(E(Y|X)) = \frac{1}{2}E(X) + \frac{b}{2} = \frac{a + 3b}{4}. \qquad \blacktriangle$$

Definition 3.11. Let X be an integrable random variable and let $Y_1, \ldots, Y_n$ be random variables, $n \in \mathbb{N}^*$. The **conditional probability of X given the random variables $Y_1, \ldots, Y_n$** is defined to be:

$$P(X \in A|Y_1, \ldots, Y_n) = E(\mathbb{I}_{\{X \in A\}} | \mathcal{F}_{Y_1, \ldots, Y_n}), \ A \in \mathcal{B},$$

where $\mathcal{F}_{Y_1, \ldots, Y_n}$ denotes the σ-field generated by the random variables $Y_1, \ldots, Y_n$, see Definition 2.4. $\qquad \blacklozenge$

Example 3.8. (1) If X is an integrable random variable and $\mathbb{Y} = (Y_1, \ldots, Y_n)$ is a random vector with discrete range $\mathbb{Y}(\Omega)$, then, as in Example 3.5-(1), we have for $A \in \mathcal{B}$ a.s.

$$P(X \in A|Y_1, \ldots, Y_n) = \sum_{y \in \mathbb{Y}(\Omega)} P(X \in A|\mathbb{Y} = y) \mathbb{I}_{\{\mathbb{Y} = y\}}$$

$$= \sum_{(y_1, \ldots, y_n) \in \mathbb{Y}(\Omega)} P(X \in A|Y_1 = y_1, \ldots, Y_n = y_n) \mathbb{I}_{\{Y_1 = y_1, \ldots, Y_n = y_n\}}.$$

(2) If (X, Y) is a continuous random vector on the probability space $(\Omega, \mathcal{F}, P)$, then, as in Example 3.6, we have for $A \in \mathcal{B}$ (see also Definition 2.20):

$$P(X \in A|Y)(\omega) = h(Y(\omega)) \text{ for a.e. } \omega \in \Omega,$$

where $h(y) = P(X \in A|Y = y)$ for $y \in \mathbb{R}$. In particular, for every $x \in \mathbb{R}$ we have

$$P(X \leq x|Y)(\omega) = h(Y(\omega)) \text{ for a.e. } \omega \in \Omega,$$

where $h(y) = F_{X|Y}(x|y)$ for $y \in \mathbb{R}$.

(3) If X and Y are independent random variables on the probability space $(\Omega, \mathcal{F}, P)$, then for every $x \in \mathbb{R}$ we have

$$P(X + Y \leq x|Y)(\omega) = h(Y(\omega)) \text{ for a.e. } \omega \in \Omega,$$

where $h(y) = P(X + y \leq x)$ for $y \in \mathbb{R}$. The result follows from Theorem 3.3-(8) with $f : \mathbb{R}^2 \to \mathbb{R}$ given by

$$f(u, v) = \begin{cases} 1, & \text{if } u + v \leq x \\ 0, & \text{otherwise.} \end{cases}$$

▲

3.3.1 Solved Problems

Problem 3.3.1. Let $\Omega = \{1, 2, 3, 4, 5, 6, 7, 8, 9, 10\}$, the probability be $P(\{i\}) = \frac{1}{10}$ for each $i \in \{1, \ldots, 10\}$, the σ-field $\mathcal{F} = \mathcal{P}(\Omega)$ (the set of all subsets of Ω). Consider

$$A = \{1, 2, 4, 10\}, \ B = \{2, 4, 6, 8\}$$

$$\mathcal{A}_1 = \{B\} \subset \mathcal{F}, \ \mathcal{A}_2 = \big\{\{1, 2, 3, 4\}, \{5, 6, 7, 8\}\big\} \subset \mathcal{F}.$$

Compute :
1) $E(\mathbb{I}_A | \mathcal{G}_1)$, where $\mathcal{G}_1 = \sigma(\mathcal{A}_1)$ is the σ-field generated by $\mathcal{A}_1$;
2) $E(\mathbb{I}_A | \mathcal{G}_2)$, where $\mathcal{G}_2 = \sigma(\mathcal{A}_2)$ is the σ-field generated by $\mathcal{A}_2$.

Solution 3.3.1: 1) We have $\bar{B} = \{1, 3, 5, 7, 9, 10\}$ and the σ-field generated by $\mathcal{A}_1$ is $\mathcal{G}_1 = \sigma(\mathcal{A}_1) = \{\emptyset, \Omega, B, \bar{B}\}$. By Example 3.3-(1), we have

$$E(\mathbb{I}_A | \mathcal{G}_1) = \frac{E(\mathbb{I}_A \cdot \mathbb{I}_B)}{P(B)} \mathbb{I}_B + \frac{E(\mathbb{I}_A \cdot \mathbb{I}_{\bar{B}})}{P(\bar{B})} \mathbb{I}_{\bar{B}} \quad \text{a.s.}$$

Therefore,

$$E(\mathbb{I}_A | \mathcal{G}_1)(\omega) = \begin{cases} \dfrac{P(A \cap B)}{P(B)}, & \text{if } \omega \in B \\ \dfrac{P(A \cap \bar{B})}{P(\bar{B})}, & \text{if } \omega \in \bar{B} \end{cases} = \begin{cases} \dfrac{1}{2}, & \text{if } \omega \in \{2, 4, 6, 8\} \\ \dfrac{1}{3}, & \text{if } \omega \in \{1, 3, 5, 7, 9, 10\}. \end{cases}$$

2) The σ-field generated by $\mathcal{A}_2$ is

$$\mathcal{G}_2 = \sigma(\mathcal{A}_2) = \{\emptyset, \Omega, \{1, 2, 3, 4\}, \{5, 6, 7, 8\}, \{9, 10\},$$
$$\{1, 2, 3, 4, 5, 6, 7, 8\}, \{5, 6, 7, 8, 9, 10\}, \{1, 2, 3, 4, 9, 10\}\}.$$

Denote

$$A_1 = \{1, 2, 3, 4\}, \ A_2 = \{5, 6, 7, 8\}, \ A_3 = \{9, 10\}.$$

and consider $\{A_1, A_2, A_3\}$, which is a partition of Ω and

$$\mathcal{G}_2 = \sigma(\mathcal{A}_2) = \sigma(\{A_1, A_2, A_3\}).$$

By Example 3.3-(2) we have

$$E(\mathbb{I}_A|\mathcal{G}_2) = \sum_{i=1}^{3} \frac{E(\mathbb{I}_A \cdot \mathbb{I}_{A_i})}{P(A_i)}\mathbb{I}_{A_i} \quad \text{a.s.}$$

This implies

$$E(\mathbb{I}_A|\mathcal{G}_2)(\omega) = \begin{cases} \dfrac{P(A \cap A_1)}{P(A_1)}, & \text{if } \omega \in A_1 \\ \dfrac{P(A \cap A_2)}{P(A_2)}, & \text{if } \omega \in A_2 \\ \dfrac{P(A \cap A_3)}{P(A_3)}, & \text{if } \omega \in A_3 \end{cases} = \begin{cases} \dfrac{3}{4}, & \text{if } \omega \in \{1,2,3,4\} \\ 0, & \text{if } \omega \in \{5,6,7,8\} \\ \dfrac{1}{2}, & \text{if } \omega \in \{9,10\}. \end{cases}$$

Problem 3.3.2. Let $g : [0,1] \to (0,\infty)$ be continuous function and let $\mathcal{G} = \{(x,y) : x \in [0,1], y \in [0,g(x)]\}$. Let (X,Y) be a random vector uniformly distributed on $\mathcal{G}$, i.e., $f_{X,Y} : \mathbb{R}^2 \to \mathbb{R}$ defined by

$$f_{X,Y}(x,y) = \begin{cases} \left(\int_0^1 g(t)dt\right)^{-1}, & \text{if } (x,y) \in \mathcal{G} \\ 0, & \text{otherwise} \end{cases}$$

is a joint density function of (X,Y). Find $E(Y|X)$.

Solution 3.3.2: We note that f_X given by

$$f_X(x) = \int_{\mathbb{R}} f_{X,Y}(x,y)dy = \int_0^{g(x)} \left(\int_0^1 g(t)dt\right)^{-1} dy = \frac{g(x)}{\displaystyle\int_0^1 g(t)dt},$$

if $x \in [0,1]$, and $f_X(x) = 0$, otherwise, is a density function of X. Since for every $x \in [0,1]$ we have $f_X(x) > 0$ and

$$E(Y|X = x) = \int_{\mathbb{R}} y f_{Y|X}(y|x)dy = \frac{g(x)}{2} \text{ for } x \in [0,1],$$

we deduce, by Example 3.6, that $E(Y|X) = \dfrac{1}{2}g(X)$ a.s.

Problem 3.3.3. Consider $\Omega = [0,1]$, $\mathcal{F} = \mathcal{B}([0,1])$, P the Lebesgue measure on $[0,1]$ and the random variables given by

$$X(\omega) = 2\omega + 1 \text{ for each } \omega \in \Omega$$

and

$$Y(\omega) = \begin{cases} 0, & \text{if } \omega \in [0,0.5] \\ 1, & \text{if } \omega \in (0.5,1]. \end{cases}$$

Compute: 1) $\mathcal{F}_Y$; 2) $E(X|Y)$.

Solution 3.3.3: 1) We denote

$$A_1 = Y^{-1}(\{0\}) = [0, 0.5], A_2 = Y^{-1}(\{1\}) = (0.5, 1].$$

$\{A_1, A_2\}$ is a partition of Ω. By Example 2.3-(2), we have that

$$\mathcal{F}_Y = \sigma(\{A_1, A_2\}) = \{\emptyset, [0, 1], [0, 0.5], (0.5, 1]\}.$$

2) By Example 3.5, we have

$$E(X|Y)(\omega) = \begin{cases} E(X|Y = 0), & \text{if } \omega \in [0, 0.5] \\ E(X|Y = 1), & \text{if } \omega \in (0.5, 1]. \end{cases}$$

We use Definition 3.8 to write

$$E(X|Y = y) = \int_{\mathbb{R}} x \, dF_{X|Y}(x|y) dx \quad \text{for } y \in \{0, 1\},$$

where

$$F_{X|Y}(x|y) = P(X \le x|Y = y) = \frac{P(X \le x, Y = y)}{P(Y = y)} \quad \text{for } y \in \{0, 1\}.$$

Computations yield to

$$F_{X|Y}(x|0) = \begin{cases} 0, & \text{if } x < 1 \\ x - 1, & \text{if } 1 \le x < 2 \\ 1, & \text{if } 2 \le x \end{cases}$$

and

$$F_{X|Y}(x|1) = \begin{cases} 0, & \text{if } x < 2 \\ x - 2, & \text{if } 2 \le x < 3 \\ 1, & \text{if } 3 \le x, \end{cases}$$

as well as

$$E(X|Y = 0) = \int_1^2 x \, dx = 1.5,$$

$$E(X|Y = 1) = \int_2^3 x \, dx = 2.5.$$

Finally, we have for a.e. $\omega \in [0, 1]$

$$E(X|Y)(\omega) = \begin{cases} 1.5, & \text{if } \omega \in [0, 0.5] \\ 2.5, & \text{if } \omega \in (0.5, 1]. \end{cases}$$

Problem 3.3.4. Consider the following algorithm:

- **generate a random vector** $(X, Y) \sim MVUnif([0,1] \times [0,1])$;
- **if** $X \leq Y$, **then** $(U, V) = (X, Y)$;
  else $(U, V) = (Y, X)$.

Prove that $E(U|V) = \frac{1}{2}V$ a.s. and find $E(U)$.

Solution 3.3.4: Let $F_{U,V}$ be the joint distribution function of (U, V). Then

$$
\begin{aligned}
F_{U,V}(u, v) &= P\big(\min\{X, Y\} \leq u, \max\{X, Y\} \leq v\big) \\
&= P\big(\max\{X, Y\} \leq v\big) - P\big(\min\{X, Y\} > u, \max\{X, Y\} \leq v\big) \\
&= \big(P(X \leq v)\big)^2 - \big(P(u < X \leq v)\big)^2, \ u, v \in \mathbb{R},
\end{aligned}
$$

where we use the fact that X and Y are independent and identically distributed. Hence,

$$
F_{U,V}(u, v) = \begin{cases}
v^2 - (u - v)^2, & \text{if } 0 < u < v < 1 \\
1 - (1 - u)^2, & \text{if } 0 < u < 1 \leq v \\
v^2, & \text{if } 0 < v < 1, v \leq u \\
1, & \text{if } u, v \geq 1 \\
0, & \text{otherwise.}
\end{cases}
$$

By Theorem 2.8-(1), we deduce that

$$
f_{U,V}(u, v) = \begin{cases}
2, & \text{if } 0 < u < v < 1 \\
0, & \text{otherwise}
\end{cases}
$$

is a joint density function of (U, V). Hence,

$$
f_V(v) = \int_{\mathbb{R}} f_{U,V}(u, v) = \begin{cases}
\int_0^v 2du, & \text{if } v \in (0,1) \\
0, & \text{otherwise}
\end{cases} = \begin{cases}
2v, & \text{if } v \in (0,1) \\
0, & \text{otherwise}
\end{cases}
$$

is a density function of V, and thus for $v \in (0,1)$ we have $f_V(v) > 0$ and

$$
E(U|V = v) = \int_{\mathbb{R}} u f_{U|V}(u|v) du = \frac{v}{2}.
$$

By Example 3.6, we have $E(U|V) = \frac{1}{2}V$ a.s. By Theorem 3.3-(2), we have

$$
E(U) = E(E(U|V)) = \frac{1}{2}E(V) = \frac{1}{2}\int_0^1 2v^2 dv = \frac{1}{3}.
$$

3.4   Variance and Other Numerical Characteristics

Definition 3.12. Let X be a random variable and let $E(X)$ be its expectation. The **variance** (or **dispersion**) of X is the number

$$V(X) = E\left((X - E(X))^2\right),$$

if $(X - E(X))^2$ is integrable. The value $\sigma = \sqrt{V(X)}$ is called the **standard deviation** of X.   ♦

Theorem 3.4. *If X and Y are random variables, then the following properties hold:*

(1) $V(X) = E(X^2) - E^2(X)$.
(2) $V(aX + b) = a^2 V(X)$ *for all* $a, b \in \mathbb{R}$.
(3) *If X and Y are independent, then*

$$V(X + Y) = V(X) + V(Y)$$

   and

$$V(X \cdot Y) = V(X)V(Y) + E^2(X)V(Y) + E^2(Y)V(X)$$
$$= E(X^2)E(Y^2) - E^2(X)E^2(Y).$$

Definition 3.13. Let X and Y be two random variables and let $E(X)$ and $E(Y)$ be their expectations. The **covariance** of the random variables X and Y is (if it exists) the number

$$\text{cov}(X, Y) = E\left((X - E(X))(Y - E(Y))\right).$$

The **correlation coefficient** of X and Y is defined by

$$\rho(X, Y) = \frac{\text{cov}(X, Y)}{\sqrt{V(X)V(Y)}},$$

if $\text{cov}(X, Y), V(X), V(Y)$ exist and $V(X) \neq 0, V(Y) \neq 0$.   ♦

Theorem 3.5. *If X, Y and Z are random variables, then the following properties hold (whenever the numerical characteristics exist):*

(1) $\text{cov}(X, X) = V(X)$.
(2) $\text{cov}(X, Y) = E(X \cdot Y) - E(X)E(Y)$.
(3) *If X and Y are independent, then* $\text{cov}(X, Y) = \rho(X, Y) = 0$, *i.e.,* X *and Y are **uncorrelated**.*
(4) $V(aX + bY) = a^2 V(X) + b^2 V(Y) + 2ab\,\text{cov}(X, Y)$ *for all* $a, b \in \mathbb{R}$.
(5) $\text{cov}(aX + bY, Z) = a\text{cov}(X, Z) + b\text{cov}(Y, Z)$ *for all* $a, b \in \mathbb{R}$.

Example 3.9. Let $X_1, \ldots, X_n$ be random variables, $n \in \mathbb{N}$, $n \geq 2$.
(1) Then

$$V\left(\sum_{k=1}^n X_k\right) = \sum_{k=1}^n V(X_k) + 2 \sum_{\substack{i,j=1 \\ i<j}}^n \operatorname{cov}(X_i, X_j).$$

(2) If $X_1, \ldots, X_n$ be pairwise uncorrelated (i.e., X_i and X_j are uncorrelated for $i, j \in \{1, \ldots, n\}$, $i \neq j$), then

$$V(X_1 + \ldots + X_n) = V(X_1) + \ldots + V(X_n).$$

In particular, this property holds if the involved random variables are pairwise independent.

Proof: (1) We proceed by induction. Clearly, the equality holds for $n = 2$, by Theorem 3.5-(4). Suppose the relation is true for any $n - 1$ random variables, for some fixed $n \in \mathbb{N}$, $n \geq 3$. Using Theorem 3.5-(4)-(5) and the induction hypothesis, we get

$$V\left(\sum_{k=1}^n X_k\right) = V\left(\sum_{k=1}^{n-1} X_k\right) + V(X_n) + 2 \sum_{i=1}^{n-1} \operatorname{cov}(X_i, X_n)$$

$$= \sum_{k=1}^{n-1} V(X_k) + 2 \sum_{\substack{i,j=1 \\ i<j}}^{n-1} \operatorname{cov}(X_i, X_j) + V(X_n) + 2 \sum_{i=1}^{n-1} \operatorname{cov}(X_i, X_n)$$

$$= \sum_{k=1}^n V(X_k) + 2 \sum_{\substack{i,j=1 \\ i<j}}^n \operatorname{cov}(X_i, X_j).$$

(2) The result follows from (1). ▲

Example 3.10. Let $n \in \mathbb{N}^*$ and $p \in (0, 1)$. An electronic system has n components that function independently and each component is functional with probability p. Let X be the number of functional components. We compute the corresponding expected value $E(X)$ and variance $V(X)$.
Solution: For every $i \in \{1, \ldots, n\}$ let $X_i \sim Ber(p)$ such that $X_i = 1$, if the ith component is functional, and $X_i = 0$, otherwise. Moreover, we consider $X_1, \ldots, X_n$ to be independent random variables. Observe that $X = X_1 + \ldots + X_n$ and $X \sim Bino(n, p)$. We have

$$E(X) = E(X_1 + \cdots + X_n) = E(X_1) + \cdots + E(X_n) = np$$

and, by the independence of $X_1, \ldots, X_n$, we have

$$V(X_1 + \cdots + X_n) = V(X_1) + \cdots + V(X_n) = np(1 - p) = np(1 - p). \quad \blacktriangle$$

Definition 3.14. Let $\mathbb{X} = (X_1, ..., X_n)$ be a random vector, where each component has variance. The $n \times n$ dimensional matrix

$$V(\mathbb{X}) = E\Big((\mathbb{X} - E(\mathbb{X}))^T (\mathbb{X} - E(\mathbb{X}))\Big) = \big(\mathrm{cov}(X_i, X_j)\big)_{i,j=\overline{1,n}}$$

is called the **covariance matrix** of the random vector $\mathbb{X}$.   ◆

Example 3.11. **(1)** In what follows we show that the covariance matrix of a random vector $\mathbb{X}$ is a positive semidefinite matrix. By definition, $V(\mathbb{X})$ is a symmetric matrix. Let $\mathrm{x} \in \mathbb{R}^n$. Observe that $\mathbb{X}\mathrm{x}^T$ is a random variable. We compute

$$\mathrm{x}V(\mathbb{X})\mathrm{x}^T = \mathrm{x}E\Big((\mathbb{X} - E(\mathbb{X}))^T (\mathbb{X} - E(\mathbb{X}))\Big)\mathrm{x}^T$$

$$= E\Big((\mathbb{X}\mathrm{x}^T - E(\mathbb{X}\mathrm{x}^T))(\mathbb{X}\mathrm{x}^T - E(\mathbb{X}\mathrm{x}^T))\Big) = V(\mathbb{X}\mathrm{x}^T) \geq 0.$$

(2) Let $\mathbb{X}$ be a random vector with expectation $E(\mathbb{X}) = \mathrm{m} \in \mathbb{R}^n$ and covariance matrix $V(\mathbb{X}) = A \in \mathcal{M}_{n \times n}(\mathbb{R})$. Consider $B \in \mathcal{M}_{n \times n}(\mathbb{R})$ to be a matrix and $\mathrm{b} \in \mathbb{R}^n$, then the random vector $\mathbb{Y} = \mathbb{X}B + \mathrm{b}$, has the following properties:

$$E(\mathbb{Y}) = E(\mathbb{X}B + \mathrm{b}) = E(\mathbb{X})B + \mathrm{b} = \mathrm{m}B + \mathrm{b}$$

and

$$V(\mathbb{Y}) = V(\mathbb{X}B + \mathrm{b}) = B^T V(\mathbb{X})B = B^T A B.$$ ▲

Definition 3.15. Let $k \in \mathbb{N}$ and let X be a random variable. The number $\nu_k = E(X^k)$ (if it exists) is called **the moment of order k of X**, while the number $E\Big((X - E(X))^k\Big)$ (if it exists) is called the **central moment of order k of X** and it is denoted by μ_k.   ◆

Example 3.12. Let $k \in \mathbb{N}^*$, $m \in \mathbb{R}$ and $\sigma > 0$. We compute the expected value, the variance and the central moment of order k of a random variable X having normal distribution $N(m, \sigma^2)$.

Solution: We shall use the fact that the integral of an odd function on a symmetric (with respect to the origin) interval is zero and some properties of the Gamma function (see Proposition A.1).

By the change of variable $t = \dfrac{x - m}{\sigma}$, we deduce that

$$E(X) = \frac{1}{\sqrt{2\pi}\sigma} \int_{-\infty}^{\infty} x \exp\left\{-\frac{(x-m)^2}{2\sigma^2}\right\} dx$$

$$= \frac{\sigma}{\sqrt{2\pi}} \int_{-\infty}^{\infty} t \exp\left\{-\frac{t^2}{2}\right\} dt + m\sqrt{\frac{2}{\pi}} \int_{0}^{\infty} \exp\left\{-\frac{t^2}{2}\right\} dt$$

$$= \frac{m}{\sqrt{\pi}} \Gamma\left(\frac{1}{2}\right) = m.$$

By the same change of variable, we deduce also that

$$\mu_k = \frac{1}{\sqrt{2\pi}\sigma} \int_{-\infty}^{\infty} (x-m)^k \exp\left\{-\frac{(x-m)^2}{2\sigma^2}\right\} dx$$

$$= \begin{cases} 0, & k \text{ is odd}, \\ \dfrac{2\sigma^k}{\sqrt{2\pi}} \displaystyle\int_0^{\infty} t^k \exp\left\{-\dfrac{t^2}{2}\right\} dt, & k \text{ is even.} \end{cases}$$

For k even, using the change of variable $s = \dfrac{t^2}{2}$, we get

$$\mu_k = \frac{2^{\frac{k}{2}}\sigma^k}{\sqrt{\pi}} \int_0^{\infty} s^{\frac{k-1}{2}} e^{-s} ds = \frac{2^{\frac{k}{2}}\sigma^k}{\sqrt{\pi}} \Gamma\left(\frac{k+1}{2}\right)$$

$$= 2^{\frac{k}{2}}\sigma^k \prod_{j=1}^{k/2} \frac{2j-1}{2} = \sigma^k (k-1)!!$$

In particular, we have $V(X) = \mu_2 = \sigma^2$. ▲

Example 3.13. We consider $\mathbb{X} = (X_1, .., X_n) \sim MVN(\mathrm{m}, A)$, where $\mathrm{m} \in \mathbb{R}^n$, $A \in \mathcal{M}_{n \times n}(\mathbb{R})$ is a positive definite matrix (see Section 2.5.4). Then, by Example 2.24, we have

$$\mathbb{X} \sim MVN(\mathrm{m}, A) \Longleftrightarrow \mathbb{Y} \sim MVN(0_n, I_n), \text{ where } \mathbb{Y} = (\mathbb{X} - \mathrm{m})A^{-\frac{1}{2}}.$$

Denote $\mathbb{Y} = (Y_1, \ldots, Y_n)$. Then the components $Y_1, \ldots, Y_n$ have standard normal distribution $N(0,1)$ and they are independent (see Problem 2.5.1). Therefore, $E(\mathbb{Y}) = 0_n$ and $V(\mathbb{Y}) = I_n$. So, by Example 3.11 we obtain $E(\mathbb{X}) = E(\mathbb{Y}A^{\frac{1}{2}} + \mathrm{m}) = \mathrm{m}$ and the covariance matrix is

$$V(\mathbb{X}) = V(\mathbb{Y}A^{\frac{1}{2}} + \mathrm{m}) = A.$$ ▲

Theorem 3.6. *Let X be a random variable with expectation $E(X)$ and let $a > 0$. Then the following inequalities are true:*

(1) **Markov's inequality:** $P\left(|X| \geq a\right) \leq \dfrac{1}{a} E(|X|)$.

(2) **Chebyshev's inequality:** $P\left(|X - E(X)| \geq a\right) \leq \dfrac{1}{a^2} V(X)$.

Example 3.14. Let X be a random variable with expectation $m = 2$ and standard deviation $\sigma = 1$. Then $P(1 < X < 5) \geq 0.5$.

Proof: Let $a = 1, b = 5$. We rewrite the probability, use Markov's inequality and obtain

$$P(a < X < b) = P\left(-\frac{b-a}{2} < X - \frac{a+b}{2} < \frac{b-a}{2} \right)$$

$$= 1 - P\left(\left| X - \frac{a+b}{2} \right| \geq \frac{b-a}{2} \right) \geq 1 - \frac{4}{(b-a)^2} E\left(\left(X - \frac{a+b}{2} \right)^2 \right)$$

$$= 1 - \frac{4}{(b-a)^2} \left(\sigma^2 + \left(m - \frac{a+b}{2} \right)^2 \right) = 0.5. \qquad \blacktriangle$$

Theorem 3.7. (*Jensen's Inequality*) *Let* $I \subseteq \mathbb{R}$ *be an interval (finite or infinite). Consider* $g : I \to \mathbb{R}$ *to be function and* X *a random variable with* $X \in I$ *a.s. such that* $E(X)$ *and* $E(g(X))$ *exist.*
(1) *If* g *is a convex function, then* $g(E(X)) \leq E(g(X))$. *If* g *is strictly convex, then equality holds if and only if* $X = E(X)$ *a.s.*
(2) *If* g *is a concave function, then* $g(E(X)) \geq E(g(X))$. *If* g *is strictly concave, then equality holds if and only if* $X = E(X)$ *a.s.*

Example 3.15. Let X and Y be positive random variables. For $\alpha \geq 1$ we compare $E\left(\max\{X^\alpha, Y^\alpha\} \right)$ with $\left(E(X) \right)^\alpha$ and $\left(E(Y) \right)^\alpha$. For $\alpha \in [0,1]$ we compare $E\left(\min\{X^\alpha, Y^\alpha\} \right)$ with $\left(E(X) \right)^\alpha$ and $\left(E(Y) \right)^\alpha$. We assume that all expectations considered above exist.
Solution: Let $\alpha \geq 0$. Note that

$$E(X^\alpha) \begin{cases} \leq \left(E(X) \right)^\alpha, & \text{if } \alpha \in (0,1) \\ \geq \left(E(X) \right)^\alpha, & \text{if } \alpha > 1 \\ = \left(E(X) \right)^\alpha, & \text{if } \alpha = 0 \text{ or } \alpha = 1, \end{cases}$$

where we use Jensen's inequality (see Theorem 3.7) for $g(x) = x^\alpha$, $x \geq 0$, which is convex for $\alpha \in (1, \infty)$ and concave for $\alpha \in (0,1)$. It is easy to check by the monotonicity of the expectation that

$$\max\left\{ E(X), E(Y) \right\} \leq E\left(\max\{X, Y\} \right), \quad E\left(\min\{X, Y\} \right) \leq \min\left\{ E(X), E(Y) \right\}$$

Then we obtain for $\alpha \geq 1$

$$E\left(\max\{X^\alpha, Y^\alpha\} \right) = E\left(\left(\max\{X, Y\} \right)^\alpha \right) \geq \left(E(\max\{X, Y\}) \right)^\alpha$$

$$\geq \left(\max\{E(X), E(Y)\} \right)^\alpha = \max\left\{ \left(E(X) \right)^\alpha, \left(E(Y) \right)^\alpha \right\}$$

and for $\alpha \in [0,1]$

$$E\left(\min\{X^\alpha, Y^\alpha\} \right) = E\left(\left(\min\{X, Y\} \right)^\alpha \right) \leq \left(E(\min\{X, Y\}) \right)^\alpha$$

$$\leq \left(\min\left\{ E(X), E(Y) \right\} \right)^\alpha = \min\left\{ \left(E(X) \right)^\alpha, \left(E(Y) \right)^\alpha \right\}. \qquad \blacktriangle$$

Theorem 3.8. (*Hölder's Inequality*) *Let* $p > 1$ *and* $q = \dfrac{p}{p-1}$. *Consider two random variables* X *and* Y *such that* $E(|X|^p) < \infty$ *and* $E(|Y|^q) < \infty$. *Then*

$$E(|XY|) \leq \left(E(|X|^p)\right)^{\frac{1}{p}} \left(E(|Y|^q)\right)^{\frac{1}{q}}.$$

For $p = q = 2$ *the above result is called the* **Cauchy–Schwarz inequality** *for random variables.*

3.4.1   Solved Problems

Problem 3.4.1. Let X be a random variable with $E(|X|) = 0$. Prove that $P(X = 0) = 1$. Deduce that a random variable has zero variance if and only if it is almost surely a constant.

Solution 3.4.1: Using Markov's inequality (see Theorem 3.6), it follows that $P(|X| \geq a) = 0$ for all $a > 0$. We take now successively $a = \dfrac{1}{n}$, $n \in \mathbb{N}^*$, and we get

$$P(X = 0) = P\left(\bigcap_{n \in \mathbb{N}^*} \left\{|X| < \frac{1}{n}\right\}\right) = \lim_{n \to \infty} P\left(|X| < \frac{1}{n}\right)$$

$$= 1 - \lim_{n \to \infty} P\left(|X| \geq \frac{1}{n}\right) = 1.$$

We have

$$E\big((X - E(X))^2\big) = 0 \Leftrightarrow P\big((X - E(X))^2 = 0\big) = 1 \Leftrightarrow P\big(X = E(X)\big) = 1.$$

Problem 3.4.2. The number of calls per hour coming to the call center of a company is modeled by using a Poisson random variable with mean 4.
1) Find the probability that no call comes in a certain hour.
2) Compute the variance of the number of calls.

Solution 3.4.2: Let X denote the number of calls per hour coming to the call center with $X \sim Poiss(\lambda)$. In order to compute λ, we calculate $E(X)$. Recall the expansion $\displaystyle\sum_{j=0}^{\infty} \frac{x^j}{j!} = e^x$ for each $x \in \mathbb{R}$. Then we write

$$E(X) = \sum_{k=0}^{\infty} k \frac{\lambda^k}{k!} e^{-\lambda} = \lambda e^{-\lambda} \sum_{k=1}^{\infty} \frac{\lambda^{k-1}}{(k-1)!} = \lambda e^{-\lambda} \sum_{j=0}^{\infty} \frac{\lambda^j}{j!} = \lambda.$$

1) So, $\lambda = 4$. We get $P(X = 0) = e^{-4} \approx 0.0183$.

2) We compute

$$E(X^2) = \sum_{k=0}^{\infty} k^2 \frac{\lambda^k}{k!} e^{-\lambda} = e^{-\lambda} \sum_{k=1}^{\infty} (k^2 - k + k) \frac{\lambda^k}{k!}$$

$$= \lambda^2 e^{-\lambda} \sum_{k=2}^{\infty} \frac{\lambda^{k-2}}{(k-2)!} + E(X) = \lambda^2 e^{-\lambda} \sum_{j=0}^{\infty} \frac{\lambda^j}{j!} + \lambda = \lambda^2 + \lambda.$$

Therefore, $V(X) = \lambda = 4$.

Problem 3.4.3. 1) Let $X \sim Gamma(a, b)$, $a, b > 0$. Compute the moments of order $k \in \mathbb{N}^*$.

2) A company studied the operating time of a certain machine before breaking down and observed that it is Gamma distributed with a mean of 24 hours and a standard deviation of 2 hours. Using Matlab,

• compute the probability that the machine functions properly for at least 18 hours before breaking down;

• if the machine has already been operating 18 hours, find the probability that it will function properly another 12 hours.

Solution 3.4.3: 1) Let $k \in \mathbb{N}^*$. We have

$$\nu_k = E(X^k) = \frac{1}{\Gamma(a)b^a} \int_0^{\infty} x^{a+k-1} \exp\left\{\frac{-x}{b}\right\} dx.$$

Making the change of variables $x = by$ and using the properties of the Gamma function (see Proposition A.1), we obtain

$$\nu_k = E(X^k) = \frac{b^k}{\Gamma(a)} \int_0^{\infty} y^{a+k-1} e^{-y} dy = \frac{b^k}{\Gamma(a)} \Gamma(a+k)$$

$$= a(a+1)\dots(a+k-1)b^k.$$

In particular, have $E(X) = ab$, $V(X) = ab^2$.

2) Denote by X the operating time of the machine before breaking down. Using 1), it follows that $a = 144, b = \dfrac{1}{6}$ and $X \sim Gamma\left(144, \dfrac{1}{6}\right)$. We have

$$p_1 = P(X \geq 18) = P(X > 18) = 1 - F_X(18)$$

and thus

$$p_2 = P(X \geq 30 | X \geq 18) = \frac{1 - F_X(30)}{p_1},$$

where F_X is the cumulative disitribution function of X.

>>p_1=1-**gamcdf**(18,144,1/6); p_2=(1-**gamcdf**(30,144,1/6))/p_1;
>>**fprintf**('p_1=%5.4f, p_2=%5.4f \n', p_1,p_2)
p_1=0.9994, p_2=0.0025

Problem 3.4.4. Suppose that $(x_1, \ldots, x_n)$ is a random permutation of $(1, \ldots, n)$. If $x_i = i$, $i \in \{1, \ldots, n\}$, then i is said to be a fixed point of the permutation. Denote by Z_n the number of fixed points of a random permutation. Compute the expected number of fixed points $E(Z_n)$ and their variance $V(Z_n)$.

Solution 3.4.4: For each $i \in \{1, \ldots, n\}$ let

$$X_i = \begin{cases} 1, & \text{if } x_i = i \\ 0, & \text{otherwise.} \end{cases}$$

We have

$$P(X_i = 1) = \frac{(n-1)!}{n!} = \frac{1}{n}, \ P(X_i = 0) = \frac{n-1}{n}.$$

Then

$$E(X_i) = \frac{1}{n} \text{ and } V(X_i) = E(X_i^2) - E^2(X_i) = \frac{n-1}{n^2}.$$

We have $E(Z_n) = E(X_1) + \cdots + E(X_n) = 1$.

When we compute the variance, we must take into account that $X_1, \ldots, X_n$ are not independent. In view of Example 3.9-(1),

$$V(Z_n) = \sum_{k=1}^{n} V(X_k) + 2 \sum_{\substack{i,k=1 \\ i<k}}^{n} \text{cov}(X_i, X_k).$$

We compute $\text{cov}(X_i, X_k)$ for $i, k \in \{1, \ldots, n\}$ with $i < k$

$$\text{cov}(X_i, X_k) = E(X_i X_k) - E(X_i)E(X_k)$$

$$= 0 \cdot P(X_i X_k = 0) + 1 \cdot P(X_i X_k = 1) - \frac{1}{n^2}$$

$$= P(X_i = 1 \cap X_k = 1) - \frac{1}{n^2} = \frac{(n-2)!}{n!} - \frac{1}{n^2} = \frac{1}{n^2(n-1)}.$$

Then

$$V(Z_n) = \frac{n-1}{n} + 2 \cdot \frac{n(n-1)}{2} \cdot \frac{1}{n^2(n-1)} = 1.$$

Problem 3.4.5. Let $(X_1, \ldots, X_n)$ be a random vector with multivariate normal distribution. If the components $X_1, \ldots, X_n$ are pairwise uncorrelated, then they are independent.

Solution 3.4.5: Since $X_1, \ldots, X_n$ are pairwise uncorrelated, it follows that their covariance matrix has diagonal form, the elements on the diagonal are $\sigma_i^2 = V(X_i)$, $i \in \{1, \ldots, n\}$. A joint density function of $\mathbb{X}$ is given for $\mathbb{x} = (x_1, \ldots, x_n) \in \mathbb{R}^n$ by

$$f(\mathbb{x}) = \frac{1}{(2\pi)^{n/2}\sigma_1 \ldots \sigma_n} \exp\left\{ -\frac{1}{2} \sum_{i=1}^{n} \frac{(x_i - \mu_i)^2}{\sigma_i^2} \right\},$$

where $\mu_i = E(X_i)$, $i \in \{1, \ldots, n\}$. We write

$$f(\mathbb{x}) = \prod_{i=1}^{n} \frac{1}{(2\pi)^{1/2}\sigma_i} \exp\left\{ -\frac{1}{2} \frac{(x_i - \mu_i)^2}{\sigma_i^2} \right\}.$$

By Theorem 2.13-(2) it follows that $X_1, \ldots, X_n$ are independent random variables.

Problem 3.4.6. Let $X \sim N(0,1)$ and $W \sim Ber(0.5)$ be independent random variables. Then the random variable $Y = (2W - 1)X$ has the following properties: 1) $Y \sim N(0,1)$; 2) X and Y are uncorrelated.
Express the joint distribution function of (X, Y) by using the distribution function F of the standard normal distribution.
Remark: Even if X and Y are normally distributed and uncorrelated, they are not necessarily independent. This is not in contradiction with Problem 3.4.5, since (X, Y) are not jointly normally distributed.

Solution 3.4.6: X and W are independent, hence

$$P(X \le y | W = a) = P(X \le y) \text{ for } a \in \{0, 1\} \text{ and any } y \in \mathbb{R}.$$

By using Theorem 1.8, we have for each $y \in \mathbb{R}$

$$
\begin{aligned}
F_Y(y) &= P(Y \le y) \\
&= P(Y \le y | W = 0)P(W = 0) + P(Y \le y | W = 1)P(W = 1) \\
&= \frac{1}{2}\Big(P(-X \le y | W = 0) + P(X \le y | W = 1) \Big) \\
&= \frac{1}{2}\Big(P(-X \le y) + P(X \le y) \Big) = F_X(y),
\end{aligned}
$$

where F_X, F_Y are the distribution functions of X, respectively Y. So, $Y \sim N(0,1)$. Since $2W - 1$ and X^2 are also independent (see Theorem 2.12), we have

$$
\begin{aligned}
\operatorname{cov}(X, Y) &= E(XY) - E(X)E(Y) = E(XY) \\
&= E\big((2W - 1)X^2\big) = E(2W - 1)E(X^2) = 0.
\end{aligned}
$$

Therefore, X and Y are uncorrelated.

The joint distribution function of (X, Y) is

$$
\begin{aligned}
F_{X,Y}(x,y) &= P(X \le x, Y \le y) \\
&= P(X \le x, Y \le y|W=0)P(W=0) + P(X \le x, Y \le y|W=1)P(W=1) \\
&= \frac{1}{2}P(\{X \le x\} \cap \{-y \le X\}) + \frac{1}{2}P(\{X \le x\} \cap \{X \le y\}).
\end{aligned}
$$

Then

$$
F_{X,Y}(x,y) = \begin{cases} \dfrac{1}{2}\left(F(\min\{x,y\}) + F(x) - F(-y)\right), & \text{if } x + y \ge 0 \\ \dfrac{1}{2}F(\min\{x,y\}), & \text{if } x + y < 0. \end{cases}
$$

Note that one can find values $x, y \in \mathbb{R}$ such that $F_{X,Y}(x,y) \ne F_X(x)F_Y(y)$ (e.g., $x \in \mathbb{R}^*, y = -x$), hence X and Y are not independent.

Problem 3.4.7. Suppose that two random variables X and Y have the following joint density function:

$$
f_{X,Y}(x,y) = \begin{cases} x - y, & \text{if } 0 \le x \le 1 \text{ and } -1 \le y \le 0 \\ 0, & \text{otherwise.} \end{cases}
$$

Find $E(X)$, $E(Y)$ and $\rho(X,Y)$.

Solution 3.4.7: We have

$$
E(X) = \int_{-\infty}^{\infty} \int_{-\infty}^{\infty} x f_{X,Y}(x,y)dxdy = \int_0^1 \int_{-1}^0 x(x-y)\,dxdy = \frac{7}{12}.
$$

Similarly, we obtain $E(Y) = -\dfrac{7}{12}$. We compute

$$
\begin{aligned}
E(X^2) &= \int_{-\infty}^{\infty} \int_{-\infty}^{\infty} x^2 f_{X,Y}(x,y)dxdy \\
&= \int_0^1 \int_{-1}^0 x^2(x-y)\,dxdy = \frac{5}{12}.
\end{aligned}
$$

and, similarly, $E(Y^2) = \dfrac{5}{12}$. Hence, $V(X) = V(Y) = \dfrac{11}{144}$. We have

$$
E(XY) = \int_{-\infty}^{\infty} \int_{-\infty}^{\infty} xy f_{X,Y}(x,y)dxdy = \int_0^1 \int_{-1}^0 xy(x-y)\,dxdy = -\frac{1}{3}
$$

and thus $\operatorname{cov}(X,Y) = \dfrac{1}{144}$ and $\rho(X,Y) = \dfrac{1}{11}$.

3.5   Moment Generating Function

Definition 3.16. The **moment generating function** of a random variable X is the function $M_X : D \subseteq \mathbb{R} \to \mathbb{R}$ defined by

$$M_X(t) = E\big(\exp\{tX\}\big) \text{ for } t \in D,$$

where D is the set of all $t \in \mathbb{R}$ for which the above expectation exists.   ♦

Theorem 3.9. *Let X and Y be random variables. Then the following properties hold:*

(1) $M_X(0) = 1$.
(2) *If $Y = aX + b$ for fixed $a, b \in \mathbb{R}$, then $M_Y(t) = e^{tb} M_X(at)$ for all $b \in \mathbb{R}$ and $t, a \in \mathbb{R}$ for which $M_X(at)$ is defined.*
(3) *If X and Y are independent random variables, then*

$$M_{X+Y}(t) = M_X(t) \cdot M_Y(t)$$

for all $t \in \mathbb{R}$ for which these moment generating functions are defined.

Theorem 3.10. *If there exists $\delta > 0$ such that M_X is defined on $(-\delta, \delta)$, then for $t \in (-\delta, \delta)$*

$$M_X(t) = E\big(\exp\{tX\}\big) = \sum_{k=0}^{\infty} \frac{t^k}{k!} E(X^k)$$

and for $n \in \mathbb{N}$

$$E(X^n) = M_X^{(n)}(0),$$

where $M_X^{(n)}(0)$ is the derivative of order n of M_X at 0.

Theorem 3.11. *Let X and Y be random variables for which there exists $\delta > 0$ such that $M_X(t) = M_Y(t)$ for each $t \in (-\delta, \delta)$. Then $F_X = F_Y$, i.e., X and Y have the same distribution.*

Remark 3.5. There is an one-to-one correspondence between the probability distribution of a random variable and the moment generating function, if this function is defined on an (open) interval containing 0; this means that the distribution of a random variable is uniquely determined by its moment generating function and the moment generating function of a random variable is (by definition) uniquely determined by the distribution of the random variable.   ▼

Example 3.16. (1) If the random variable X has a binomial distribution $Bino(n, p)$, $n \in \mathbb{N}^*$, $p \in (0, 1)$, then its moment generating function is given for each $t \in \mathbb{R}$ by

$$M_X(t) = \sum_{k=0}^{n} e^{tk} C(n, k) p^k (1 - p)^{n-k} = (e^t p + 1 - p)^n.$$

(2) If $X_k \in Bino(n_k, p)$ (with $n_k \in \mathbb{N}^*$, $p \in (0, 1)$), $k \in \{1, \ldots, m\}$, $m \in \mathbb{N}^*$, are independent random variables, then $X_1 + \ldots X_m \sim Bino\left(n_1 + \cdots + n_m, p\right)$, because, by Theorem 3.9-(3), we have

$$M_{X_1 + \cdots + X_m}(t) = M_{X_1}(t) \cdot \ldots \cdot M_{X_m}(t) = (e^t p + 1 - p)^{n_1 + \cdots + n_m} \text{ for } t \in \mathbb{R},$$

which is the moment generating function of a random variable having $Bino\left(n_1 + \cdots + n_m, p\right)$ distribution and, by Theorem 3.11, it follows that

$$X_1 + \cdots + X_m \sim Bino\left(n_1 + \cdots + n_m, p\right).$$

(3) If $X \sim Gamma(a, b)$, $a, b > 0$, then the moment generating function of X is defined for $t < \dfrac{1}{b}$ by

$$M_X(t) = \int_0^\infty \frac{1}{\Gamma(a) b^a} x^{a-1} \exp\left\{ tx - \frac{x}{b} \right\} dx = \frac{1}{(1 - bt)^a}.$$

(4) For $X \sim \chi^2(n, \sigma)$, $n \in \mathbb{N}^*$, $\sigma > 0$, since $Gamma\left(\dfrac{n}{2}, 2\sigma^2\right) = \chi^2(n, \sigma)$, we obtain

$$M_X(t) = \frac{1}{(1 - 2\sigma^2 t)^{\frac{n}{2}}} \text{ for } t < \frac{1}{2\sigma^2}.$$

(5) Let $X_1, \ldots, X_5 \sim Unif[a, b]$, $a < b$, be independent random variables. Is $S_5 = X_1 + \ldots + X_5$ uniformly distributed?

Solution: For every $i \in \{1, \ldots 5\}$ let f_{X_i} be density function of X_i and write

$$M_{X_i}(t) = \int_\mathbb{R} e^{tx} f_{X_i}(x) dx = \frac{e^{bt} - e^{at}}{t(b - a)} \text{ for } t \in \mathbb{R}^*.$$

Obviously, $M_{X_i}(0) = 1$. Then for $t \in \mathbb{R}^*$ we have

$$M_{S_5}(t) = M_{X_1}(t) \cdot \ldots \cdot M_{X_5}(t) = \left(\frac{e^{bt} - e^{at}}{t(b - a)} \right)^5$$

and $M_{S_5}(0) = 1$.

Assume that there exist real numbers $A < B$ such that

$$\left(\frac{e^{bt} - e^{at}}{t(b-a)}\right)^5 = \frac{e^{Bt} - e^{At}}{t(B-A)} \text{ for each } t \in \mathbb{R}^*. \tag{3.4}$$

Then

$$\frac{(e^{bt} - e^{at})^5}{t^4(e^{Bt} - e^{At})} = \frac{(b-a)^5}{B-A} \text{ for each } t \in \mathbb{R}^*.$$

We write

$$\lim_{t \to \infty} \frac{(e^{bt} - e^{at})^5}{t^4(e^{Bt} - e^{At})} = \lim_{t \to \infty} \left(\frac{e^{5bt}}{t^4 e^{Bt}} \cdot \frac{(1 - \exp\{(a-b)t\})^5}{1 - \exp\{(A-B)t\}}\right)$$

$$= \lim_{t \to \infty} \frac{\exp\{(5b - B)t\}}{t^4} = \begin{cases} 0, & \text{for } 5b \leq B, \\ \infty, & \text{for } 5b > B, \end{cases} \neq \frac{(b-a)^5}{B-A}.$$

So, (3.4) does not hold. Hence, M_{S_5} is not the moment generating function of a uniform distribution and, by Theorem 3.11, it follows that S_5 is not uniformly distributed.   ▲

Remark 3.6. The characteristic function of a random variable X is the function $\Phi_X : \mathbb{R} \to \mathbb{C}$ defined by

$$\Phi_X(t) = E\Big(\exp\{itX\}\Big) = E\big(\cos(tX)\big) + iE\big(\sin(tX)\big) \text{ for } t \in \mathbb{R}.$$

Note that in comparison to the moment generating function, this is a complex valued function and it is defined for all $t \in \mathbb{R}$. There is an one-to-one correspondence between the distribution of a random variable and the characteristic function, i.e., the distribution of a random variable is uniquely determined by its characteristic function and the characteristic function of a random variable is (by definition) uniquely determined by the distribution of the random variable.

For various properties of the characteristic function, we refer the reader to [Gut (2005), Chapter 4].   ▼

3.5.1   Solved Problems

Problem 3.5.1. Let X be a random variable that has the moment generating function

$$M_X(t) = \frac{e^t(1 + e^{2t})^3}{8}, \quad t \in \mathbb{R}.$$

Find the probability mass function of X.

Solution 3.5.1: We have
$$M_X(t) = \frac{1}{8}e^t + \frac{3}{8}e^{3t} + \frac{3}{8}e^{5t} + \frac{1}{8}e^{7t}, \quad t \in \mathbb{R}.$$
In view of Theorem 3.11, we deduce that
$$P(X = 1) = P(X = 7) = \frac{1}{8} \text{ and } P(X = 3) = P(X = 5) = \frac{3}{8}.$$

Problem 3.5.2. 1) Find the moment generating function, the mean and the variance of $X \sim Poiss(\lambda)$, $\lambda > 0$.
2) Let X and Y be independent random variables with $X \sim Poiss(\lambda)$ and $Y \sim Poiss(\mu)$, $\lambda, \mu > 0$. Prove that $X + Y \sim Poiss(\lambda + \mu)$.

Solution 3.5.2: 1) Since $P(X = k) = \dfrac{\lambda^k}{k!}e^{-\lambda}$ for $k \in \mathbb{N}$, we get for all $t \in \mathbb{R}$

$$M_X(t) = \sum_{k=0}^{\infty} e^{kt}\frac{\lambda^k}{k!}e^{-\lambda} = e^{-\lambda}\sum_{k=0}^{\infty}\frac{(\lambda e^t)^k}{k!} = \exp\left\{\lambda(e^t - 1)\right\}.$$

Then $M'_X(0) = \lambda$ and $M''_X(0) = \lambda(1 + \lambda)$. So, $E(X) = \lambda$ and $E(X^2) = \lambda(1 + \lambda)$, which implies $V(X) = \lambda(1 + \lambda) - \lambda^2 = \lambda$.
2) Since X and Y are independent, it follows by Theorem 3.9 and by 1) that

$$M_{X+Y}(t) = M_X(t) \cdot M_Y(t) = \exp\left\{(\lambda + \mu)(e^t - 1)\right\}, t \in \mathbb{R}.$$

This is the moment generating function of a random variable having a Poisson distribution with parameter $\lambda + \mu$. By Theorem 3.11, it follows that $X + Y \sim Poiss(\lambda + \mu)$.

Problem 3.5.3. Suppose that the numbers of misprints per page in a document of 200 pages are independent and they are modeled by a Poisson distribution with parameter $\lambda = 0.025$.
1) If 40 pages are inspected, what is the probability that the total number of misprints will be at most 5?
2) If all pages are inspected, what is the probability that the total number of misprints will be at least 15?

Solution 3.5.3: If the number of misprints of a page has a Poisson distribution with parameter λ, then the number of misprints in 40 pages has a Poisson distribution with parameter $40\lambda = 1$ (see Problem 3.5.2).
1) The probability that the total number of misprints in 40 inspected pages is at most 5 is $e^{-1}\sum_{k=0}^{5}\dfrac{1}{k!}$.

`>>poisscdf(5,1)`
ans = 0.9994

2) The number of misprints in 200 pages has a Poisson distribution with parameter $200\lambda = 5$ (see Problem 3.5.2). The probability that the total number of misprints in all pages will be at least 15 is

$$e^{-5} \sum_{k=15}^{\infty} \frac{5^k}{k!} = 1 - e^{-5} \sum_{k=0}^{14} \frac{5^k}{k!}.$$

`>>1-poisscdf(14,5)`
ans = 0.0002

Problem 3.5.4. 1) Let $n \in \mathbb{N}^*$ and let $X_i \sim N(m_i, \sigma_i^2)$, $m_i \in \mathbb{R}$, $\sigma_i > 0$, $i \in \{1, \ldots, n\}$, be independent random variables. For $c_1, \ldots, c_n \in \mathbb{R}$ prove that $S_n = c_1 X_1 + \ldots + c_n X_n$ has $N\left(\sum_{i=1}^{n} c_i m_i, \sum_{i=1}^{n} c_i^2 \sigma_i^2\right)$ distribution.

2) If

$$m_1 = 2, m_2 = 1, m_3 = 0, m_4 = 0, m_5 = 1, m_6 = 2,$$
$$\sigma_1^2 = 4, \sigma_2^2 = 1, \sigma_3^2 = 4, \sigma_4^2 = 2, \sigma_5^2 = 1, \sigma_6^2 = 1,$$
$$c_1 = 0.5, c_2 = 1, c_3 = 0.5, c_4 = 1, c_5 = 2, c_6 = 1,$$

compute the moment of order 4 for S_6 by using the Matlab Symbolic Math Toolbox.

Solution 3.5.4: 1) The moment generating function of X_i is for each $t \in \mathbb{R}$

$$M_{X_i}(t) = \frac{1}{\sqrt{2\pi}\sigma_i} \int_{\mathbb{R}} \exp\left\{tx - \frac{(x - m_i)^2}{2\sigma_i^2}\right\} dx = \exp\left\{m_i t + \frac{1}{2}\sigma_i^2 t^2\right\}.$$

Then, for $t \in \mathbb{R}$

$$M_{S_n}(t) = M_{X_1}(c_1 t) \cdot \ldots \cdot M_{X_n}(c_n t) = \exp\left\{mt + \frac{1}{2}\sigma^2 t^2\right\}$$

where $m = \sum_{i=1}^{n} c_i m_i$, $\sigma^2 = \sum_{i=1}^{n} c_i^2 \sigma_i^2$. This is the moment generating function of the normal distribution $N(m, \sigma^2)$, hence, by Theorem 3.11, $S_n \sim N(m, \sigma^2)$.

2) $M_{S_6}(t) = \exp\{6t + 5t^2\}$ for $t \in \mathbb{R}$ and the moment of order 4 is, by Theorem 3.10,

$$E(S_6^4) = \left(\exp\{6t + 5t^2\}\right)^{(4)}\Big|_{t=0}.$$

The computations in Matlab indicate $E(S_6^4) = 3756$.

m=[2, 1, 0, 0, 1, 2] *%means*
s=[4, 1, 4, 2, 1, 1] *% variances*
c=[0.5, 1, 0.5, 1, 2, 1] *% coefficients*
syms M(t)
M(t)=**exp**(t*c*m'+0.5*t^2*c.^2*s');
mom=**subs**(**diff**(M,4),t,0); *%fourth order derivative at 0*
fprintf('The moment of order 4 is %s. \n ', char(mom))
>> The moment of order 4 is 3756.

3.6 Joint Moment Generating Function

Definition 3.17. The **joint moment generating function** of a random vector $\mathbb{X} = (X_1, \ldots, X_n)$ is the function $M_{\mathbb{X}} : D \subseteq \mathbb{R}^n \to \mathbb{R}^n$ defined by

$$M_{\mathbb{X}}(\mathbb{t}) = E\big(\exp\{t_1 X_1 + \ldots + t_n X_n\} \big) = E\big(\exp\{\mathbb{t}\mathbb{X}^T\} \big)$$

for $\mathbb{t} = (t_1, \ldots, t_n) \in D$, where D is the set of all $\mathbb{t} \in \mathbb{R}^n$ for which the above expectation exists. ♦

Notation: We denote the Euclidean norm of a vector $\mathbb{t} = (t_1, \ldots, t_n) \in \mathbb{R}^n$ by

$$\|\mathbb{t}\| = (t_1^2 + \ldots + t_n^2)^{\frac{1}{2}}.$$

Theorem 3.12. [DasGupta (2010), Theorem 11.7, p. 264] *If there exists $\delta > 0$ such that $M_{\mathbb{X}}$ is defined for* $\mathbb{t}= (t_1, \ldots, t_n) \in \mathbb{R}^n$ *with* $\|\mathbb{t}\| < \delta$, *then*

$$E(X_1^{k_1} \cdot \ldots \cdot X_n^{k_n}) = \frac{\partial^{k_1 + \ldots + k_n} M_{\mathbb{X}}(t_1, \ldots, t_n)}{\partial t_1^{k_1} \ldots \partial t_n^{k_n}}\bigg|_{\mathbb{t}=0_n}.$$

Theorem 3.13. [DasGupta (2010), Theorem 11.8, p. 265] *Let $\mathbb{X}$ and $\mathbb{Y}$ be random vectors for which there exists $\delta > 0$ such that $M_{\mathbb{X}}(\mathbb{t}) = M_{\mathbb{Y}}(\mathbb{t})$ for each $\mathbb{t} \in \mathbb{R}^n$ with $\|\mathbb{t}\| < \delta$. Then $F_{\mathbb{X}} = F_{\mathbb{Y}}$, i.e., $\mathbb{X}$ and $\mathbb{Y}$ have the same distribution.*

Example 3.17. (1) Similar to Theorem 3.9-(2), we have, in the case of random vectors, the following property: Let $\mathbb{X}$ be a random vector with values in $\mathbb{R}^n$, let $B \in \mathcal{M}_{n \times n}(\mathbb{R})$ be a matrix and let $b \in \mathbb{R}^n$ be a vector. Then the random vector $\mathbb{Y} = \mathbb{X}B + \mathbb{b}$ has the joint moment generating function

$$M_{\mathbb{Y}}(\mathbb{t}) = e^{\mathbb{t}\mathbb{b}^T} M_{\mathbb{X}}(\mathbb{t}B^T),$$

for the values $\mathbb{t} \in \mathbb{R}^n$ for which $M_{\mathbb{X}}(\mathbb{t}B^T)$ exists.
(2) Let $\mathbb{X} = (X_1, \ldots, X_n) \sim MVN(0_n, I_n)$. Then, by Problem 2.5.1, we

have that $X_1, \ldots, X_n$ are independent random variables. By using Theorem 3.9 and the computations in Problem 3.5.4, we write

$$\begin{aligned} M_{\mathbb{X}}(\mathfrak{t}) &= E\big(\exp\{t_1 X_1 + \ldots + t_n X_n\}\big) \\ &= E\big(\exp\{t_1 X_1\}\big) \cdot \ldots \cdot E\big(\exp\{t_n X_n\}\big) \\ &= \exp\left\{\frac{1}{2}(t_1^2 + \ldots + t_n^2)\right\} = \exp\left\{\frac{1}{2}\mathfrak{t}\mathfrak{t}^T\right\} \end{aligned}$$

for every $\mathfrak{t} = (t_1, \ldots, t_n) \in \mathbb{R}^n$.

(3) Consider $\mathbb{X} = (X_1, \ldots, X_n) \sim MVN(0_n, I_n)$. Let $A \in \mathcal{M}_{n \times n}(\mathbb{R})$ be a positive definite matrix and $\mathfrak{m} \in \mathbb{R}^n$ be a vector. Then the random vector $\mathbb{Y} = \mathbb{X}A^{\frac{1}{2}} + \mathfrak{m}$ satisfies $\mathbb{Y} \sim MVN(\mathfrak{m}, A)$, where $A^{\frac{1}{2}}$ is the square root of the matrix A, see Proposition A.4. The joint moment generating function of $\mathbb{Y} \sim MVN(\mathfrak{m}, A)$ is by (1) and (2)

$$M_{\mathbb{Y}}(\mathfrak{t}) = \exp\left\{\mathfrak{t}\mathfrak{m}^T + \frac{1}{2}\mathfrak{t}A\mathfrak{t}^T\right\} \quad \text{for every } \mathfrak{t} \in \mathbb{R}^n.$$

(4) If $\mathbb{X} = (X_1, \ldots, X_n) \sim MVN(\mathfrak{m}, A)$, with $\mathfrak{m} = (m_1, \ldots, m_n) \in \mathbb{R}$ and $A = (a_{ij})_{i,j=\overline{1,n}}$ positive definite matrix, then for each $i \in \{1, \ldots, n\}$ and $t \in \mathbb{R}$ we have

$$M_{X_i}(t) = M_{\mathbb{X}}(0, \ldots, 0, \underbrace{t}_{\text{position } i}, 0, \ldots, 0) = \exp\left\{m_i t + \frac{1}{2}a_{ii}t^2\right\}.$$

This is the moment generating function for the $N(m_i, a_{ii})$ normal distribution (see Problem 3.5.4). Hence,

$$X_i \sim N(m_i, a_{ii}) \quad \text{for each } i \in \{1, \ldots, n\},$$

which means that each component of a multivariate normally distributed random vector is in fact a normally distributed random variable.   ▲

Remark 3.7. Let $A \in \mathcal{M}_{n \times n}(\mathbb{R})$ be a symmetric and positive semidefinite matrix. Analogously to Example 3.17-(3), we can use the joint moment generating function to define the **degenerate case of the multivariate normal distribution**. We say that a random vector $\mathbb{Y}$ has $MVN(\mathfrak{m}, A)$ distribution, if its joint moment generating function is

$$M_{\mathbb{Y}}(\mathfrak{t}) = \exp\left\{\mathfrak{t}\mathfrak{m}^T + \frac{1}{2}\mathfrak{t}A\mathfrak{t}^T\right\} \quad \text{for every } \mathfrak{t} \in \mathbb{R}^n.$$

In this case, it is not necessarily that there exists a density function. In the one dimensional case, we have the degenerate case when the variance equals to zero, i.e., the random variable is a.s. a constant value (see Problem 3.4.1). By convention we accept this degenerate case too (it could appear in some special cases of Gaussian processes, see Remark 5.7).   ▼

3.6.1 Solved Problems

Problem 3.6.1. The random vector $\mathbb{X} = (X_1, \ldots, X_n)$, $n \in \mathbb{N}^*$, has multivariate normal distribution $MVN(\mathbb{m}, A)$, with $\mathbb{m} = (m_1, \ldots, m_n) \in \mathbb{R}$ and $A \in \mathcal{M}_{n \times n}(\mathbb{R})$ positive definite matrix, if and only if any linear combination $\mathbb{b}\mathbb{X}^T = \sum_{i=1}^{n} b_i X_i$ with $\mathbb{b} = (b_1, \ldots, b_n) \in \mathbb{R}^n \setminus \{0_n\}$ has normal distribution $N(\mathbb{b}\mathbb{m}^T, \mathbb{b}A\mathbb{b}^T)$.

Solution 3.6.1: Let $\tau \in \mathbb{R}$ and $\mathbb{b} = (b_1, \ldots, b_n) \in \mathbb{R}^n \setminus \{0_n\}$ be arbitrary. Consider $\mathbb{X} = (X_1, \ldots, X_n) \sim MVN(\mathbb{m}, A)$. By using Example 3.17-(3) we have

$$M_{b_1 X_1 + \ldots + b_n X_n}(\tau) = E\big(\exp\{\tau b_1 X_1 + \ldots + \tau b_n X_n\}\big)$$

$$= M_{\mathbb{X}}(b_1 \tau, \ldots, b_n \tau) = \exp\left\{\tau \mathbb{b}\mathbb{m}^T + \frac{1}{2}\tau^2 \mathbb{b}A\mathbb{b}^T\right\},$$

which is the moment generating function for the $N(\mathbb{b}\mathbb{m}^T, \mathbb{b}A\mathbb{b}^T)$ normal distribution. Hence, $b_1 X_1 + \ldots + b_n X_n$ is a normally distributed random variable (see Problem 3.5.4).

Consider now $\mathbb{b}\mathbb{X}^T = b_1 X_1 + \ldots + b_n X_n$ to be a normally distributed random variable for each $\mathbb{b} = (b_1, \ldots, b_n) \in \mathbb{R}^n \setminus \{0_n\}$. Hence, $X_1, \ldots, X_n$ are normally distributed random variables, i.e.,

$$X_i \sim N(m_i, \sigma_i^2) \text{ for each } i \in \{1, \ldots, n\}$$

and (see Example 3.9)

$$E(b_1 X_1 + \ldots + b_n X_n) = b_1 m_1 + \ldots + b_n m_n = \mathbb{b}\mathbb{m}^T,$$

$$V(b_1 X_1 + \ldots + b_n X_n) = \sum_{i,j=1}^{n} b_i b_j \mathrm{cov}(X_i, X_j) = \mathbb{b}A\mathbb{b}^T,$$

where $\mathbb{m} = (m_1, \ldots, m_n) \in \mathbb{R}^n$ and $A = V(\mathbb{X})$. This implies

$$b_1 X_1 + \ldots + b_n X_n \sim N(\mathbb{b}\mathbb{m}^T, \mathbb{b}A\mathbb{b}^T).$$

Then we have

$$M_{\mathbb{X}}(\mathbb{b}) = E\big(\exp\{b_1 X_1 + \ldots + b_n X_n\}\big) = \exp\left\{\mathbb{b}\mathbb{m}^T + \frac{1}{2}\mathbb{b}A\mathbb{b}^T\right\}.$$

By using Example 3.17-(3) and Theorem 3.13, we have $\mathbb{X} \sim MVN(\mathbb{m}, A)$.

Problem 3.6.2. The random vector $\mathbb{X} = (X_1, \ldots, X_n)$ has multivariate uniform distribution $MVUnif([0, 1] \times \cdots \times [0, 1])$. Let $A \in \mathcal{M}_{n \times n}(\mathbb{R})$. Prove that $\mathbb{X}A$ has multivariate uniform distribution if and only if each line, respectively each column, of A has exactly one nonzero element.

Solution 3.6.2: Let $A = (a_{ij})_{i,j=\overline{1,n}}$. We note that

$$
M_{\mathbb{X}}(\mathbb{t}) = \begin{cases} \displaystyle\prod_{\substack{i=1 \\ t_i \neq 0}}^{n} \frac{e^{t_i} - 1}{t_i}, & \mathbb{t} \neq 0_n \\ 1, & \mathbb{t} = 0_n, \end{cases}
$$

where we use the notation $\mathbb{t} = (t_1, \ldots, t_n)$. We assume that

$$
\mathbb{X}A \sim MVUnif([b_1, c_1] \times \cdots \times [b_n, c_n]),
$$

where $b_j < c_j$, for every $j \in \{1, \ldots, n\}$. Fix $i \in \{1, \ldots, n\}$ and denote

$$
\mathbb{e}_i = (0, \ldots, 0, \underbrace{t}_{\text{position } i}, 0, \ldots, 0)
$$

and $J = \{j \in \{1, \ldots, n\} : a_{ji} \neq 0\}$. By Example 3.17-(1), we have

$$
M_{\mathbb{X}A}(t\mathbb{e}_i) = M_{\mathbb{X}}(t\mathbb{e}_i A^T) = \prod_{j \in J} \frac{e^{ta_{ji}} - 1}{ta_{ji}} = \frac{e^{tc_i} - e^{tb_i}}{t(c_i - b_i)}, \quad t \in \mathbb{R}^*,
$$

and $M_{\mathbb{X}A}(0_n) = 1$; observe that the above relation implies also $J \neq \emptyset$. Using the same arguments as in Example 3.16-(5) and taking into account that

$$
\frac{\prod_{j \in J} a_{ji}}{c_i - b_i} = \lim_{t \to \infty} \frac{\exp\left\{ t\left(\sum_{j \in J} a_{ji} - c_i \right) \right\}}{t^{\#J-1}},
$$

we deduce that $\#J = 1$, because we have a nonzero number on the left hand side. Hence, the ith column of A has exactly one nonzero element. Therefore, we deduce that each column of A has exactly one nonzero element.

Suppose that there is a line only with zeros. Then $\det A = \det(A^T) = 0$ and thus there exists $\mathbb{v} = (v_1, \ldots, v_n) \in \mathbb{R}^n \setminus \{0_n\}$ such that $\mathbb{v}A^T = 0_n$. Let $I_1 = \{i \in \{1, \ldots, n\} : v_i < 0\}$ and $I_2 = \{i \in \{1, \ldots, n\} : v_i > 0\}$. We note that

$$
M_{\mathbb{X}A}(t\mathbb{v}) = M_{\mathbb{X}}(t\mathbb{v}A^T) = \prod_{i \in I_1 \cup I_2} \frac{e^{tv_i c_i} - e^{tv_i b_i}}{tv_i(c_i - b_i)} = 1, \quad t \neq 0.
$$

Hence (see Example 3.16-(5)),

$$
\prod_{i \in I_1 \cup I_2} v_i(c_i - b_i) = \lim_{t \to \infty} \frac{\exp\left\{ t\left(\displaystyle\sum_{i \in I_1} v_i b_i + \sum_{i \in I_2} v_i c_i \right) \right\}}{t^{\#I_1 + \#I_2}} \in \{0, \infty\},
$$

because $\#I_1 + \#I_2 > 0$, which gives a contradiction. Since each column of A has exactly one nonzero element, A has exactly n nonzero elements. Since each line of A has at least a nonzero element, we deduce that each line of A has exactly one nonzero element.

Suppose now that each line, respectively each column, of A has exactly one nonzero element. Then there exist a permutation σ of $(1, \ldots, n)$ and $a_1, \ldots, a_n \in \mathbb{R}^*$ such that $\mathfrak{t}A^T = (a_{\sigma(1)}t_{\sigma(1)}, \ldots, a_{\sigma(n)}t_{\sigma(n)})$, for all $\mathfrak{t} = (t_1, \ldots, t_n) \in \mathbb{R}^n$. Hence

$$M_{\mathbb{X}A}(\mathfrak{t}) = M_{\mathbb{X}}(\mathfrak{t}A^T) = \begin{cases} \displaystyle\prod_{\substack{i=1 \\ t_i \neq 0}}^{n} \frac{e^{a_i t_i} - 1}{a_i t_i}, & \mathfrak{t} \neq 0_n \\ 1, & \mathfrak{t} = 0_n \end{cases}$$

and thus $\mathbb{X}A$ has multivariate uniform distribution, by Theorem 3.13.

3.7 Problems for Chapter 3

Problem 3.7.1. The probability to win a game in each trial is $p \in (0, 1)$. A gambler repeats playing a certain game until he wins for 1) the first time; 2) the third time. Compute the expected number of games played by the gambler.

Solution 3.7.1: Let X be the number of games played by the gambler.
1) $X - 1 \sim Geo(p)$. We have $E(X - 1) = \dfrac{1 - p}{p}$ from the solution of Example 3.1. We obtain $E(X) = \dfrac{1}{p}$.
2) Denote by X_1 be the number of trials until the first win is obtained, let X_2 be the number of additional trials to the second win and let X_3 be the number of additional trials to the third win. We have $X = X_1 + X_2 + X_3$, then $E(X) = \dfrac{3}{p}$.

Problem 3.7.2. Let $p_1, \ldots, p_m \in (0, 1)$ be such that $p_1 + \ldots + p_m = 1$. Also, let $c_1, \ldots, c_m$ be m distinct characters. For each position in an n characters code, one of the characters $c_1, \ldots, c_m$ is randomly and independently chosen with the corresponding probabilities $p_1, \ldots, p_m$. Let $k \in \{0, 1, \ldots, n\}$ and let C_k be the random variable that counts the number of characters that appear exactly k times in the code. Write a program that generates codes as described above and computes the arithmetic mean value of the number of characters that appear exactly k times in a code. Find the expected value of C_k and compare it with the value given by your program.

Solution 3.7.2: For every $i \in \{1, \ldots, m\}$ let

$$X_i = \begin{cases} 1, & c_i \text{ appears exactly } k \text{ times in the code} \\ 0, & \text{otherwise} \end{cases}$$

and note that $P(X_i = 1) = C(n,k)p_i^k(1-p_i)^{n-k}$. Hence

$$E(C_k) = E\left(\sum_{i=1}^m X_i\right) = \sum_{i=1}^m C(n,k)p_i^k(1-p_i)^{n-k}.$$

```
function Y=discret(x,p,N,M)
%N×M random numbers/characters are generated in the matrix Y
%according to the values in the vector x and their corresponding
%probabilities in the vector p, using the inversion method
[~,~,ind] = histcounts(rand(N,M),[0 cumsum(p)]);
Y=x(ind);
end
```

```
function [codes,Ck,ECk,E]=codegen(c,p,n,N,k)
%codes=is a set of N random codes with n characters
%chosen from c with corresponding probabilities given by p
%Ck=number of characters that appear k times in each code
%ECk=arithmetic mean of Ck
%E=the expected value
codes=discret(c,p,N,n);
Ck=zeros(1,N);
for i=1:N
    Ck(i)=sum(sum(codes(i,:)'==c)==k);
end
ECk=mean(Ck);
E=sum(binopdf(k,n,p));
end
```

```
>>[codes,Ck,ECk,E]=codegen('abcde',[0.3,0.1,0.3,0.1,0.2],6,10000,2);
>>[ECk, E]
ans = 1.0981 1.0909
```

Problem 3.7.3. A random code of $n \in \mathbb{N}^*$ digits is generated such that for each position in the code one of the following digits: 0, 2, 4, 6, 8, is randomly and independently chosen with the following corresponding probabilities: 0.3, 0.1, 0.3, 0.1, 0.2. Determine the expected value of the number of distinct digits in the code and compare this theoretical value with the corresponding arithmetic mean value given by 1000 codes generated by simulating the above procedure in a program.

Solution 3.7.3: We note that the expected value of the number of distinct digits in the code is $5 - E(C_0) = 5 - (0.7^n + 0.9^n + 0.7^n + 0.9^n + 0.8^n)$, where C_0 is the number of digits that do not appear in the code and we use the solution of Problem 3.7.2.

For the simulation, we can use the program from Problem 3.7.2. We consider $n = 5$ in the following.

```
>>[~,~,ECk,E]=codegen(0:2:8,[0.3,0.1,0.3,0.1,0.2],6,1000,0);
>>[5-E, 5-ECk]
ans = 3.4397 3.4560
```

Problem 3.7.4. (Example of universal hashing) Consider all 10 characters strings that one user can choose to register on some website. Suppose each character is represented by 7 bits, which give the corresponding ASCII code, and thus each username is represented by a column vector of 70 bits. Let us call the usernames in the binary representation keys. Let U be the set, called universe, off all keys.

Let $k \in \mathbb{N}^*$, $k < 70$, and $m = 2^k$. Let H be the set of all possible vectors of k bits. Let us call the elements of H hash codes.

The size of H should be small, when compared to the size of the universe U. The main scope of the hashing is to speed-up the search for certain keys, by using the hash codes, which necessitate less memory to be stored than the keys do.

We say that $h : U \to H$ is a hash function. For every $x, y \in U$, $x \neq y$, we say that x is in collision with y, if $h(x) = h(y)$.

We say that a finite family $\mathcal{H}$ of hash functions from U to H is universal, if choosing uniformly from $\mathcal{H}$ a random hash function h we have

$$P\big(h(x) = h(y)\big) \leq \frac{1}{m}, \quad \text{for all } x, y \in U, x \neq y,$$

i.e., any two keys of the universe are in collision with probability at most $\frac{1}{m}$, when the hash function h is drawn randomly from $\mathcal{H}$.

The main idea of the universal hashing is to reduce the number of collisions.

For every matrix A of $k \times 70$ bits, let $h_A : U \to H$ be given by

$$h_A(x) = (A \cdot x) \bmod 2, \quad x \in U.$$

Let $\mathcal{H} = \{h_A : A \text{ is a binary } k \times 70 \text{ matrix}\}$. Let $S \subsetneq U$ be a set of n keys, $n \in \mathbb{N}^*$ (S represents a set of registered usernames). Let h be a randomly drawn hash function from $\mathcal{H}$.

1) Prove that $\mathcal{H}$ is a universal family of hash functions.

2) Prove that the expected number of elements from S which are in collision with a fixed key $x \in U \setminus S$ is $\frac{n}{m}$.

3) We say that we have a false positive match in S for a key $x \in U \setminus S$, if x is in collision with at least one key from S (i.e., x has an already used hash code, even though x is a new username that has not been registered before). Prove that the probability of a false positive match in S is at most $\dfrac{n}{m}$. In particular, if $k = \lceil \log_2(100n) \rceil$, then the probability of a false positive match in S is at most 1%, where $\lceil \cdot \rceil$ is the ceiling function.

4) Generate a random matrix corresponding to h with $k = \lceil \log_2(100n) \rceil$ columns and simulate $n \in \{500, 1000, 1500\}$ registrations, by randomly generating usernames that contain only capital letters or decimal digits. For each registration store only the corresponding hash code and verify if a previous hash code has been stored before (in this case, print a message that the username is invalid).

5) Choose a username x. Generate a set S of random $n \in \{1000, 2000, 3000\}$ keys which do not contain x. Generate $N \in \{500, 1000, 1500\}$ random matrices with $k = \lceil \log_2(100n) \rceil$ columns and for each corresponding hash function check if we have a false positive match in S for the username x. Return the percentage of false positives.

Solution 3.7.4 1) Let $A = (A_{ij})_{\substack{i=\overline{1,k} \\ j=\overline{1,70}}}$ be a binary $k \times 70$ matrix. Let $x, y \in U, x \neq y$ and $z = (x - y) \bmod 2$. Since $z \neq 0$, there exists an element of z which is not 0. Without loss of generality, let us assume this element is z_1, where $z_1, \ldots, z_{70}$ are the elements of z. Hence $z_1 = 1$. In the following, we consider all the expression and operations, which involve A, to be modulo 2. In view of the law of total probability (see Theorem 1.8), we deduce that

$$P\big(h_A(x) = h_A(y)\big) = P(Az = 0) = P\bigg(\bigcap_{i=1}^{k} \Big\{ A_{i1} = \sum_{j=2}^{70} A_{ij} \cdot z_j \Big\} \bigg)$$

$$= \sum_{\substack{a_{ij} \in \{0,1\} \\ i=\overline{1,k}, j=\overline{2,70}}} P\bigg(\bigcap_{i=1}^{k} \Big\{ A_{i1} = \sum_{j=2}^{70} a_{ij} \cdot z_j \Big\} \Big| \bigcap_{i=1}^{k} \bigcap_{j=2}^{70} \{ A_{ij} = a_{ij} \} \bigg)$$

$$\cdot P\bigg(\bigcap_{i=1}^{k} \bigcap_{j=2}^{70} \{ A_{ij} = a_{ij} \} \bigg)$$

$$= \sum_{\substack{a_{ij} \in \{0,1\} \\ i=\overline{1,k}, j=\overline{2,70}}} \frac{1}{2^k} P\bigg(\bigcap_{i=1}^{k} \bigcap_{j=2}^{70} \{ A_{ij} = a_{ij} \} \bigg) = \frac{1}{2^k} = \frac{1}{m}.$$

2) For every $y \in S$ let $C_{x,y}$ be the random variable that takes the value 1, if x and y are in collision, and 0, otherwise. In view of the proof for 1), the expected number of elements from S in collision with x is equal to

$$E\left(\sum_{y\in S} C_{x,y}\right) = \sum_{y\in S} E(C_{x,y}) = \sum_{y\in S} \frac{1}{m} = \frac{n}{m}.$$

3) Recall that h is a randomly drawn hash function from $\mathcal{H}$. The probability of a false positive match in S for x is

$$P\left(\bigcup_{y\in S}\{h(x) = h(y)\}\right) \le \sum_{y\in S} P(h(x) = h(y)) = \frac{n}{m},$$

where we use Problem 1.3.1 and the property from 1). If $k \ge \log_2(100n)$, then $\frac{n}{m} \le \frac{1}{100}$.

4)

```
function UniHash(n)
k=ceil(log2(n*100));
A=binornd(1,0.5,k,70);
s=randsample(['A':'Z','0':'9'],10,true);
HS=mod(A*str2num(reshape(dec2bin(double(s'),7),70,1)),2)';
for i=2:n
    s=randsample(['A':'Z','0':'9'],10,true);
    hs=mod(A*str2num(reshape(dec2bin(double(s'),7),70,1)),2)';
    if ismember(hs,HS,'rows')
        fprintf('The username %s was not valid.\n',s);
    else
        HS=[HS; hs];
    end
end
end
```

```
>> UniHash(1000)
The username OWZSOEZ05S was not valid.
The username Q69YSCQHTV was not valid.
```

5)

```
function S=usernames(n,username)
 S=[ ];
 for i=1:n
    s=randsample(['A':'Z','0':'9'],10,true);
    while s==username
     s=randsample(['A':'Z','0':'9'],10,true);
    end
    S=[S;s];
 end
end
```

```
function UniHash_simul(N,username,S)
n=size(S,1);
k=ceil(log2(n*100));
SB=reshape(str2num(reshape(dec2bin(double(S'),7)',70*n,1)),70,n);
ub=reshape(str2num(reshape(dec2bin(double(username'),7)',70,1)),70,1);
count=0;
for j=1:N
    A=binornd(1,0.5,k,70);
    hs=mod(A*SB,2)';
    hu=mod(A*ub,2)';
    if ismember(hu,hs,'rows')
        count=count+1;
    end
end
fprintf('The percentage of false positives is %3.2f%%.\n',count/N*100);
end
```

```
>>S=usernames(3000,'ABCDE12345');
>>UniHash_simul(1000,'ABCDE12345',S)
The percentage of false positives is 0.60%.
```

For further properties of hashing, we refer to [Mitzenmacher and Upfal (2005), Section 5.5].

Problem 3.7.5. In an online multiplayer video game, each player starts on his own with an either red or green flag and then can make alliances with any other player that has an opposite color flag. The game starts with $n \in \mathbb{N}^*$ players. Suppose that there are $m \in \mathbb{N}^*$ pairs of players that are friends and are willing to establish an alliance, if possible.

1) For $n = m = 7$, identify the n players with the vertices of an undirected graph and establish m pairs of friends by connecting each corresponding pair of vertices with an edge, in Matlab. Next, randomly and independently assign to each vertex (i.e., player) a color (for the corresponding flag) and then highlight with blue each edge which represents an alliance between friends.

2) For $n = m = 7$ and the established friendships at 1), simulate $N \in \{500, 1000, 1500\}$ times the random selection of the flags and estimate the expected value of the number of alliances.

3) Prove that, if we randomly and independently assign the colors of the flags (we toss a fair coin for each player to assign the color), then the expected value of the number of alliances between friends is $\frac{m}{2}$.

4) Prove that it is possible to assign the colors of the flags such that there are at least $\frac{m}{2}$ possible alliances between friends.

Solution 3.7.5: 1)

```
clf; hold on; axis off;
E=[1 2;1 3;3 4;3 5;4 6;4 7;6 7]; G=graph(E(:,1),E(:,2));
g=plot(G,'MarkerSize',8,'NodeColor','r','EdgeColor','k','NodeFontSize',15);
flags=1:numnodes(G);
flags_green=flags(binornd(1,0.5,1,numnodes(G))==1);
flags_red=setdiff(flags,flags_green);
highlight(g,flags_green,'NodeColor','g');
for i=1:numedges(G)
   if sum(ismember(G.Edges.EndNodes(i,:),flags_red))==1
      highlight(g,G.Edges.EndNodes(i,:),'EdgeColor','b','LineWidth',2);
   end
end
```

2)

```
function video_game_simul(N)
E=[1 2;1 3;3 4;3 5;4 6;4 7;6 7]; G=graph(E(:,1),E(:,2));
flags=1:numnodes(G);
alliances=zeros(1,N);
for j=1:N
flags_green=flags(binornd(1,0.5,1,numnodes(G))==1);
flags_red=setdiff(flags,flags_green);
count=0;
for i=1:numedges(G)
   if sum(ismember(G.Edges.EndNodes(i,:),flags_red))==1
      count=count+1;
   end
end
alliances(j)=count;
end
fprintf('The estimated expected value is %3.2f.\n',mean(alliances));
end
```

```
>> video_game_simul(1000)
The estimated expected value is 3.48.
```

3) We shall consider again the identification of the players and friendships with an undirected graph. Also, we identify each alliance between friends with a blue edge.

Let

$$X_i = \begin{cases} 1, & \text{the } i\text{th edge is blue,} \\ 0, & \text{otherwise.} \end{cases}$$

We note that $X = \sum_{i=1}^{m} X_i$ is a random variable that gives the number of blue edges (i.e., alliances between friends). We have

$$E(X) = \sum_{i=1}^{m} E(X_i) = \sum_{i=1}^{m} \frac{1}{2} = \frac{m}{2}.$$

4) Let $\{x_1, \dots, x_n\}$ be the range of X. Suppose that it is not possible to assign the colors of the flags such that there are at least $\frac{m}{2}$ possible alliances between friends. Then we have $P(X \geq E(X)) = 0$ (in view of 3)). Hence, $x_i < E(X)$ for all $i \in \{1, \dots, n\}$ and thus

$$E(X) = \sum_{i=1}^{n} x_i \cdot P(X = x_i) < \sum_{i=1}^{n} E(X) \cdot P(X = x_i) = E(X),$$

which is false. Therefore, our supposition is false, thus $P(X \geq E(X)) > 0$. Hence, it is possible to assign the colors of the flags such that there are at least $\frac{m}{2}$ possible alliances between friends.

Remark: Problem 3.7.5 4) is an application of the so-called probabilistic method, regarding the Max-Cut problem, see [Mitzenmacher and Upfal (2005), Chapter 6].　　　　　　　　　　　　　　　　　　　　　　　▼

Problem 3.7.6. 1) Write a program that, for given $q_0, q_1, \dots, q_n \in (0, 1)$, $n \in \mathbb{N}$, generates a number in the following way: first, forms a binary number, by randomly and independently choosing, for the bit on position $i \in \{0, 1, \dots, n\}$, 1 with probability q_i and 0 with probability $1 - q_i$, and then converts the number to base 10. The position 0 is the last position in the binary number.

2) Consider the following game in two. The first player chooses $p_0, p_1, \dots, p_n \in (0, 1)$, $n \in \mathbb{N}$, as input for the program. The second player takes into account the input chosen by the first player and chooses his input for the program such that the expected value of his output is equal to the expected value of the output of the first player, but his variance is as large as possible. Find the input of the second player. Simulate this game by generating 1000 numbers for each player and print the corresponding estimated expected values and variances.

Solution 3.7.6: 1)

```
function Z=rndnumb(q,k) %generates k numbers using the probabilities in q
B=binornd(ones(k,length(q)),repmat(flip(q),k,1));
Z=bin2dec(num2str(B));
end
```

2) Let X be a number generated by the first player and Y be a number generated by the second player. For every $i \in \{0, 1, \ldots, n\}$ let X_i be the bit on position i corresponding to X and Y_i be the bit on position i corresponding to Y. Let $q_0, q_1, \ldots, q_n \in (0, 1)$ be the input of the second player. Then

$$E(X) = E\left(\sum_{i=0}^{n} 2^i X_i\right) = \sum_{i=0}^{n} 2^i p_i, \ V(X) = \sum_{i=0}^{n} 2^{2i} p_i(1 - p_i),$$

$$E(Y) = E\left(\sum_{i=0}^{n} 2^i Y_i\right) = \sum_{i=0}^{n} 2^i q_i, \ V(Y) = \sum_{i=0}^{n} 2^{2i} q_i(1 - q_i).$$

By the Cauchy–Schwarz inequality (for real numbers), we have

$$V^2(Y) = \left(\sum_{i=0}^{n} 2^{2i} q_i(1 - q_i)\right)^2 \leq \left(\sum_{i=0}^{n} 2^{2i} q_i^2\right) \cdot \left(\sum_{i=0}^{n} 2^{2i}(1 - q_i)^2\right)$$

and the equality holds if and only if there exists $\alpha \in \mathbb{R}$ such that $q_i = \alpha(1 - q_i)$ for each $i = \overline{0, n}$. Hence, the input of the second player satisfies the following conditions: $\sum_{i=0}^{n} 2^i q_i = \sum_{i=0}^{n} 2^i p_i$ and $q_0 = q_1 = \ldots = q_n$, and thus

$$q_0 = q_1 = \ldots = q_n = \frac{1}{2^{n+1} - 1} \sum_{i=0}^{n} 2^i p_i$$

has to be the choice of the second player.

```
>> n=4; p=rand(1,n+1), q=repmat(2.^[0:n]*p'/(2^(n+1)-1),1,n+1)
p = 0.6000 0.2500 0.9300 0.9600 0.0300
q = 0.4187 0.4187 0.4187 0.4187 0.4187
>> X=rndnumb(p,1000); Y=rndnumb(P,1000);
>> [mean(X), mean(Y); var(X,1), var(Y,1)]
ans = 12.9570 12.9920
      12.5052 81.1139
%the corresponding theoretical values
>> [2.^[0:n]*p', 2.^[0:n]*P'; 2.^[0:2:2*n]*(p.*(1-p))', 2.^[0:2:2*n]*(q.*(1-q))']
ans = 12.9800 12.9800
      11.9388 82.9966
```

For a discussion regarding this problem, see [Feller (1968), Example (b), p. 230].

Problem 3.7.7. A code of 20 symbols is transmitted through a communication channel with interference. Each symbol is correctly received with probability 0.9 and all symbols are independently affected or not. Write a program that computes the expected value of the number of received symbols until 10 consecutive symbols are correctly received or the last symbol is received. Simulate the transmission and reception of 1000 codes and estimate the desired expected value.

Solution 3.7.7: Let X_n denote the number of received symbols until 10 consecutive symbols are correctly received or the last symbol is received, if $n \in \mathbb{N}^*$ symbols are transmitted. Also, for every $i \in \{1, \ldots, 10\}$ let A_i be the event that the first incorrectly received symbol is on position i and let A be the event that the first 10 symbols are all correctly received.

Let $n \in \mathbb{N}$, $n \geq 11$, and $i \in \{1, \ldots, 10\}$. Then for $k \in \{1, \ldots, n\}$ we have $P(X_n = k | A_i) = P(X_{n-i} = k - i)$ because the event $X_n = k$ given A_i occurs if and only if $k - i$ symbols are transmitted after the incorrectly received symbol on position i until 10 consecutive symbols are correctly received or the last symbol is received (also, note that the symbols are independent). Hence,

$$E(X_n | A_i) = \sum_{k=1}^{n} k P(X_{n-i} = k - i)$$

$$= i + \sum_{k=1}^{n} (k - i) P(X_{n-i} = k - i) = i + E(X_{n-i}).$$

By Theorem 3.2, we have

$$E(X_n) = \sum_{i=1}^{10} E(X_n | A_i) P(A_i) + E(X_n | A) P(A)$$

$$= \sum_{i=1}^{10} (i + E(X_{n-i})) \cdot 0.9^{i-1} \cdot 0.1 + 10 \cdot 0.9^{10},$$

for $n \in \mathbb{N}$, $n \geq 11$. Also, we note that $E(X_n) = n$, for $n \in \mathbb{N}^*$, $n \leq 10$. In the following, we consider a recursive program that computes the value of $E(X_n)$.

```
function E=expechar(n)
if n<=10 E=n;
  else
      E=10*0.9^10;
      for i=1:10
          E=E+(i+expechar(n-i))*0.9^(i-1)*0.1;
      end
  end
end
```

Also, we consider an iterative program that computes the value of $E(X_n)$, which works faster than the recursive one.

```
function E=expechar(n)
  if n<=10
    E=n; return;
  end
  e=1:10;
  for k=11:n
    E=sum((([1:10]+e(10:-1:1)).*0.9.^[0:9])*0.1+10*0.9^10;
    e(1:9)=e(2:10); e(10)=E;
  end
end
end
```

Next, we consider the simulations. The main idea is to identify each correctly received symbol with 1 and each incorrectly received symbol with 0.

```
function expechar_sim
X=zeros(1,1000);
for i=1:1000
    c=binornd(1,0.9,1,20);
    j=1; count=0;
    while count<10&&j<=20
        if c(j)==1 count=count+1;
          else count=0;
        end
        X(i)=X(i)+1; j=j+1;
    end
end
fprintf('The estimated expected value is %5.3f.\n',mean(X));
fprintf('The expected value is %5.3f.\n',expechar(20));
end
```

```
>>expechar_sim
The estimated expected value is 14.986.
The expected value is 14.944.
```

Problem 3.7.8. Let $n \in \mathbb{N}$. Consider the following approximate counting algorithm due to Morris, given in pseudocode:

$$X_0 = 0$$
$$\text{for } i = 1, \ldots, n$$
$$X_i = X_{i-1} + 1 \text{ with probability } \frac{1}{2^{X_{i-1}}} \text{ and}$$
$$X_i = X_{i-1} \text{ with probability } 1 - \frac{1}{2^{X_{i-1}}}$$
$$\text{end}$$

1) Implement this algorithm in Matlab.

2) Prove that $E(2^{X_n}) = n + 1$ and $V(2^{X_n}) = \dfrac{n(n-1)}{2}$.

3) Prove that $E(X_n) \leq \log_2(n + 1)$.

4) Write a Matlab function that estimates the expected value and the variance of X_n. Compare the estimated expected value with the number of bits in the binary representation of n.

Solution 3.7.8: 1)

```
function X=Morris(n)
X=0;
for i=1:n
    X=X+binornd(1,1/2^X);
end
end
```

2) We prove that $E(2^{X_n}) = n + 1$ and $E(4^{X_n}) = \dfrac{n(n-1)}{2} + (n+1)^2$, by induction with respect to $n \in \mathbb{N}$. We clearly have $E(2^{X_0}) = E(4^{X_0}) = 1$. Also, we observe that $1 \leq X_n \leq n$ for $n \in \mathbb{N}^*$. Assume that $E(2^{X_{n-1}}) = n$ and $E(4^{X_{n-1}}) = \dfrac{(n-1)(n-2)}{2} + n^2$, where $n \in \mathbb{N}^*$. We note that

$$E(2^{X_n}) = \sum_{i=1}^{n} 2^i P(X_n = i) \text{ and } E(4^{X_n}) = \sum_{i=1}^{n} 4^i P(X_n = i)$$

and for every $i \in \{1, 2, \ldots\}0$

$$
\begin{aligned}
P(X_n = i) &= P\big(\{X_n = X_{n-1} + 1\} \cap \{X_{n-1} = i - 1\}\big) \\
&\quad + P\big(\{X_n = X_{n-1}\} \cap \{X_{n-1} = i\}\big) \\
&= P\big(X_n = X_{n-1} + 1 \big| X_{n-1} = i - 1\big) P(X_{n-1} = i - 1) \\
&\quad + P\big(X_n = X_{n-1} \big| X_{n-1} = i\big) P(X_{n-1} = i) \\
&= \frac{1}{2^{i-1}} \cdot P(X_{n-1} = i - 1) + \left(1 - \frac{1}{2^i}\right) \cdot P(X_{n-1} = i).
\end{aligned}
$$

Hence,

$$E(2^{X_n}) = \sum_{i=1}^{n-1} 2^i P(X_{n-1} = i) + \sum_{i=1}^{n} P(X_{n-1} = i) = E(2^{X_{n-1}}) + 1 = n + 1$$

and

$$E(4^{X_n}) = \sum_{i=1}^{n} 4^i P(X_{n-1} = i) + 3 \sum_{i=1}^{n} 2^i P(X_{n-1} = i)$$

$$= \frac{(n-1)(n-2)}{2} + n^2 + 3n = \frac{n(n-1)}{2} + (n+1)^2.$$

Now, it is easy to deduce also that $V(2^{X_n}) = \frac{n(n-1)}{2}$.

3) Since $\log_2$ is a concave function, we can use Jensen's inequality (see Theorem 3.7) and 2) to obtain

$$E(X_n) = E\left(\log_2(2^{X_n})\right) \leq \log_2\left(E(2^{X_n})\right) = \log_2(n+1).$$

4)

```
function Morris_simul(n,N)
X=zeros(1,N);
for i=1:N
    X(i)=Morris(n);
end
fprintf('The estimated expected value is %5.3f.\n',mean(X));
fprintf('The estimated variance is %4.3f.\n',var(X,1));
fprintf('The number of bits is %d.\n',length(dec2bin(n)));
end
```

```
>> Morris_simul(123456,100)
The estimated expected value is 16.640.
The estimated variance is 0.790.
The number of bits is 17.
```

For a detailed analysis of Morris' algorithm, see [Flajolet (1985)].

Problem 3.7.9. Consider the following function in Matlab, based on a randomized type algorithm called reservoir sampling:

```
% p ∈ (0,1) and k ∈ ℕ*
function [N,R]=reservoir(p,k)
N=k; R=1:k;
while rand<p
    N=N+1;
    i=unidrnd(N);
    if i<=k
        R(i)=N;
    end
end
end
```

1) Let X_n be the total number of changes in the vector R in the **while** loop of the above function, given that $N = n$ (i.e., the output N equals n), where $n \in \mathbb{N}$ is such that $n \geq k$. Prove that $E(X_n) = O(\log n)$, as $n \to \infty$, i.e., $\displaystyle\limsup_{n\to\infty} \left| \frac{E(X_n)}{\log n} \right| < \infty.$

2) Let $n \in \mathbb{N}$ be such that $n \geq k$. Prove that

$$P(j \in R | N = n) = \frac{k}{n} \text{ for all } j \in \{1, \ldots, n\}.$$

3) Prove that, if $k = 1$, then $E(R) = \dfrac{2 - p}{2(1 - p)}$.

Solution 3.7.9: Before solving the tasks, we note that the above reservoir sampling function does two things at the same time: generates a number N, using the geometric distribution with parameter p, and samples, without replacement, k numbers from the set $1, \ldots, N$, while **rand** returns a value less than p.

1) Let $n \in \mathbb{N}$, $n \geq k$. For every $j \in \mathbb{N}^*$ let $I_{j,n}$ be the random variable that takes the value 1, if the assignment in the jth **if** takes place, and the value 0, otherwise, given that the output N equals n. Note that N takes the value $j + k$ in the **while** loop when the jth **if** is called and thus $E(I_{j,n}) = \dfrac{k}{j + k}$ for $j \in \mathbb{N}^*$. Then

$$E(X_n) = \sum_{j=1}^{n-k} E(I_{j,n}) = k \sum_{j=1}^{n-k} \frac{1}{j + k}$$

$$< k \sum_{j=1}^{n-k} (\log(j + k) - \log(j + k - 1)) = k(\log(n) - \log(k)),$$

where we use the inequality $\log(x) - \log(x - 1) > \dfrac{1}{x}$ for $x > 1$, which follows by the mean value theorem. Hence, $E(X_n) = O(\log n)$, as $n \to \infty$.

2) Obviously, the relation holds for $n = k$. Next, fix $n \in \mathbb{N}$, $n > k$. For every $l \in \{k, \ldots, n\}$, let $R_{l,n}$ be the vector R after the $(l - k)$th iteration in the **while** loop, given that the output N equals n. We clearly have $P(j \in R | N = n) = P(j \in R_{n,n})$ for all $j \in \{1, \ldots, n\}$.

We prove by induction that $P(j \in R_{l,n}) = \dfrac{k}{l}$ for all $j \in \{1, \ldots, l\}$, with respect to $l \in \{k, \ldots, n\}$. We clearly have $P(j \in R_{k,n}) = 1$, for all $j \in \{1, \ldots, k\}$. Suppose we have $P(j \in R_{l,n}) = \dfrac{k}{l}$ for all $j \in \{1, \ldots, l\}$, where $l \in \{k, \ldots, n - 1\}$. We note that $l + 1 \in R_{l+1,n}$ if and only if $i = \textbf{unidrnd}(N) \leq k$ in the $(l + 1 - k)$th iteration of the **while** loop and thus, since i is a uniformly distributed random number in the set $\{1, \ldots, l + 1\}$, we have $P(l + 1 \in R_{l+1,n}) = \dfrac{k}{l + 1}$. For every $j \in \{1, \ldots, l\}$, we have $j \in R_{l+1,n}$ if and only if $j \in R_{l,n}$ and $i = \textbf{unidrnd}(N)$ is not equal to the index of j in the vector R in the $(l + 1 - k)$th iteration of the **while** loop. So, using the induction hypothesis, $P(j \in R_{l+1,n}) = \dfrac{k}{l} \cdot \dfrac{l}{l+1} = \dfrac{k}{l+1}$. Hence, we conclude that

$$P(j \in R | N = n) = P(j \in R_{n,n}) = \frac{k}{n} \text{ for all } j \in \{1, \ldots, n\}.$$

3) Using the law of total probability (see Theorem 1.8), the result from 1) and some basic results in analysis (see [Schinazi (2012), p. 215–216]), we deduce that

$$E(R) = \sum_{i=1}^{\infty} i P(R = i) = \sum_{i=1}^{\infty} i \sum_{n=i}^{\infty} P(R = i | N = n) \cdot P(N = n)$$

$$= \sum_{i=1}^{\infty} i \sum_{n=i}^{\infty} \frac{1}{n} p^{n-1}(1-p) = \sum_{i=1}^{\infty} i \cdot \frac{1-p}{p} \sum_{n=i}^{\infty} \frac{p^n}{n}$$

$$= \frac{1-p}{p} \sum_{i=1}^{\infty} i \cdot \sum_{n=i}^{\infty} \int_0^p x^{n-1} dx = \frac{1-p}{p} \sum_{i=1}^{\infty} i \cdot \int_0^p x^{i-1} \sum_{n=0}^{\infty} x^n dx$$

$$= \frac{1-p}{p} \sum_{i=1}^{\infty} i \cdot \int_0^p x^{i-1} \cdot \frac{1}{1-x} dx = \frac{1-p}{p} \int_0^p \sum_{i=1}^{\infty} i x^{i-1} \cdot \frac{1}{1-x} dx$$

$$= \frac{1-p}{p} \int_0^p \left(\frac{1}{1-x} \right)' \cdot \frac{1}{1-x} dx = \frac{2-p}{2(1-p)}.$$

Problem 3.7.10. Let $m, n, k \in \mathbb{N}$ be such that $m + n > 0$ and $k \leq m + n$. A vector v contains n negative numbers and m strictly positive numbers. k numbers are randomly chosen without replacement from v and copied in a vector w. Let $X(m, n, k)$ be the number of negative numbers in w.
1) Simulate 1000 times the above experiment and estimate the expected value of $X(10, 8, 7)$.
2) Prove that

$$E\big(X(m,n,k)\big) = \frac{n}{m+n}\Big(1 + E\big(X(m, n-1, k-1)\big)\Big)$$
$$+ \frac{m}{m+n} E\big(X(m-1, n, k-1)\big)$$

for $m, n, k \geq 1$, $k \leq m + n$.
3) Write a recursive function in Matlab that computes $E\big(X(m, n, k)\big)$. Print the value of $E\big(X(10, 8, 7)\big)$ and compare it with the value obtained for 1).
4) Prove by induction that

$$E\big(X(m, n, k)\big) = \frac{kn}{m+n} \text{ for } m, n, k \in \mathbb{N}, \ m + n > 0, \ k \leq m + n. \quad (3.5)$$

Solution 3.7.10: 1)

```
>> mean(hygernd(10+8,8,7,1,1000))
ans = 3.1150
```

2) Let A be the event that the first number in w is a negative number. By Theorem 3.2, we deduce that

$$E\big(X(m,n,k)\big) = E\big(X(m,n,k)|A\big) \cdot P(A) + E\big(X(m,n,k)|\bar{A}\big) \cdot P(\bar{A})$$

$$= E\big(X(m,n-1,k-1)+1\big) \cdot \frac{n}{m+n} + E\big(X(m-1,n,k-1)\big) \cdot \frac{m}{m+n}.$$

3)

```
function E=rechyge(m,n,k)
  if m==0 E=k;
   elseif n==0 E=0;
    else E=(n*(1+rechyge(m,n-1,k-1))+m*rechyge(m-1,n,k-1))/(m+n);
   end
end
>> rechyge(10,8,7)
ans = 3.1111
```

4) If $m = 0$ in (3.5), then the formula $E\big(X(0,n,k)\big) = k$ holds for $k \in \{0,1,\ldots,n\}$. If $n = 0$ in (3.5), then the formula $E\big(X(m,0,k)\big) = 0$ holds for $k = 0$. Moreover, if $k = 0$, then $E\big(X(m,n,0)\big) = 0$ for all $m,n \in \mathbb{N}$.

Let $m,n \in \mathbb{N}^*$ be fixed. We prove by induction with respect to $k \in \{1,\ldots,m+n\}$ that

$$E\big(X(i,j,k)\big) = \frac{kj}{i+j}$$

for each $i \in \{0,\ldots,m\}$, $j \in \{0,\ldots,n\}$ with $i+j \geq k$. Obviously,

$$E\big(X(i,j,1)\big) = \frac{j}{i+j} \text{ for each } i \in \{0,\ldots,m\}, j \in \{0,\ldots,n\}, i+j \geq 1.$$

Assume that $E\big(X(i,j,k)\big) = \dfrac{kj}{i+j}$ for each $i \in \{0,\ldots,m\}, j \in \{0,\ldots,n\}$ with $i+j \geq k$, where $k \in \{1,\ldots,m+n\}$ is fixed.

For $i \in \{0,\ldots,m\}, j \in \{0,\ldots,n\}$ with $i+j \geq k+1$, from the recursive relation obtained at 2) and the induction hypothesis, we get

$$E\big(X(i,j,k+1)\big) = E\big(X(i,j-1,k)+1\big) \cdot \frac{j}{i+j} + E\big(X(i-1,j,k)\big) \cdot \frac{i}{i+j}$$

$$= \left(1 + \frac{k(j-1)}{i+j-1}\right) \cdot \frac{j}{i+j} + \frac{kj}{i-1+j} \cdot \frac{i}{i+j} = \frac{(k+1)j}{i+j}.$$

Obviously, if $i = 0$ or $j = 0$ the formula $E\big(X(i,j,k+1)\big) = \dfrac{(k+1)j}{i+j}$ holds, in view of the discussion at beginning of the solution for 4).

Problem 3.7.11. Let $U_1, U_2 \sim Unif[0,1]$ be independent random variables and let $a \in (0,1)$.

1) Find a density function of $Z = aU_1 + (1-a)U_2$.

2) Compute the expectation and variance of Z.

3) Determine the value of a, if $V(Z) = \dfrac{1}{24}$.

Solution 3.7.11: 1) First, we assume that $a \in \left(0, \dfrac{1}{2}\right]$. Then $a \leq 1 - a$, which we shall use below. Clearly, we have $aU_1 \sim Unif[0, a]$ and $(1-a)U_2 \sim Unif[0, 1-a]$. Let f_{aU_1}, $f_{(1-a)U_2}$ be density functions of aU_1, respectively $(1-a)U_2$. Since

$$\begin{cases} u \in (0, a) \\ z - u \in (0, 1-a) \end{cases} \Leftrightarrow \begin{cases} z \in (0, 1) \\ u \in (\max\{0, z-(1-a)\}, \min\{a, z\}), \end{cases}$$

the following function given by (see Example 2.15)

$$f_Z(z) = \int_{\mathbb{R}} f_{aU_1}(u) f_{(1-a)U_2}(z-u)du = \int_{\max\{0, z-(1-a)\}}^{\min\{a, z\}} \frac{1}{a(1-a)} du$$

$$= \begin{cases} \dfrac{z}{a(1-a)}, & \text{if } z \in (0, a) \\ \dfrac{1}{1-a}, & \text{if } z \in [a, 1-a] \\ \dfrac{1-z}{a(1-a)}, & \text{if } z \in (1-a, 1) \end{cases}$$

and $f_Z(z) = 0$, if $z \notin (0, 1)$, is a density function of Z.

If $a \in \left(\dfrac{1}{2}, 1\right)$, then one can use the above proof by replacing a with $\alpha = 1 - a$, since, in this case, we have $Z = \alpha U_2 + (1-\alpha)U_1$ with $\alpha \in \left(0, \dfrac{1}{2}\right)$. Hence, the function

$$f_Z(z) = \begin{cases} \dfrac{z}{(1-a)a}, & \text{if } z \in (0, 1-a) \\ \dfrac{1}{a}, & \text{if } z \in [1-a, a] \\ \dfrac{1-z}{(1-a)a}, & \text{if } z \in (a, 1) \\ 0, & \text{if } z \notin (0, 1) \end{cases}$$

is a density function of Z.

2) We have $E(Z) = aE(U_1) + (1-a)E(U_2) = \dfrac{1}{2}$ and

$$V(Z) = a^2 V(U_1) + (1-a)^2 V(U_2) = \frac{2a^2 - 2a + 1}{12}.$$

3) Using 2), it follows $a = \dfrac{1}{2}$.

Problem 3.7.12. If $X_1, \ldots, X_n$ are independent random variables with normal distribution $N(0, \sigma^2)$, where $n \in \mathbb{N}^*$ and $\sigma > 0$, then $X_1^2 + \cdots + X_n^2$ has $\chi^2(n, \sigma)$ distribution.

Solution 3.7.12: We write for each $t < \dfrac{1}{2\sigma^2}$

$$M_{X_1^2}(t) = \frac{1}{\sqrt{2\pi}\sigma} \int_{\mathbb{R}} \exp\left\{ -\frac{x^2(1 - 2\sigma^2 t)}{2\sigma^2} \right\} dx$$

$$= \frac{1}{(1 - 2\sigma^2 t)^{\frac{1}{2}}} \int_{\mathbb{R}} \frac{1}{\sqrt{2\pi}\sigma} \exp\left\{ -\frac{u^2}{2\sigma^2} \right\} du = \frac{1}{(1 - 2\sigma^2 t)^{\frac{1}{2}}}.$$

Then for each $t < \dfrac{1}{2\sigma^2}$

$$M_{X_1^2 + \cdots + X_n^2}(t) = M_{X_1^2}(t) \cdot \cdots \cdot M_{X_n^2}(t) = \frac{1}{(1 - 2\sigma^2 t)^{\frac{n}{2}}}.$$

By Theorem 3.11 and Example 3.16-(4), we have $X_1^2 + \cdots + X_n^2 \sim \chi^2(n, \sigma)$.
Remark: The property contained in this problem can be proved without moment generating functions by using cumulative distribution functions and mathematical induction. ▼

Problem 3.7.13. Let X and Y be independent random variables having normal distribution $N(0, 1)$. Determine the expectation of the following random variable

$$Z = \exp\left\{ \frac{X^2 + Y^2}{4} \right\} (1 + X^2 + Y^2).$$

Solution 3.7.13: Using Problem 3.7.12, it follows that $U = X^2 + Y^2$ has $\chi^2(2, 1) = Exp\left(\dfrac{1}{2}\right)$ distribution (see Sections 2.5.6 and 2.5.7). Therefore, $Z = \exp\left\{\dfrac{U}{4}\right\}(1 + U)$ and

$$E(Z) = \int_{\mathbb{R}} \exp\left\{ \frac{u}{4} \right\} (1 + u) f_U(u) du = \int_0^\infty \exp\left\{ -\frac{u}{4} \right\} \frac{1 + u}{2} du = 10,$$

where f_U is a density function of U and we use the result from Example 3.2.

Problem 3.7.14. A lottery winner decides to spend his price of 4 million euros, each year, in independent amounts (of million of euros) that follow the $Unif[0,2]$ distribution.

1) Estimate the expected number of years before the winner spends half of his price by using simulations in Matlab.

2) Find the expected number of years before the winner spends half of his price.

Solution 3.7.14: 1)

```
function myears=lotteryprize(N)
myears=0;
for i=1:N
    years=0;
    money=4;
    while 1
        money=money-unifrnd(0,2);
        if money<2
            break;
        else
            years=years+1;
        end
    end
    myears=myears+years;
end
myears=myears/N;
end
```

```
>> lotteryprize(10000)
ans = 1.7188
```

2) Let $Z_l \sim Unif[0,2]$ be the amount of money spent in the lth year, for $l \in \mathbb{N}^*$, and

$$Y = \sup\{l \in \mathbb{N}^* : Z_1 + \ldots + Z_l < 2\}.$$

Then Y is a random variable which indicates the number of years before the lottery winner spends half of his price. We note that

$$Y = \sup\{l \in \mathbb{N}^* : X_0 + \ldots + X_{l-1} < 1\},$$

where $X_j = \dfrac{1}{2} Z_{j+1} \sim Unif[0,1]$, $j \in \mathbb{N}$. In view of Problem 2.9.28-3), we have

$$P(Y = k + 1) = \frac{1}{(k+1)!} - \frac{1}{(k+2)!}, \quad k \in \mathbb{N}.$$

Hence,

$$E(Y) = \sum_{k=0}^{\infty} (k+1) \left(\frac{1}{(k+1)!} - \frac{1}{(k+2)!} \right) = \sum_{k=0}^{\infty} \left(\frac{1}{k!} - \frac{1}{(k+1)!} + \frac{1}{(k+2)!} \right)$$

$$= \sum_{k=0}^{\infty} \frac{1}{k!} + \sum_{k=0}^{\infty} \left(\frac{1}{(k+2)!} - \frac{1}{(k+1)!} \right) = e - 1 \approx 1.7183 \text{ (years)}.$$

Problem 3.7.15. Let $k \in \mathbb{N}^*$, $p_1, \ldots, p_n \in (0, 1)$, $n \in \mathbb{N}^*$, be such that $p_1 + \ldots + p_n = 1$ and $\lambda_1, \ldots, \lambda_n > 0$. A job arrives in the queueing system represented below — Fig. 3.1. The job is routed through one of the n networks, with corresponding probabilities $p_1, \ldots, p_n$. Each network consists of k independent servers connected in series, represented by circles in the picture. The response time of each server is given by a random variable that follows the exponential distribution with a corresponding parameter λ_i, as specified in the picture. Determine the expected value and the variance of the total response time of the queueing system.

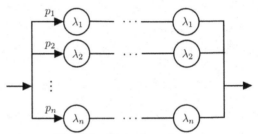

Fig. 3.1: Queueing system

Solution 3.7.15: Let $i \in \{1, \ldots, n\}$. Let A_i be the event that the job is routed through the ith network. Let X_i be the response time of the ith network. By Problem 2.9.26 1), X_i follows the Erlang distribution given by the density function

$$f_i(x) = \begin{cases} 0, & x \leq 0 \\ \dfrac{\lambda_i^k x^{k-1} e^{-\lambda_i x}}{(k-1)!}, & x > 0. \end{cases}$$

By computations one can easily deduce that $E(X_i) = \dfrac{k}{\lambda_i}$ and $E(X_i^2) = \dfrac{k(k+1)}{\lambda_i^2}$.

Let X be the total response time of the system. For every $x \in \mathbb{R}$, we have by the law of total probability (see Theorem 1.8)

$$P(X \leq x) = \sum_{i=1}^{n} P(X \leq x | A_i) \cdot P(A_i) = \sum_{i=1}^{n} p_i P(X_i \leq x).$$

Hence

$$f_X(x) = \sum_{i=1}^{n} p_i f_i(x), \quad x \in \mathbb{R}$$

is a density function of X. We mention that the distribution of X is called the hyper-Erlang distribution.

We deduce that

$$E(X) = k \sum_{i=1}^{n} \frac{p_i}{\lambda_i}$$

and

$$V(X) = k(k+1) \sum_{i=1}^{n} \frac{p_i}{\lambda_i^2} - k^2 \left(\sum_{i=1}^{n} \frac{p_i}{\lambda_i} \right)^2.$$

Problem 3.7.16. 1) For given $n, k \in \mathbb{N}^*$, $k \leq n$, generate 1000 sets, each of k numbers randomly chosen (without replacement) from the set $\{1, \ldots, n\}$. For each generated set, put in a vector x the lowest number of the set. Plot in Matlab the empirical cumulative distribution function and print the mean value of the vector x.

2) Compute the cumulative distribution function and the expected value of the random variable X representing the lowest number in a set of k random numbers generated as described in 1).

3) Plot the cumulative distribution function of X in the figure of the empirical cumulative distribution function and compare the expected value with the mean value obtained from 1) for $n = 10$ and $k = 5$.

Solution 3.7.16: 1)

```
function lowestnumb_ecdf(n,k)
x=zeros(1,1000);
for i=1:1000
    x(i)=min(randsample(n,k));
end
fprintf('The mean value of x is %5.4f.\n',mean(x));
[EF,t]=ecdf(x); t(1)=0;
clf; hold on;
for i=1:length(t)-1
    plot([t(i) t(i+1)],[EF(i) EF(i)],'-b','LineWidth',2);
    plot([t(i+1) t(i+1)],[EF(i) EF(i+1)],':k');
end
end
```

2) For every $m \in \{0, 1, \ldots, n - k\}$ we have

$$P(X > m) = \frac{C(n - m, k) \cdot C(m, 0)}{C(n, k)},$$

since we can use the probabilistic model of sampling without replacement with two states: "the number is strictly larger than m" and "the number is less than or equal to m" (see Remark 2.12). Then

$$F_X(x) = P(X \le x)$$

$$= \begin{cases} 0, & x < 1 \\ 1 - \dfrac{C(n - m, k)}{C(n, k)}, & m \le x < m + 1, \ m \in \{1, \ldots, n - k\} \\ 1, & x \ge n - k + 1. \end{cases}$$

Next, we note that for every $m \in \{1, \ldots, n - k + 1\}$ we have

$$P(X = m) = \frac{C(n - m, k - 1) \cdot C(1, 1) \cdot C(m - 1, 0)}{C(n, k)},$$

since we can use the probabilistic model of sampling without replacement with three states: "the number is strictly larger than m", "the number is equal to m" and "the number is strictly less than m" (see Section 2.4.6). Hence

$$E(X) = \sum_{m=1}^{n-k+1} m \cdot \frac{C(n - m, k - 1)}{C(n, k)}$$

$$= \frac{n + 1}{C(n, k)} \sum_{m=1}^{n-k+1} C(n - m, k - 1) - \frac{1}{C(n, k)} \sum_{m=1}^{n-k+1} (n - m + 1) C(n - m, k - 1)$$

$$= \frac{n + 1}{C(n, k)} \sum_{j=k-1}^{n-1} C(j, k - 1) - \frac{k}{C(n, k)} \sum_{j=k}^{n} C(j, k). \tag{3.6}$$

For every $K, N \in \mathbb{N}^*$ with $K \le N$ we have

$$\sum_{j=K}^{N} C(j, K) = C(N + 1, K + 1), \tag{3.7}$$

by using the idea that the $\displaystyle\sum_{j=K}^{N} C(j, K)$ is the coefficient of x^K in the polynomial expansion

$$q(x) = (1 + x)^K + \cdots + (1 + x)^N = x^{-1}(1 + x)^{N+1} - x^{-1}(1 + x)^K.$$

In view of (3.6) and (3.7), we deduce that

$$E(X) = \frac{(n+1)C(n,k)}{C(n,k)} - \frac{kC(n+1,k+1)}{C(n,k)} = \frac{n+1}{k+1}.$$

3) We call the Matlab function from 1) with $n = 10$, $k = 5$, then we return the expected value of X and we plot the cumulative distribution function of X:

```
function lowestnumb_cdf(n,k)
fprintf('The expected value of X is %5.4f.\n',(n+1)/(k+1));
for i=0:max(t)-1
    if i<n-k F=1-nchoosek(n-i,k)/nchoosek(n,k);
    else F=1; end
    plot([i i+1],[F F],'--r','LineWidth',2);
end
end
```

```
>> lowestnumb_ecdf(10,5);
The mean value of x is 1.8410.
>> hold on; lowestnumb_cdf(10,5);
The expected value of X is 1.8333.
```

Problem 3.7.17. Let $c_1, \ldots, c_m$ be m distinct characters, $m \in \mathbb{N}^*$. In an urn there are b_i balls with the character c_i printed on them, for $i \in \{1, \ldots, m\}$. For each position in an n characters code, $n \le B = b_1 + \ldots + b_m$, a ball from the urn is randomly extracted without replacement. Let $k \in \{0, 1, \ldots, n\}$ and let Y_k be the random variable that counts the number of characters that appear exactly k times in the code. Write a Matlab program that generates codes as described above and computes the arithmetic mean value of the number of characters that appear exactly k times in a code. Find the expected value of Y_k and compare it with the value given by your program.

Solution 3.7.17: Let

$$X_i = \begin{cases} 1, & c_i \text{ appears exactly } k \text{ times in the code} \\ 0, & \text{otherwise}, \quad i \in \{1, \ldots, m\}. \end{cases}$$

Then we have (see Remark 2.12)

$$P(X_i = 1) = \frac{C(b_i,k) \cdot C(B - b_i, n - k)}{C(B,n)},$$

if $b_i \ge k$ and $B - b_i \ge n - k$, and $P(X_i = 1) = 0$, otherwise. Hence,

$$E(Y_k) = E\Big(\sum_{i=1}^{n} X_i\Big) = \sum_{\substack{i=1 \\ b_i \ge k \\ B-b_i \ge n-k}}^{n} \frac{C(b_i,k) \cdot C(B - b_i, n - k)}{C(B,n)}.$$

```
function [codes,Yk,EYk,E]=codegen2(c,b,n,N,k)
%codes=is a set of N random codes with n characters
%chosen from an urn with b(i) balls with the character c(i)
%Yk=number of characters that appear k times in each code
%EYk=arithmetic mean value of Yk
%E=the expected value
urn=repelem(c,b); Yk=zeros(1,N);
codes=repmat(c(1),N,n);
for i=1:N
    codes(i,:)=randsample(urn,n);
    Yk(i)=sum(sum(codes(i,:)'==c)==k);
end
EYk=mean(Yk);
E=0; B=sum(b);
for i=1:length(b)
 if b(i)>=k && B-b(i)>=n-k
  E=E+hygepdf(k,B,b(i),n);
 end
end
end
```

```
>>[codes,Yk,EYk,E]=codegen2('abcde',[3,1,3,1,2],6,1000,2);
>>[EYk,E]
ans = 1.3610 1.3333
```

Problem 3.7.18. There are 30, 10, 30, 10, 20 balls in an urn with the letters a, b, c, d, respectively e, printed on them. A code is generated by randomly extracting n balls, $n \leq 70$, without replacement. Determine the expected number of distinct letters in the code and compare this theoretical value with the corresponding arithmetic mean value given by 1000 codes generated by simulating the extractions in a program.

Solution 3.7.18: We note that the expected number of distinct letters in the code is

$$5 - E(Y_0) = 5 - \frac{C(70,n) + C(90,n) + C(70,n) + C(90,n) + C(80,n)}{C(B,n)},$$

where Y_0 is the number of letters that do not appear in the code and we use Problem 3.7.17.

For the simulation, we can use the program from Problem 3.7.17, where $n = 50$.

```
>>[~,~,EYk,E]=codegen2('abcde',[30,10,30,10,20],50,1000,1);
>>[5-E, 5-EYk]
ans = 4.9855 4.9870
```

Problem 3.7.19. Let $X \sim Exp(1)$ and $Y \sim Unif[0,1]$ be independent random variables that show the lifetime of two electronic components. Let

U and W be the random variables that show the lifetime of the system formed by connecting the electric components in parallel, respectively in series. Find $E(U)$, $E(W)$, $V(U)$, $V(W)$, $E(U+W)$, $V(U+W)$, $E(U \cdot W)$ and $V(U \cdot W)$. Estimate these values in a program. Are U and W independent?

Solution 3.7.19: We clearly have $U = \max\{X, Y\}$ and $W = \min\{X, Y\}$. If $u \in [0, 1)$, then

$$F_U(u) = P(\max\{X, Y\} \le u) = P(X \le u) \cdot P(Y \le u) = (1 - e^{-u})u.$$

Hence,

$$F_U(u) = \begin{cases} 0, & \text{if } u < 0 \\ (1 - e^{-u})u, & \text{if } 0 \le u < 1 \\ 1 - e^{-u}, & \text{if } 1 \le u \end{cases}$$

and thus

$$f_U(u) = \begin{cases} 0, & \text{if } u < 0 \\ e^{-u}(u - 1) + 1, & \text{if } 0 \le u < 1 \\ e^{-u}, & \text{if } 1 \le u \end{cases}$$

is a density function of U. Therefore,

$$E(U) = \int_0^1 e^{-u}(u^2 - u) + u\,du + \int_1^\infty e^{-u}u\,du = \frac{3}{2} - \frac{1}{e},$$

$$E(U^2) = \int_0^1 e^{-u}(u^3 - u^2) + u^2\,du + \int_1^\infty e^{-u}u^2\,du = \frac{13}{3} - \frac{6}{e}$$

and

$$V(U) = E(U^2) - E^2(U) = \frac{25}{12} - \frac{3}{e} - \frac{1}{e^2}.$$

If $w \in [0, 1)$, then

$$F_W(w) = P(\min\{X, Y\} \le w) = 1 - P(X \ge w) \cdot P(Y \ge w) = 1 - e^{-w}(1 - w).$$

Hence

$$F_W(w) = \begin{cases} 0, & \text{if } w < 0 \\ 1 - e^{-w}(1 - w), & \text{if } 0 \le w < 1 \\ 1, & \text{if } 1 \le w, \end{cases}$$

and thus

$$f_W(w) = \begin{cases} e^{-w}(2 - w), & \text{if } 0 \le w < 1 \\ 0, & \text{if } w \notin [0, 1) \end{cases}$$

is a density function of W. Therefore,

$$E(W) = \int_0^1 e^{-w}(2w - w^2)dw = \frac{1}{e},$$

$$E(W^2) = \int_0^1 e^{-w}(2w^2 - w^3)dw = \frac{6}{e} - 2$$

and

$$V(W) = E(W^2) - E^2(W) = \frac{6}{e} - 2 - \frac{1}{e^2}.$$

We easily deduce that $E(U + W) = E(U) + E(W) = \frac{3}{2}$. Since we do not know whether or not U and W are independent, the key-idea to compute the other values is to observe that

$$U + W = X + Y \quad \text{and} \quad U \cdot W = X \cdot Y.$$

Hence,

$$V(U + W) = V(X) + V(Y) = 1 + \frac{1}{12} = \frac{13}{12},$$

$$E(U \cdot W) = E(X) \cdot E(Y) = \frac{1}{2}$$

and (see Theorem 3.4-(3))

$$V(U \cdot W) = V(X \cdot Y) = V(X) \cdot V(Y) + E^2(X) \cdot V(Y) + E^2(Y)V(X) = \frac{5}{12}.$$

Since $E(U \cdot W) \neq E(U) \cdot E(W)$, U and W are not independent.

```
>> X=exprnd(1,1,1000); Y=unifrnd(0,1,1,1000);
>> U=max([X;Y]); W=min([X;Y]);
>> [mean(U),mean(W);var(U,1),var(W,1)]
ans = 1.1452 0.3759
       0.8213 0.0723
>> [3/2-1/exp(1),1/exp(1);25/12-3/exp(1)-1/exp(1)^2,6/exp(1)-2-1/exp(1)^2]
ans = 1.1321 0.3679
       0.8444 0.0719
>> [mean(U+W),mean(U.*W);var(U+W,1),var(U.*W,1)]
ans = 1.5211 0.5108
       1.0542 0.4084
>> [3/2,1/2;13/12,5/12]
ans = 1.5000 0.5000
       1.0833 0.4167
```

Problem 3.7.20. n batteries function independently with the following corresponding probabilities $p_1, \ldots, p_n$ and have the following corresponding strictly positive voltages $v_1, \ldots, v_n$, where $n \in \mathbb{N}^*$. Suppose that the batteries are connected into a system that always chooses to use one battery from the functional ones, if any, that has:
1) the highest voltage;
2) the lowest voltage.
If no battery is functional, the system is not functional and thus has 0 voltage. Write a program that prints the expected value and the variance of the voltage of the system, for each situation, and estimate these values by simulations.

Solution 3.7.20: For every $i \in \{1, \ldots, n\}$, let V_i be the random variable that is equal to the voltage of the ith battery (i.e., V_i equals v_i, if the battery is working, and equals 0, if the battery is not working). Also, let X be the voltage of the system.

Now, let $\{v_{i_1}, \ldots, v_{i_m}\}$ be the set of distinct voltages of the batteries, $m \in \mathbb{N}^*$, $m \leq n$. The only possible values of X, in each situation, are $v_{i_1}, \ldots, v_{i_m}$ and 0. Let $j \in \{1, \ldots, m\}$. We denote

$$I_j = \{i \in \{1, \ldots, n\} : v_i > v_{i_j}\},$$
$$S_j = \{i \in \{1, \ldots, n\} : v_i < v_{i_j}\},$$
$$K_j = \{k \in \{1, \ldots, n\} : v_k = v_{i_j}\}.$$

Also, let

$$P_j = \prod_{i \in I_j}(1 - p_i), \text{ if } I_j \neq \emptyset, \ P_j = 1, \text{ if } I_j = \emptyset,$$

$$R_j = \prod_{i \in S_j}(1 - p_i), \text{ if } S_j \neq \emptyset, \ R_j = 1, \text{ if } S_j = \emptyset,$$

$$Q_j = P\left(\bigcup_{k \in K_j} \{V_k = v_{i_j}\} \right).$$

The value of Q_j can be computed by using the inclusion-exclusion principle (see Theorem 1.5) and the independence of the batteries. The following function in Matlab, performs recursively the desired computations.

```
function Q=incl_excl(p)
if length(p)==1 Q=p;
else
    q=incl_excl(p(1:end-1)); Q=q+p(end)-q*p(end);
end
end
```

Note that P_j is the probability that all batteries with voltage strictly higher than v_{i_j} do not work, R_j is the probability that all batteries with voltage strictly lower than v_{i_j} do not work and Q_j is the probability that at least one battery with voltage v_{i_j} works.

1) In this case, $X = \max\{V_1, \ldots, V_n\}$. Hence,

$$P(X = v_{i_j}) = P\big(\max\{V_1, \ldots, V_n\} = v_{i_j}\big) = Q_j \cdot P_j,$$

since $\max\{V_1, \ldots, V_n\} = v_{i_j}$ occurs if and only if all the batteries with voltages strictly higher than v_{i_j} do not work and at least one battery with voltage v_{i_j} works. So, $E(X) = \displaystyle\sum_{j=1}^{m} v_{i_j} Q_j P_j$ and

$$V(X) = \sum_{j=1}^{m} v_{i_j}^2 Q_j P_j - \left(\sum_{l=1}^{m} v_{i_j} Q_j P_j \right)^2.$$

2) In this case,

$$X = \begin{cases} 0, & \text{if } V_1 = \ldots = V_n = 0 \\ \min\big\{V_i : V_i \neq 0, i \in \{1, \ldots, n\}\big\}, & \text{otherwise.} \end{cases}$$

Hence,

$$P(X = v_{i_j}) = P\big(\min\{V_1, \ldots, V_n\} = v_{i_j}\big) = Q_j \cdot R_j,$$

since $\min\{V_1, \ldots, V_n\}$ occurs if and only if all the batteries with voltages strictly lower than v_{i_j} do not work and at least one battery with voltage v_{i_j} works. So, $E(X) = \displaystyle\sum_{j=1}^{m} v_{i_j} Q_j R_j$ and

$$V(X) = \sum_{j=1}^{m} v_{i_j}^2 Q_j R_j - \left(\sum_{l=1}^{m} v_{i_j} Q_j R_j \right)^2.$$

```
function voltages_simul(p,v,N)
w=unique(v);m=length(w);
P=ones(m,1);Q=P;R=P;
for j=1:m
    P(j)=prod(1-p(v>w(j)));
    R(j)=prod(1-p(v<w(j)));
    Q(j)=incl_excl(p(v==w(j)));
end
V=v.*binornd(ones(N,length(p)),repmat(p,N,1));
X=max(V,[ ],2);
fprintf('The estimated expected value of the maximum voltage is %4.3f.\n',mean(X));
fprintf('The estimated variance of the maximum voltage is %4.3f.\n',var(X,1));
```

fprintf('The expected value of the maximum voltage is %4.3f.\n',w*(Q.*P));
fprintf('The variance of the maximum voltage is %4.3f.\n',w.^2*(Q.*P)-(w*(Q.*P))^2);
fprintf('\n');
V(V==0)=**NaN**; X=**min**(V,[ ],2); X(**isnan**(X))=0;
fprintf(['The estimated expected value of the minimum voltage is %4.3f.\n',**mean**(X));
fprintf('The estimated variance of the minimum voltage is %4.3f.\n',**var**(X,1));
fprintf('The expected value of the minimum voltage is %4.3f.\n',w*(Q.*R));
fprintf('The variance of the minimum voltage is %4.3f.\n',w.^2*(Q.*R)-(w*(Q.*R))^2);
end

>>n=10;p=**rand**(1,n);v=**unidrnd**(5,1,n);N=1000;
>>voltages_simul(p,v,N)
The estimated expected value of the maximum voltage is 4.895.
The estimated variance of the maximum voltage is 0.168
The expected value of the maximum voltage is 4.890.
The variance of the maximum voltage is 0.167.

The estimated expected value of the minimum voltage is 1.161.
The estimated variance of the minimum voltage is 0.145.
The expected value of the minimum voltage is 1.162.
The variance of the minimum voltage is 0.149.

Problem 3.7.21. Let $p \in (0,1)$ and $\mu > 0$. An incoming email is sent either to a server A with probability p or to a server B with probability $1 - p$. The number Z of incoming emails (in the time interval of 1 hour) is modeled by a Poisson distribution with parameter μ. Determine the expectation of the difference between the number of emails received by server A and the number of emails received by server B.

Solution 3.7.21: Let X be the number of emails received by server A and Y be the number of emails received by server B. We clearly have $Z = X + Y$. For $n \in \mathbb{N}$ we have

$$P(X = k|Z = n) = \begin{cases} C(n,k)p^k(1-p)^{n-k}, & \text{if } k \in \{0,\dots,n\} \\ 0, & \text{otherwise.} \end{cases}$$

By the law of total probability (see Theorem 1.8), we deduce that

$$P(X = k) = \sum_{n=0}^{\infty} P(X = k|Z = n) \cdot P(Z = n)$$

$$= \sum_{n=k}^{\infty} C(n,k)p^k(1-p)^{n-k}\frac{e^{-\mu}\mu^n}{n!} = \frac{(\mu p)^k}{k!}e^{-\mu p},$$

for all $k \in \mathbb{N}$. Hence, X has a Poisson distribution with parameter μp. Its expectation is μp (see Problem 3.5.2-1)). We have to determine $E(X - Y)$. We have

$$E(X - Y) = E(X - (Z - X)) = 2E(X) - E(Z) = \mu(2p - 1).$$

Problem 3.7.22. Let $X_1 \sim Unif[0,1]$ and $X_2 \sim Exp(1)$ be independent. Let $Y_1 = \min\{X_1, X_2\}$ and $Y_2 = \max\{X_1, X_2\}$. Find $E(Y_2|Y_1)$.

Solution 3.7.22: Since $X_1 \sim Unif[0,1]$ and $X_2 \sim Exp(1)$ are independent, the function

$$f_{X_1,X_2}(x_1, x_2) = \begin{cases} e^{-x_2}, & \text{if } x_1 \in [0,1], x_2 > 0 \\ 0, & \text{otherwise} \end{cases}$$

is a joint density function of (X_1, X_2). By Example 2.25-(3), the following

$$f_{Y_1,Y_2}(y_1, y_2 | X_1 < X_2) = \begin{cases} \dfrac{f_{X_1,X_2}(y_1, y_2)}{P(X_1 < X_2)}, & \text{if } y_1 \in (0,1), y_1 < y_2 \\ 0, & \text{otherwise} \end{cases}$$

$$= \begin{cases} \dfrac{e^{-y_2}}{P(X_1 < X_2)}, & \text{if } y_1 \in (0,1), y_1 < y_2 \\ 0, & \text{otherwise} \end{cases}$$

and

$$f_{Y_1,Y_2}(y_1, y_2 | X_2 < X_1) = \begin{cases} \dfrac{f_{X_1,X_2}(y_2, y_1)}{P(X_2 < X_1)}, & \text{if } 0 < y_1 < y_2 < 1 \\ 0, & \text{otherwise} \end{cases}$$

$$= \begin{cases} \dfrac{e^{-y_1}}{P(X_2 < X_1)}, & \text{if } 0 < y_1 < y_2 < 1 \\ 0, & \text{otherwise} \end{cases}$$

are conditional density functions. By Example 2.25-(2), we have the following joint density function of (Y_1, Y_2)

$$f_{Y_1,Y_2}(y_1, y_2) = \begin{cases} e^{-y_2}, & \text{if } y_1 \in (0,1), y_1 < y_2 \\ 0, & \text{otherwise} \end{cases}$$

$$+ \begin{cases} e^{-y_1}, & \text{if } 0 < y_1 < y_2 < 1 \\ 0, & \text{otherwise.} \end{cases}$$

So, the following function is a density function of Y_1

$$f_{Y_1}(y_1) = \int_{\mathbb{R}} f_{Y_1,Y_2}(y_1, y_2) dy_2 = \int_{y_1}^{\infty} e^{-y_2} dy_2 + \int_{y_1}^{1} e^{-y_1} dy_2 = e^{-y_1}(2 - y_1),$$

if $y_1 \in (0,1)$, and $f_{Y_1}(y_1) = 0$, otherwise. For $y_1 \in (0,1)$ we note that $f_{Y_1}(y_1) > 0$ and

$$E(Y_2|Y_1 = y_1) = \int_{\mathbb{R}} y_2 f_{Y_2|Y_1}(y_2|y_1) dy_2$$

$$= \frac{1}{e^{-y_1}(2-y_1)} \int_{y_1}^{\infty} y_2 e^{-y_2} dy_2 + \frac{1}{2-y_1} \int_{y_1}^{1} y_2 dy_2$$

$$= \frac{1+y_1}{2-y_1} + \frac{1-y_1^2}{2(2-y_1)} = \frac{(y_1+1)(3-y_1)}{2(2-y_1)}$$

and thus, by Example 3.6, we have

$$E(Y_2|Y_1) = \frac{(Y_1+1)(3-Y_1)}{2(2-Y_1)} \text{ a.s.}$$

Note that $E(Y_2|Y_1)$ can be found also by using similar arguments to those in the solution of Problem 3.3.4.

Problem 3.7.23. Let $A = \begin{pmatrix} 1 & -1 \\ -1 & 2 \end{pmatrix}$ and let $\mathbb{X} = (X_1, X_2) \sim MVN(0_2, A)$. Prove that $E(X_1|X_2) = -\frac{1}{2}X_2$ a.s.

Solution 3.7.23: The function

$$f_{\mathbb{X}}(\mathbb{x}) = \frac{1}{2\pi} \exp\left\{-\frac{1}{2}(2x_1^2 + x_2^2 + 2x_1x_2)\right\}, \quad \mathbb{x} = (x_1, x_2)^T \in \mathbb{R}^2$$

is a joint density function of $\mathbb{X}$. Hence,

$$f_{X_2}(x_2) = \int_{-\infty}^{\infty} \frac{1}{2\pi} \exp\left\{-\frac{1}{2}(2x_1^2 + x_2^2 + 2x_1x_2)\right\} dx_1 = \frac{1}{2\sqrt{\pi}} \exp\left\{-\frac{x_2^2}{4}\right\}$$

for $x_2 \in \mathbb{R}$ is a density function of X_2. So, for $x_2 \in \mathbb{R}$ we have

$$f_{X_1|X_2}(x_1|x_2) = \frac{1}{\sqrt{\pi}} \exp\left\{-\left(x_1 + \frac{x_2}{2}\right)^2\right\}, \quad x_1 \in \mathbb{R}.$$

Therefore,

$$E(X_1|X_2 = x_2) = \int_{-\infty}^{\infty} \frac{x_1}{\sqrt{\pi}} \exp\left\{-\left(x_1 + \frac{x_2}{2}\right)^2\right\} dx_1 = -\frac{x_2}{2}, \ x_2 \in \mathbb{R},$$

and thus, by Example 3.6, $E(X_1|X_2) = -\frac{1}{2}X_2$ a.s.

Problem 3.7.24. Let $X \sim Exp(\alpha)$ and $Y \sim Exp(\beta)$ be independent random variables, where $\alpha, \beta > 0$.

1) Prove that $E(X + Y|X - Y) = |X - Y| + \frac{2}{\alpha + \beta}$ a.s.

2) Find $E(|X - Y|)$.

Solution 3.7.24: Let $B = \begin{pmatrix} 1 & 1 \\ 1 & -1 \end{pmatrix}$ and $U = X + Y$, $V = X - Y$. We note that $B^{-1} = \dfrac{1}{2}B$ and $(U, V) = (X, Y) \cdot B$. By Example 2.14, we have a joint density function of (U, V) given by

$$f_{U,V}(u, v) = \frac{1}{|\det B|} f_{X,Y}\big((u, v) \cdot B^{-1}\big) = \frac{1}{2} f_X \Big(\frac{u + v}{2}\Big) f_Y \Big(\frac{u - v}{2}\Big)$$

$$= \begin{cases} \dfrac{1}{2} \exp\left\{ -\dfrac{(\alpha + \beta)u}{2} - \dfrac{(\alpha - \beta)v}{2} \right\}, & \text{if } u > |v| \\ 0, & \text{otherwise,} \end{cases}$$

where f_X, f_Y are density functions of X, respectively Y. Hence, we have a density function of V given by for $v \in \mathbb{R}$

$$f_V(v) = \int_{|v|}^{\infty} \frac{1}{2} \exp\left\{ -\frac{(\alpha + \beta)u}{2} - \frac{(\alpha - \beta)v}{2} \right\} du$$

$$= \frac{1}{\alpha + \beta} \exp\left\{ -\frac{(\alpha + \beta)|v|}{2} - \frac{(\alpha - \beta)v}{2} \right\}$$

and thus

$$f_{U|V}(u|v) = \begin{cases} \dfrac{\alpha + \beta}{2} \exp\left\{ \dfrac{(\alpha + \beta)(|v| - u)}{2} \right\}, & \text{if } u > |v| \\ 0, & \text{otherwise.} \end{cases}$$

So, for $v \in \mathbb{R}$

$$E(U|V = v) = \int_{|v|}^{\infty} \frac{\alpha + \beta}{2} u \exp\left\{ \frac{(\alpha + \beta)(|v| - u)}{2} \right\} du = |v| + \frac{2}{\alpha + \beta}.$$

Therefore, $E(X + Y|X - Y) = |X - Y| + \dfrac{2}{\alpha + \beta}$ a.s. and

$$\frac{1}{\alpha} + \frac{1}{\beta} = E(X) + E(Y) = E(X + Y)$$

$$= E\big(E(X + Y|X - Y)\big) = E(|X - Y|) + \frac{2}{\alpha + \beta},$$

which implies $E(|X - Y|) = \dfrac{1}{\alpha + \beta} \Big(\dfrac{\alpha}{\beta} + \dfrac{\beta}{\alpha}\Big)$.

Problem 3.7.25. Let $n \in \mathbb{N}^*$ and $X \sim Unif[1, n + 1]$. Prove that $E(\lfloor X \rfloor | X) = \lfloor X \rfloor$ a.s. and $E(X|\lfloor X \rfloor) = \lfloor X \rfloor + \dfrac{1}{2}$ a.s., where $\lfloor \cdot \rfloor$ is the floor function.

Solution 3.7.25 Let $(\Omega, \mathcal{F}, P)$ be the probability space on which X is defined. Since

$$\lfloor X \rfloor = \sum_{k=1}^{n} k \cdot \mathbb{I}_{X^{-1}([k,k+1))},$$

we have (see Examples 2.2 and 2.5) that $\lfloor X \rfloor$ is a discrete random variable on $(\Omega, \mathcal{F}, P)$. Clearly, the σ-field $\mathcal{F}_X$ generated by X is contained in $\mathcal{F}$ and the σ-field $\mathcal{F}_{\lfloor X \rfloor}$ generated by $\lfloor X \rfloor$ is contained in $\mathcal{F}_X$, because $X^{-1}([k, k+1)) \in \mathcal{F}_X$ for $k \in \{1, \ldots, n\}$. By Theorem 3.3-(4), we have a.s. $E(\lfloor X \rfloor \,|\, X) = \lfloor X \rfloor$. By Example 3.5-(1), we have

$$E(X - \lfloor X \rfloor \,|\, \lfloor X \rfloor) = \sum_{k=1}^{n} E(X - \lfloor X \rfloor \,|\, \lfloor X \rfloor = k) \mathbb{I}_{X^{-1}([k,k+1))}.$$

By Definition 3.6, for every $k \in \{1, \ldots, n\}$ we have

$$E(X - \lfloor X \rfloor \,|\, \lfloor X \rfloor = k) = \frac{E\Big((X - \lfloor X \rfloor) \cdot \mathbb{I}_{\{X \in [k,k+1)\}}\Big)}{P(X \in [k, k+1))}$$

$$= n \int_{k}^{k+1} \frac{x - k}{n} dx = \frac{1}{2}.$$

Hence, $E(X - \lfloor X \rfloor \,|\, \lfloor X \rfloor) = \dfrac{1}{2}$ a.s. and, by Theorem 3.3-(4)-(5), we deduce the desired relation.

Problem 3.7.26. We consider a random experiment in which the events A and $\bar{A}$ occur with probabilities $P(A) = p$ and $P(\bar{A}) = 1 - p$, $p \in (0, 1)$. A negative binomial distributed random variable X can also be described by $X = \sum_{j=1}^{n} X_j$, where $X_1, \ldots, X_n \sim Geo(p)$, $n \in \mathbb{N}^*$, are independent random variables. This relation is clear, since X_j is the number of occurrences of the event $\bar{A}$ before the event A occurs the first time, while $X_1 + \ldots + X_j$ is the number of occurrences of $\bar{A}$ before the event A occurs the jth time, where $j \in \{1, \ldots, n\}$.

1) Find the moment generating functions for X_j, $j \in \{1, \ldots, n\}$, and X.
2) Compute the expectation and the variance of a random variable X having negative binomial distribution.

Solution 3.7.26: 1) Let $j \in \{1, \ldots, n\}$. By the definition of a geometric distribution (see Section 2.4.9) we get for the moment generating function

$$M_{X_j}(t) = \sum_{k=0}^{\infty} e^{tk}(1-p)^k p = p \sum_{k=0}^{\infty} ((1-p)e^t)^k = \frac{p}{1 - (1-p)e^t},$$

if $(1 - p)e^t < 1$, that is, $t < -\log(1 - p)$. Obviously, M_{X_j} is defined in a neighborhood of 0. It follows, from Theorem 3.9, that

$$M_X(t) = \left(\frac{p}{1 - (1 - p)e^t} \right)^n \quad \text{for } t < -\log(1 - p).$$

2) We use Theorem 3.10 and get

$$E(X) = M_X'(0) = \frac{np^n(1 - p)e^t}{(1 - (1 - p)e^t)^{n+1}} \Bigg|_{t=0} = \frac{n(1 - p)}{p}$$

and by elementary calculations

$$E(X^2) = M_X''(0) = \frac{n(1 - p)(n - np + 1)}{p^2}.$$

Consequently, we get the variance by

$$V(X) = E(X^2) - E^2(X) = \frac{n(1 - p)}{p^2}.$$

Problem 3.7.27. Let $X \sim NBin(n_1, p)$ and $Y \sim NBin(n_2, p)$ be independent random variables, $n_1, n_2 \in \mathbb{N}^*$ and $p \in (0, 1)$. Prove that $X + Y \sim NBin(n_1 + n_2, p)$.

Solution 3.7.27: Since X and Y are independent we get with Theorem 3.9 and Problem 3.7.26

$$M_{X+Y}(t) = M_X(t)M_Y(t) = \left(\frac{p}{1 - (1 - p)e^t} \right)^{n_1} \cdot \left(\frac{p}{1 - (1 - p)e^t} \right)^{n_2}$$

$$= \left(\frac{p}{1 - (1 - p)e^t} \right)^{n_1 + n_2}$$

for $t < -\log(1 - p)$. We obtained the moment generating function of the $NBin(n_1 + n_2, p)$ distribution and, by Remark 3.5, it follows that

$$X + Y \sim NBin(n_1 + n_2, p).$$

Problem 3.7.28. The random variable X has a moment generating function that satisfies

$$M_X(t) = \left(\frac{1}{3 - 2e^t} \right)^3 \quad \text{for } t \in (-\delta, \delta),$$

where $\delta > 0$ is fixed. Compute $P(X > 1)$.

Solution 3.7.28: Observe that

$$\frac{1}{3 - 2e^t} = \sum_{k=0}^{\infty} \frac{1}{3} \left(\frac{2}{3}\right)^k e^{kt} \text{ for } t < \log\left(\frac{3}{2}\right).$$

This is the moment generating function of a random variable having geometric distribution $Geo\left(\frac{1}{3}\right)$. By adding up 3 independent random variables, each one being $Geo\left(\frac{1}{3}\right)$ distributed, we obtain a random variable Y having $NBin\left(3, \frac{1}{3}\right)$ distribution and

$$M_Y(t) = \left(\frac{1}{3 - 2e^t}\right)^3 \text{ for } t < \log\left(\frac{3}{2}\right).$$

Since, $0 \in \left(-\infty, \log\left(\frac{3}{2}\right)\right)$, it follows, by Theorem 3.11, that X and Y have the same distribution. Hence, $X \sim NBin\left(3, \frac{1}{3}\right)$ and

$$P(X > 1) = 1 - P(X = 0) - P(X = 1) = \frac{8}{9}.$$

Problem 3.7.29. 1) Let X and Y be independent random variables such that $X \sim Unif[a, b]$ and Y is a discrete random variable with

$$P(Y = a) = P(Y = b) = 0.5,$$

where $a < b$. Prove that $X + Y \sim Unif[2a, 2b]$.
2) Let $X \sim Unif[0, 1]$ and $Y \sim Unid(n)$ be independent random variables, where $n \in \mathbb{N}^*$. Prove that $X + Y \sim Unif[1, n + 1]$.

Solution 3.7.29: 1) In view of Example 3.16-(5), we have

$$M_X(t) = \begin{cases} \dfrac{e^{tb} - e^{ta}}{t(b - a)}, & \text{if } t \neq 0 \\ 1, & \text{if } t = 0. \end{cases}$$

The moment generating function of Y is

$$M_Y(t) = \frac{e^{tb} + e^{ta}}{2}, \quad t \in \mathbb{R}.$$

Since X and Y are independent, we have

$$M_{X+Y}(t) = M_X(t) \cdot M_Y(t) = \begin{cases} \dfrac{e^{2tb} - e^{2ta}}{2t(b - a)}, & \text{if } t \neq 0 \\ 1, & \text{if } t = 0. \end{cases}$$

In view of the uniqueness of the moment generating function (see Theorem 3.11) and the expression of the moment generating function of a continuous uniformly distributed random variable, we deduce that $X + Y \sim Unif[2a, 2b]$.

2) We have

$$M_X(t) = \begin{cases} \dfrac{e^t - 1}{t}, & \text{if } t \neq 0 \\ 1, & \text{if } t = 0. \end{cases}$$

We determine the moment generating function of Y:

$$M_Y(t) = \begin{cases} \dfrac{e^{t(n+1)} - e^t}{n(e^t - 1)}, & \text{if } t \neq 0 \\ 1, & \text{if } t = 0. \end{cases}$$

Hence,

$$M_{X+Y}(t) = M_X(t) \cdot M_Y(t) = \begin{cases} \dfrac{e^{t(n+1)} - e^t}{nt}, & \text{if } t \neq 0 \\ 1, & \text{if } t = 0, \end{cases}$$

and thus $X + Y \sim Unif[1, n + 1]$.

Problem 3.7.30. Let X be a discrete random variable such that

$$P(X = n^2) = \frac{1}{2^n} \text{ for } n \in \mathbb{N}^*.$$

1) Compute the expectation of X.

2) Show that there is no $\delta > 0$ such that the moment generating function of X is defined on $(0, \delta)$.

Solution 3.7.30: 1) It holds $\displaystyle\sum_{n=1}^{\infty} \frac{n^2}{2^n} = \sum_{n=1}^{\infty} n^2 x^n \Big|_{x = \frac{1}{2}}$. For $|x| < 1$ define

$G(x) = \dfrac{x}{1 - x}$. By the property of the geometric series we have for $|x| < 1$

$G(x) = \displaystyle\sum_{n=1}^{\infty} x^n$ and by [Schinazi (2012), Power series and differentiation, p. 216] we have

$$\sum_{n=1}^{\infty} n^2 x^n = x(xG'(x))' \quad \text{for } |x| < 1.$$

Then for $x = \dfrac{1}{2}$ we get $\displaystyle\sum_{n=1}^{\infty} \frac{n^2}{2^n} = 6$. Therefore, $E(X) = 6$.

2) Let $\delta > 0$ and $t \in (0, \delta)$. Because $\displaystyle\lim_{n \to \infty} \frac{e^{tn^2}}{2^n} = \infty$, we have $\displaystyle\sum_{n=1}^{\infty} \frac{e^{tn^2}}{2^n} = \infty$.

which implies that M_X is not defined on $(0, \delta)$. Note that M_X is defined on $(-\infty, 0]$, because

$$M_X(t) \leq \sum_{n=1}^{\infty} \frac{1}{2^n} = 1, \ t \leq 0.$$

Problem 3.7.31. Let $X_1, X_2, \ldots, X_m$ $(m \in \mathbb{N}^*)$ be independent random variables and denote $S_m = X_1 + \cdots + X_m$. Prove that:
1) if $X_k \in Gamma(a_k, b)$, $a_k > 0$, $k \in \{1, \ldots, m\}$, $b > 0$, then

$$S_m \sim Gamma\left(a_1 + \cdots + a_m, b\right);$$

2) if $X_k \in \chi^2(n_k, \sigma)$, $n_k \in \mathbb{N}^*$, $k \in \{1, \ldots, m\}$, $\sigma > 0$, then

$$S_m \sim \chi^2\left(n_1 + \cdots + n_m, \sigma\right);$$

3) if $X_k \in Exp(\lambda)$, $k \in \{1, \ldots, m\}$, $\lambda > 0$, then $S_m \sim Gamma\left(m, \frac{1}{\lambda}\right)$.

Solution 3.7.31: We use Example 3.16 and Theorem 3.11:
1) For each $t < \frac{1}{b}$ it holds

$$M_{S_m}(t) = M_{X_1}(t) \cdot \ldots \cdot M_{X_m}(t) = \frac{1}{(1 - bt)^{a_1 + \cdots + a_m}}.$$

Then $S_m \sim Gamma\left(a_1 + \cdots + a_m, b\right)$.
2) We note that $Gamma\left(\frac{n_k}{2}, 2\sigma^2\right) = \chi^2(n_k, \sigma)$, $k \in \{1, \ldots, m\}$, and we use 1).
3) We note that $Exp(\lambda) = Gamma\left(1, \frac{1}{\lambda}\right)$ and we use 1).

Problem 3.7.32. Let $X \sim Gamma(a_1, b)$ and $Y \sim Gamma(a_2, b)$ be independent random variables, where $a_1, a_2, b > 0$. Prove that $\frac{X}{X + Y} \sim Beta(a_1, a_2)$ and that $\frac{X}{X + Y}$ is independent of $X + Y$, which has $Gamma(a_1 + a_2, b)$ distribution.

Solution 3.7.32: Observe that $X > 0$ and $Y > 0$ a.s. Denote $U = \frac{X}{X + Y}$ and $V = X + Y$. Hence $U \in (0, 1)$ and $V > 0$ a.s. We write the distribution function of U

$$F_U(u) = P\left(U \leq u\right) = \begin{cases} 0, & \text{if } u < 0 \\ P\left(\frac{X}{Y} \leq \frac{u}{1-u}\right), & \text{if } 0 \leq u < 1 \\ 1, & \text{if } 1 \leq u, \end{cases}$$

which implies that

$$
f_U(u) = \begin{cases} f_{\frac{X}{Y}}\left(\dfrac{u}{1-u}\right) \cdot \dfrac{1}{(1-u)^2}, & \text{if } u \in [0,1) \\ 0, & \text{if } u \notin [0,1) \end{cases}
$$

is a density function of U. Consider f_X and f_Y to be density functions of the independent random variables X and Y. Then we can write a density function of $\dfrac{X}{Y}$ for $y > 0$ by (see Example 2.15)

$$
\begin{aligned}
f_{\frac{X}{Y}}(y) &= \int_{\mathbb{R}} |v| f_X(yv) f_Y(v) dv \\
&= \frac{y^{a_1-1}}{\Gamma(a_1)\Gamma(a_2)b^{a_1+a_2}} \int_0^\infty v^{a_1+a_2-1} \exp\left\{ -\frac{(y+1)v}{b} \right\} dv \\
&= \frac{y^{a_1-1}}{(y+1)^{a_1+a_2}} \cdot \frac{1}{\Gamma(a_1)\Gamma(a_2)} \int_0^\infty t^{a_1+a_2-1} e^{-t} dt \\
&= \frac{y^{a_1-1}}{(y+1)^{a_1+a_2}} \cdot \frac{\Gamma(a_1+a_2)}{\Gamma(a_1)\Gamma(a_2)}.
\end{aligned}
$$

By the properties of the Beta function, we can write a density function of U by

$$
f_U(u) = \frac{1}{B(a_1,a_2)} \cdot u^{a_1-1} \cdot (1-u)^{a_2-1} \text{ for } u \in [0,1)
$$

and $f_U(u) = 0$ for $u \notin [0,1)$. Therefore, $U \sim Beta(a_1,a_2)$.

Since X and Y are independent, we deduce by Problem 3.7.31-1) that $V = X + Y \sim Gamma(a_1+a_2, b)$. The function

$$
f_V(y) = \begin{cases} 0, & \text{if } v \le 0 \\ \dfrac{v^{a_1+a_2-1}}{\Gamma(a_1+a_2)b^{a_1+a_2}} \exp\left\{ -\dfrac{v}{b} \right\}, & \text{if } v > 0 \end{cases}
$$

is a density function of V.

We have $X = UV, Y = (1-U)V$ and, denoting

$$
h \colon (0,1) \times (0,\infty) \to (0,\infty) \times (0,\infty)
$$
$$
h(u,v) = \big(uv, (1-u)v\big), u \in (0,1), v > 0,
$$

we see that $(U,V) = g(X,Y)$, where $g = h^{-1}$ (note that h is bijective). Let J_h be the Jacobian matrix of h. Then $\det(J_h(u,v)) = v$. By Theorem 2.14, we have a joint density function of (U,V) given by

$$
f_{U,V}(u,v) = f_{X,Y}\big(h(u,v)\big)|\det(J_h(u,v))| = f_{X,Y}\big(uv, (1-u)v\big)v
$$

if $u \in (0,1)$, $v > 0$, and $f_{U,V}(u,v) = 0$, otherwise; a joint density function of (X,Y) is given for $x, y > 0$ by

$$f_{X,Y}(x,y) = f_X(x) f_Y(y) = \frac{x^{a_1-1} y^{a_2-1}}{\Gamma(a_1)\Gamma(a_2) b^{a_1+a_2}} \exp\left\{ -\frac{x+y}{b} \right\}.$$

This implies

$$f_{U,V}(u,v) = \underbrace{\frac{u^{a_1-1}(1-u)^{a_2-1} \Gamma(a_1+a_2)}{\Gamma(a_1)\Gamma(a_2)}}_{=f_U(u)} \cdot \underbrace{\frac{v^{a_1+a_2-1}}{\Gamma(a_1+a_2) b^{a_1+a_2}} \exp\left\{ -\frac{v}{b} \right\}}_{=f_V(v)}$$

for $u \in (0,1)$ and $v > 0$. If $u \notin (0,1)$ or $v \le 0$, obviously $f_{U,V}(u,v) = 0 = f_U(u) f_V(v)$. Hence, U and V are independent random variables.

Problem 3.7.33. 1) Let $m, n \in \mathbb{N}^*$ such that $m < n$ and $\lambda > 0$. John Doe works in an insurance company. Suppose that one day the company serves n clients and m of these clients are served by John Doe. The time intervals (in minutes) spent by the clients in the company are independent and follow the $Exp(\lambda)$ distribution. Find the distribution of the total amount of time spent by the clients in the company and the distribution of the proportion of time spent by the clients served by John Doe. Prove that this proportion of time is independent of the total amount of time.

2) For $\lambda \in \{15, 20, 25\}$ simulate in Matlab 500 days in the insurance company, as described at 1), with the same number of clients given by the input values m and n. Plot on the same figure the empirical cumulative distribution function and the cumulative distribution function for the proportion of time spent by the clients served by John Doe.

Solution 3.7.33: 1) Let X be the amount of time spent by the clients served by John Doe and Y be the amount of time spent by the clients served by the other employees of the company. In view of Problem 3.7.31-3), we have $X \sim Gamma\left(m, \frac{1}{\lambda}\right)$ and $Y \sim Gamma\left(n - m, \frac{1}{\lambda}\right)$. Also, we note that $X + Y$ represents the total amount of time spent by the clients in the company and $\frac{X}{X+Y}$ represents the proportion of time spent by the clients served by John Doe. From Problem 3.7.32 we have that $X + Y \sim Gamma\left(n, \frac{1}{\lambda}\right)$ and $\frac{X}{X+Y} \sim Beta(m, n - m)$ and these two random variables are independent. Moreover, we note that the latter does not depend on λ.

2)

```
function timecomp(m,n,lambda)
times=exprnd(1/lambda,n,500);
X=sum(times(1:m,:)); Y=sum(times(m+1:n,:));
clf; hold on; axis([0 1 0 1]); grid on;
ecdf(X./(X+Y)); fplot(@(t) betacdf(t,m,n-m),[0,1]);
legend('ECDF','CDF');
end

>> timecomp(5,30,20);
```

Problem 3.7.34. 1) Let $X \sim N(-1,1)$ and $Y \sim Poiss(1)$ be independent random variables. Let $Z = X + Y$ and for every $z < 0$ denote by $\phi_z : (-\infty, 0) \to (0, \infty)$ the function given by $\phi_z(t) = e^{-tz} M_Z(t)$, $t < 0$. Prove for every $z < 0$ that ϕ_z has a global minimum point in $(-\infty, 0)$ and that the cumulative distribution function of Z satisfies

$$F_Z(z) \leq \min_{t<0} \phi_z(t).$$

2) Generate, in Matlab, 1000 numbers that follow the distribution of Z and plot, on the same figure, the empirical cumulative distribution function of these numbers and the function $z \mapsto \min_{t<0} \phi_z(t)$ for $z < 0$.

Solution 3.7.34: 1) Since $M_X(t) = \exp\left\{-t + \frac{t^2}{2}\right\}$, $t \in \mathbb{R}$ (see Problem 3.5.4), and $M_Y(t) = \exp\{e^t - 1\}$, $t \in \mathbb{R}$ (see Problem 3.5.2-1)), we have

$$M_Z(t) = \exp\left\{e^t + \frac{t^2}{2} - t - 1\right\}, \quad t \in \mathbb{R}.$$

Let $z < 0$. In order to prove that ϕ_z has a global minimum point in $(-\infty, 0)$, it suffices to prove that the function given by

$$\varphi_z(t) = e^t + \frac{t^2}{2} - t - 1 - tz, \quad t < 0$$

has a global minimum point in $(-\infty, 0)$. We note that $\varphi_z'(t) = e^t + t - 1 - z$, $t < 0$. Hence, φ_z is convex and has a unique critical point. Thus, φ_z has a global minimum point in $(-\infty, 0)$. By Markov's inequality (see Theorem 3.6), for $t < 0$ we get $F_Z(z) = P(Z \leq z) = P(e^{tZ} \geq e^{tz}) \leq e^{-tz} M_Z(t)$, hence, $F_Z(z) \leq \min_{t<0} \phi_z(t)$. We mention that the right term of this inequality is called the Chernoff bound of the cumulative distribution function of Z. For more details, see [Mitzenmacher and Upfal (2005), Chapter 4].

2)

```
function chernoff(k)
X=normrnd(-1,1,1,k); Y=poissrnd(1,1,k);
Z=X+Y;
clf; hold on; axis equal; axis([min(Z) 0 0 1]);
ecdf(Z,'censoring',Z>=0)
z=min(Z):0.1:0; phi=zeros(size(z));
for i=1:length(z)
    t0=fzero(@(t) exp(t)+t-1-z(i),-1);
    phi(i)=exp(exp(t0)+t0^2/2-t0-1-z(i)*t0);
end
plot(z,phi); legend('Empirical CDF','Chernoff bound');
end
```

$>>$ chernoff(1000)

Problem 3.7.35. 1) Let $a < b$ and X be a random variable such that $E(X) = 0$ and $a \leq X \leq b$ a.s. Prove the following inequality

$$M_X(t) \leq \exp\left\{\frac{(b-a)^2t^2}{8}\right\}, \quad t \in \mathbb{R}.$$

This result is called Hoeffding's lemma (see [Hoeffding (1963)]).

2) Let $X \sim Bino\left(n, \dfrac{1}{n+1}\right)$ and $Y \sim Beta(n,1)$ be independent random variables and denote $Z = X - Y$, for $n \in \mathbb{N}^*$. Prove that

$$M_Z(t) \leq \exp\left\{\frac{(n+1)^2t^2}{8}\right\}, \quad t \in \mathbb{R}.$$

Generate, in Matlab, 1000 numbers that follow the distribution of Z, for some given $n \in \mathbb{N}^*$ as input, and plot on the same figure the empirical moment generating function of these numbers and the above function that bounds the moment generating function of Z, on the interval $[-1, 1]$.

Solution 3.7.35: 1) Let $t \in \mathbb{R}$. We have

$$e^{tx} \leq \frac{b-x}{b-a}e^{ta} + \frac{x-a}{b-a}e^{tb}, \quad x \in [a,b],$$

in view of the convexity of the function $x \mapsto e^{tx}$ on $[a, b]$. Hence, taking into account the above inequality and that X satisfies $E(X) = 0$ and $a \leq X \leq b$ a.s., we get by Theorem 3.1-(4)

$$M_X(t) \leq \frac{be^{ta} - ae^{tb}}{b-a}.$$

Let $g : \mathbb{R} \to \mathbb{R}$ be given by

$$g(y) = \log\left(\frac{be^{ya} - ae^{yb}}{b - a}\right), \quad y \in \mathbb{R}.$$

In view of the following

$$g(y) = ya + \log\left(b - a\exp\{y(b - a)\}\right) - \log(b - a), \quad y \in \mathbb{R},$$

we deduce that g is a well defined C^2 function that satisfies

$$g(0) = g'(0) = 0$$

and

$$g''(y) = (b - a)^2 \cdot \frac{b(-a)\exp\{y(b - a)\}}{\left(b - a\exp\{y(b - a)\}\right)^2} \leq (b - a)^2 \cdot \frac{1}{4}, \quad y \in \mathbb{R}.$$

Hence, applying Taylor's theorem, there exists ξ between 0 and t such that $g(t) = \dfrac{g''(\xi)}{2}t^2$ and thus $M_X(t) \leq \exp\{g(t)\} \leq \exp\left\{\dfrac{(b - a)^2 t^2}{8}\right\}$.

2) Using the properties of the Beta function given in Proposition A.2, it can be easily deduced that $E(Y) = \dfrac{n}{n + 1}$. Also, we have $E(X) = \dfrac{n}{n + 1}$ (see Example 3.10). Moreover, we have $0 \leq X \leq n$ and $0 \leq Y \leq 1$ a.s. and thus $-1 \leq X - Y \leq n$ a.s. By 1), we deduce the desired inequality.

```
function empmomgen(n)
X=binornd(n,1/(n+1),1,1000);
Y=betarnd(n,1,1,1000);
clf; hold on; grid on;
t=[-1:0.1:1]'; xlim([-1 1]);
plot(t,mean(exp(t*(X-Y)),2),'r');
plot(t,exp((n+1)*t.^2./8),'b');
legend('Empirical moment generating function','Hoeffding bound');
end
```

```
>> empmomgen(3)
```

Problem 3.7.36. 1) Let $X_1, \cdots, X_n$ be independent random variables such that $0 \leq X_i \leq 1$ a.s., $i \in \{1, \ldots, n\}$, $n \in \mathbb{N}^*$, and denote

$$\bar{X} = \frac{1}{n}(X_1 + \ldots + X_n).$$

Prove Hoeffding's inequality:

$$P(\bar{X} - E(\bar{X}) \geq t) \leq \exp\{-2nt^2\}, \quad t > 0.$$

2) Let $a, b > 0$ and $n \in \mathbb{N}^*$. Let $X_1, \cdots, X_n \sim Beta(a, b)$ be independent random variables and denote

$$\bar{X} = \frac{1}{n}(X_1 + \ldots + X_n).$$

Prove that the cumulative distribution function of $\bar{X}$ satisfies

$$F_{\bar{X}}(t) \geq 1 - \exp\left\{-2n\left(t - \frac{a}{a+b}\right)^2\right\} \quad \text{for } t > \frac{a}{a+b}.$$

Generate, in Matlab, 1000 numbers that follow the distribution of $\bar{X}$ and plot on the same figure the empirical cumulative distribution function of these numbers and the above function, which bounds the cumulative distribution function of $\bar{X}$, for some values of a, b and n.

Solution 3.7.36: 1) Let $t > 0$. Reasoning as in the solution of Problem 3.7.34-1) (with the help of Markov's inequality, see Theorem 3.6), we can obtain the following Chernoff bound

$$P\big(\bar{X} - E(\bar{X}) \geq t\big) \leq \min_{s>0} e^{-st} M_{\bar{X}-E(\bar{X})}(s).$$

Letting $a_i = -E(X_i)$ and $b_i = 1 - E(X_i)$, we have $a_i \leq X_i - E(X_i) \leq b_i$ a.s. and thus the Hoeffding lemma (see Problem 3.7.35-1)) yields

$$M_{X_i-E(X_i)}(s) \leq \exp\left\{\frac{(b_i - a_i)^2 s^2}{8}\right\} = \exp\left\{\frac{s^2}{8}\right\}, \quad s \in \mathbb{R},$$

for each $i \in \{1, \ldots, n\}$. Hence, using the properties of the moment generating function (see Theorem 3.9), we get

$$M_{\bar{X}-E(\bar{X})}(s) = \prod_{i=1}^{n} M_{X_i-E(X_i)}\left(\frac{s}{n}\right) \leq \exp\left\{\frac{s^2}{8n}\right\}, \quad s > 0.$$

Therefore,

$$P(\bar{X} - E(\bar{X}) \geq t) \leq \exp\left\{\min_{s>0}\left(\frac{s^2}{8n} - st\right)\right\} = \exp\{-2nt^2\}.$$

For more details regarding the Hoeffding inequality, see [Hoeffding (1963)]. 2) We note that $E(X_i) = \dfrac{B(a+1,b)}{B(a,b)} = \dfrac{a}{a+b}$, $i = \{1, \ldots, n\}$ (see the properties of the Beta function given in Proposition A.2). Hence, $E(\bar{X}) = \dfrac{a}{a+b}$. The Hoeffding inequality from 1) implies that

$$P\left(\bar{X} - \frac{a}{a+b} \geq t\right) \leq \exp\{-2nt^2\}, \quad t > 0.$$

So,

$$P\left(\bar{X} \geq t\right) \leq \exp\left\{-2n\left(t - \frac{a}{a+b}\right)^2\right\} \quad \text{for } t > \frac{a}{a+b}.$$

From here we can easily deduce the desired inequality.

```
function hoeffding(a,b,n)
X=mean(betarnd(a,b,n,1000),1);
clf; hold on; axis([0 1 0 1]); grid on;
ecdf(X);
t=0:0.01:1;
plot(t,1-exp(-2*n*(t-a/(a+b)).^2));
plot([a/(a+b) a/(a+b)],[0 1],'--k');
legend('Empirical cumulative distribution function','Hoeffding bound','x=a/(a+b)');
end
```

```
>> hoeffding(1,2,3)
```

Chapter 4

Sequences of Random Variables

4.1 Convergence of Sequences of Random Variables

In this chapter the random variables of a given sequence are studied on the same probability space $(\Omega, \mathcal{F}, P)$.

Definition 4.1. A sequence $(X_n)_{n \in \mathbb{N}}$ of random variables **converges in probability** to a random variable X, if

$$\lim_{n \to \infty} P\Big(|X_n - X| \le \varepsilon\Big) = 1 \quad \text{for every } \varepsilon > 0$$

or, equivalently, if

$$\lim_{n \to \infty} P\Big(|X_n - X| > \varepsilon\Big) = 0 \quad \text{for every } \varepsilon > 0.$$

This convergence is denoted by $X_n \xrightarrow{P} X$. ♦

Definition 4.2. A sequence $(X_n)_{n \in \mathbb{N}}$ of random variables converges **almost surely** (or **almost everywhere**, or **with probability 1**) to a random variable X, if

$$P\Big(\lim_{n \to \infty} X_n = X\Big) = 1.$$

This convergence is denoted by $X_n \xrightarrow{\text{a.s.}} X$. ♦

Definition 4.3. A sequence $(X_n)_{n \in \mathbb{N}}$ of random variables converges **in mean of order r** (with $r > 0$) to a random variable X, if

$$\lim_{n \to \infty} E\big(|X_n - X|^r\big) = 0.$$

This convergence is denoted by $X_n \xrightarrow{L^r} X$.

In the special case $r = 2$ this convergence is called **mean square convergence**. ♦

Definition 4.4. A sequence $(X_n)_{n\in\mathbb{N}}$ of random variables converges **in distribution** to a random variable X, if

$$\lim_{n\to\infty} F_{X_n}(x) = F_X(x)$$

at each continuity point x of F_X, where the above functions are the corresponding distribution functions of X_n, respectively X. This convergence is denoted by $X_n \xrightarrow{d} X$.   ◆

Remark 4.1. In the literature there exist results in which it is explicitly mentioned that the involved random variables are not necessarily defined on the same probability space, e.g., Skorohod's Theorem, see [Gut (2005), Theorem 13.1, p. 258].   ▼

Theorem 4.1. [Gut (2005), Proposition 1.2, p. 206] *Let $(X_n)_{n\in\mathbb{N}}$ be a sequence of random variables and let X be a random variable. Then the following statements are equivalent:*

(1) $X_n \xrightarrow{a.s.} X$.

(2) $\displaystyle\lim_{n\to\infty} P\Big(\bigcup_{k\geq n}\big\{|X_k - X| > \varepsilon\big\}\Big) = 0$ *for every $\varepsilon > 0$.*

(3) $\displaystyle\lim_{n\to\infty} P\Big(\sup_{k\geq n}|X_k - X| > \varepsilon\Big) = 0$ *for every $\varepsilon > 0$.*

(4) $\displaystyle P\Big(\limsup_{n\to\infty}\big\{|X_k - X| > \varepsilon\big\}\Big) = 0$ *for every $\varepsilon > 0$.*

The following theorem is a consequence of Theorem 4.1-(4) and the Borel–Cantelli Lemma (see Theorem 1.10).

Theorem 4.2. *Let $(X_n)_{n\in\mathbb{N}^*}$ be a sequence of random variables and let X be a random variable such that*

$$\sum_{n=1}^{\infty} P\big(|X_n - X| > \varepsilon\big) < \infty   \text{ for every } \varepsilon > 0.$$

Then $X_n \xrightarrow{a.s.} X$.

Theorem 4.3. [Hajek (2015), Proposition 2.7, p. 47] *Let $(X_n)_{n\in\mathbb{N}}$ be a sequence of random variables and let X be a random variable. Then the following implications are true:*

(1) *If $X_n \xrightarrow{a.s.} X$, then $X_n \xrightarrow{P} X$.*

(2) *If $r > 0$ and $X_n \xrightarrow{L^r} X$, then $X_n \xrightarrow{P} X$.*

(3) *If $X_n \xrightarrow{P} X$, then $X_n \xrightarrow{d} X$.*

Theorem 4.4. [Hajek (2015), Corollary 2.10, p. 53] *Let* $(X_n)_{n\in\mathbb{N}}$ *be a sequence of random variables and let* X *be a random variable such that* $X_n \xrightarrow{P} X$. *Then there exists a (deterministic) subsequence* $(n_k)_{k\in\mathbb{N}}$ *of indices such that* $X_{n_k} \xrightarrow{a.s.} X$.

Theorem 4.5. [Hajek (2015), Proposition 2.7, p. 47] *If* $X_n \xrightarrow{P} X$ *and* $X_n \xrightarrow{P} Y$, *then* $P(X = Y) = 1$. *The same property also holds in the case of almost sure convergence and convergence in mean of order* $r > 0$. *Moreover, if* $X_n \xrightarrow{d} X$ *and* $X_n \xrightarrow{d} Y$, *then* X *and* Y *have the same distribution.*

Theorem 4.6. [Van der Vaart (1998), Theorem 2.3, p. 7] *Let* $G : \mathbb{R}^2 \to \mathbb{R}$ *be a continuous function. Then*

$$X_n \xrightarrow{a.s.} X, \ Y_n \xrightarrow{a.s.} Y \implies G(X_n, Y_n) \xrightarrow{a.s.} G(X, Y),$$

while

$$X_n \xrightarrow{P} X, \ Y_n \xrightarrow{P} Y \implies G(X_n, Y_n) \xrightarrow{P} G(X, Y).$$

Theorem 4.7. [Gut (2005), Theorem 11.4, p. 249] *Let* $(X_n)_{n\in\mathbb{N}}$, $(Y_n)_{n\in\mathbb{N}}$ *be sequences of random variables such that* $X_n \xrightarrow{d} X$ *and* $Y_n \xrightarrow{P} c$, *where* X *is a random variable and* $c \in \mathbb{R}$ *is a constant. Then*

$$X_n + Y_n \xrightarrow{d} X + c, \quad X_n \cdot Y_n \xrightarrow{d} X_n \cdot c$$

and

$$\frac{X_n}{Y_n} \xrightarrow{d} \frac{X}{c}, \ \text{if } c \neq 0 \text{ and for each } n \in \mathbb{N} \ Y_n \neq 0 \ a.s.$$

Theorem 4.8. [Billingsley (1986), Corollary 1, p. 344] *Let* $I \subseteq \mathbb{R}$. *Let* $(X_n)_{n\in\mathbb{N}}$ *be a sequence of random variables and* X *be a random variable that take values in* I *a.s. and satisfy* $X_n \xrightarrow{d} X$. *If* $h : I \to \mathbb{R}$ *is a continuous function, then* $h(X_n) \xrightarrow{d} h(X)$.

Theorem 4.9. [Gut (2005), Theorem 9.5, p. 242] *Let* $(X_n)_{n\in\mathbb{N}}$ *be a sequence of random variables and let* X *be a random variable such that* $\lim_{n\to\infty} M_{X_n}(t) = M_X(t)$ *for all* t *in a neighborhood of zero on which the involved moment generating functions are well defined. Then* $X_n \xrightarrow{d} X$.

The following are the Cauchy criteria for convergence of random variables, see [Hajek (2015), Proposition 2.9, p. 52].

Theorem 4.10. *Let* $(X_n)_{n\in\mathbb{N}}$ *be a sequence of random variables.*
(1) $(X_n)_{n\in\mathbb{N}}$ *converges a.s. to some random variable if and only if*

$$P\Big(\lim_{m,n\to\infty} |X_m - X_n| = 0\Big) = 1.$$

(2) $(X_n)_{n\in\mathbb{N}}$ *converges in mean square to some random variable if and only if* $(X_n)_{n\in\mathbb{N}}$ *is a Cauchy sequence in the* L^2 *sense, meaning* $E(X_n^2) < \infty$ *for all* $n \in \mathbb{N}$ *and*

$$\lim_{m,n\to\infty} E\big((X_m - X_n)^2\big) = 0.$$

(3) $(X_n)_{n\in\mathbb{N}}$ *converges in probability to some random variable if and only if for every* $\varepsilon > 0$

$$\lim_{m,n\to\infty} P\big(|X_m - X_n| > \varepsilon\big) = 0.$$

Theorem 4.11. (Dominated Convergence Theorem) [Gut (2005), Theorem 5.3, p. 57] *Let* $(X_n)_{n\in\mathbb{N}}$ *be a sequence of random variables and let* X *and* Y *be random variables such that* $E(Y) < \infty$, $|X_n| \leq Y$ *a.s. for all* $n \in \mathbb{N}$ *and* $X_n \xrightarrow{a.s.} X$. *Then* $E\big(|X_n|\big) < \infty$ *for each* $n \in \mathbb{N}$, $E\big(|X|\big) < \infty$ *and* $E\big(|X_n - X|\big) \to 0$. *In particular,* $E(X_n) \to E(X)$.

Definition 4.5. A sequence $(X_n)_{n\in\mathbb{N}}$ of random variables is called uniformly integrable, if

$$\lim_{c\to\infty} \sup_{n\in\mathbb{N}} E\big(|X_n|\mathbb{I}_{\{|X_n|>c\}}\big) = 0. \qquad \blacklozenge$$

Theorem 4.12. [Gut (2005), Theorem 5.4, p. 221] *Let* $r > 0$ *and* $(X_n)_{n\in\mathbb{N}}$ *be a sequence of random variables with* $E\big(|X_n|^r\big) < \infty$ *for all* $n \in \mathbb{N}$ *that converges in probability to a random variable* X. *Then the following statements are equivalent:*
(1) $\big(|X_n|^r\big)_{n\in\mathbb{N}}$ *is uniformly integrable.*
(2) $X_n \xrightarrow{L^r} X$.
(3) $E\big(|X_n|^r\big) \to E\big(|X|^r\big)$.
Moreover, if $r \geq 1$ *and one of the above holds, then* $E(X_n) \to E(X)$.

Theorem 4.13. [Billingsley (1968), Theorem 5.4, p. 32] *If* $(X_n)_{n\in\mathbb{N}}$ *is a sequence of uniformly integrable random variables such that* $X_n \xrightarrow{d} X$, *then* $E\big(|X|\big) < \infty$ *and* $E(X_n) \to E(X)$. *Also, if* X *and* X_n *are nonnegative and integrable for all* $n \in \mathbb{N}$ *such that* $X_n \xrightarrow{d} X$ *and* $E(X_n) \to E(X)$, *then* $(X_n)_{n\in\mathbb{N}}$ *is uniformly integrable.*

Theorem 4.14. (Scheffé's Theorem) [Gut (2005), Theorem 6.3, p. 227]
Let $(X_n)_{n\in\mathbb{N}}$ be a sequence of continuous random variables and let $(f_n)_{n\in\mathbb{N}}$ be a sequence of corresponding probability density functions such that $f_n(x) \to f(x)$, as $n \to \infty$, for a.e. $x \in \mathbb{R}$, where f is a probability density function of a continuous random variable X. Then $X_n \xrightarrow{d} X$.

4.1.1 Solved Problems

Problem 4.1.1. Let $(X_n)_{n\in\mathbb{N}^*}$ be a sequence of random variables such that for each $n \in \mathbb{N}^*$ we have $P(X_n = 1) = p_n$ and $P(X_n = 0) = 1 - p_n$, where $p_n \in [0,1]$. Let $r > 0$. Prove the following equivalences:

$$X_n \xrightarrow{P} 0 \iff \lim_{n\to\infty} p_n = 0 \iff X_n \xrightarrow{L^r} 0.$$

Moreover, if $(X_n)_{n\in\mathbb{N}^*}$ is a sequence of independent random variables and $p_n = \dfrac{1}{n}$ for each $n \in \mathbb{N}^*$, then $(X_n)_{n\in\mathbb{N}^*}$ does not converge a.s. to zero.

Solution 4.1.1: We have

$$X_n \xrightarrow{P} 0 \iff \lim_{n\to\infty} P(|X_n| > \varepsilon) = 0 \text{ for all } \varepsilon > 0.$$

But for each $n \in \mathbb{N}^*$

$$P(|X_n| > \varepsilon) = \begin{cases} P(X_n = 1) = p_n, & \text{if } \varepsilon < 1 \\ P(\emptyset) = 0, & \text{if } \varepsilon \geq 1. \end{cases}$$

Hence,

$$X_n \xrightarrow{P} 0 \iff \lim_{n\to\infty} p_n = 0.$$

For the convergence in mean of order r we compute

$$E(|X_n|^r) = 1^r \cdot p_n + 0^r \cdot (1 - p_n) = p_n,$$

therefore

$$X_n \xrightarrow{L^r} 0 \iff \lim_{n\to\infty} p_n = 0.$$

For each $n \in \mathbb{N}^*$ we consider the events $A_n = \{\omega \in \Omega : |X_n(\omega)| > \varepsilon\}$, where $\varepsilon \in (0,1)$, which are independent and satisfy

$$\sum_{n=1}^{\infty} P(A_n) = \sum_{n=1}^{\infty} \frac{1}{n} = \infty.$$

By the Borel–Cantelli Lemma (see Theorem 1.10), it follows that

$$P(\limsup_{n\to\infty} A_n) = 1.$$

This implies

$$1 = \lim_{n\to\infty} P\Big(\bigcup_{k\geq n} A_k\Big) = \lim_{n\to\infty} P(\sup_{k\geq n}|X_k| > \varepsilon).$$

By using Theorem 4.1, we obtain that $(X_n)_{n\in\mathbb{N}^*}$ does not converge a.s. to zero.

Problem 4.1.2. Let $(X_n)_{n\in\mathbb{N}}$ be a sequence of independent normally distributed random variables with $E(X_n) = 0$ for all $n \in \mathbb{N}$ which converges in probability to a random variable X, i.e., $X_n \xrightarrow{P} X$. Prove that $X_n \xrightarrow{L^2} X$.

Solution 4.1.2: We denote $\Delta_{nm} = X_n - X_m$ for each $n, m \in \mathbb{N}$. By the convergence in probability and by Theorem 4.10 it follows that for each $\varepsilon > 0$ it holds

$$\lim_{n,m\to\infty} P(|\Delta_{nm}| > \varepsilon) = 0.$$

For all $m, n \in \mathbb{N}$ with $m \neq n$, since X_n and X_m are independent and normally distributed random variables with zero mean, by Problem 3.5.4 we have $X_n - X_m \sim N(0, \sigma_{nm}^2)$, where $\sigma_{nm}^2 = E(\Delta_{nm}^2) = V(X_n) + V(X_m)$, and thus

$$P(|\Delta_{nm}| > \varepsilon) = 2\left(1 - F\left(\frac{\varepsilon}{\sigma_{nm}}\right)\right),$$

where F is the distribution function of the standard normal distribution $N(0,1)$. Therefore, for each $\varepsilon > 0$

$$\lim_{\substack{n,m\to\infty \\ n\neq m}} F\left(\frac{\varepsilon}{\sigma_{nm}}\right) = 1.$$

But F is a strictly increasing and continuous function whose limit to ∞ is 1, therefore the above limit implies

$$\lim_{n,m\to\infty} \sigma_{nm} = 0,$$

which leads to

$$\lim_{n,m\to\infty} E(|X_n - X_m|^2) = 0.$$

By the Cauchy criterion from Theorem 4.10 it follows that $(X_n)_{n\in\mathbb{N}}$ converges in mean square. This convergence must be to the random variable X, by Theorem 4.5 (uniqueness of the limit for the convergence in probability) and by Theorem 4.3 (mean square convergence implies convergence in probability).

Problem 4.1.3. Let $(X_n)_{n\in\mathbb{N}}$ be a sequence of random variables such that

$$P\big(X_n \leq X_{n+1} \text{ for each } n \in \mathbb{N}\big) = 1. \tag{4.1}$$

Prove that $X_n \xrightarrow{a.s.} X$ if and only if $X_n \xrightarrow{P} X$. Note that the same result holds in the case

$$P\big(X_n \geq X_{n+1} \text{ for each } n \in \mathbb{N}\big) = 1.$$

Solution 4.1.3: $X_n \xrightarrow{a.s.} X$ implies $X_n \xrightarrow{P} X$, by Theorem 4.3.

Assume $X_n \xrightarrow{P} X$ and that (4.1) holds. Theorem 4.4 yields the existence of a subsequence $(n_k)_{k \in \mathbb{N}}$ of indices such that $X_{n_k} \xrightarrow{a.s.} X$. Denote

$$\Omega_1 = \{\omega \in \Omega : \lim_{k \to \infty} X_{n_k}(\omega) = X(\omega)\}$$

and

$$\Omega_2 = \{\omega \in \Omega : X_n(\omega) \leq X_{n+1}(\omega) \text{ for each } n \in \mathbb{N}\}.$$

Then $P(\Omega_1 \cap \Omega_2) = 1$. Let $\omega \in \Omega_1 \cap \Omega_2$ and $\varepsilon > 0$. Then there exists $k_0 \in \mathbb{N}$ such that

$$|X(\omega) - X_{n_k}(\omega)| \leq \varepsilon \text{ for all } k \in \mathbb{N} \text{ with } k \geq k_0.$$

Moreover, for every $n \in \mathbb{N}$ with $n \geq n_{k_0}$ there exists $k \in \mathbb{N}$ with $k \geq k_0$ such that $n_k \leq n < n_{k+1}$ and thus

$$0 \leq X(\omega) - X_{n_{k+1}}(\omega) \leq X(\omega) - X_n(\omega) \leq X(\omega) - X_{n_k}(\omega) \leq \varepsilon.$$

Hence, for each $\omega \in \Omega_1 \cap \Omega_2$ we have $\lim_{n \to \infty} X_n(\omega) = X(\omega)$, i.e., $X_n \xrightarrow{a.s.} X$.

Problem 4.1.4. Let $(X_n)_{n \in \mathbb{N}^*}$ be a sequence of independent random variables with $Unif[a, b]$ distribution. Define for each $n \in \mathbb{N}^*$

$$Y_n = \max\{X_1, \ldots, X_n\} \text{ and } Z_n = \min\{X_1, \ldots, X_n\}.$$

Prove that $Y_n \xrightarrow{P} b$ and $Z_n \xrightarrow{P} a$. Do these sequences converge a.s.?

Solution 4.1.4: From the definition of the uniform distribution, for each $n \in \mathbb{N}^*$, the distribution function of X_n is given by

$$F_{X_n}(x) = P(X_n \leq x) = \begin{cases} 0, & \text{if } x < a \\ \dfrac{x - a}{b - a}, & \text{if } a \leq x < b \\ 1, & \text{if } b \leq x. \end{cases}$$

Then, by the independence of the random variables in $(X_n)_{n \in \mathbb{N}^*}$, we have for every $x \in \mathbb{R}$

$$P(Y_n \leq x) = P(\max\{X_1, \ldots, X_n\} \leq x) = P\Big(\bigcap_{k=1}^n \{X_k \leq x\}\Big) = \big(F_{X_1}(x)\big)^n.$$

For $x = b$ we get $P(Y_n \leq b) = 1$. Moreover, we have

$$P(|Y_n - b| > \varepsilon) = P(b - Y_n > \varepsilon) = P(Y_n < b - \varepsilon)$$

$$= \begin{cases} \left(1 - \frac{\varepsilon}{b-a}\right)^n, & \text{if } 0 < \varepsilon \leq b - a \\ 0, & \text{if } b - a < \varepsilon. \end{cases}$$

Therefore, $\lim\limits_{n \to \infty} P(|Y_n - b| > \varepsilon) = 0$ for each $\varepsilon > 0$. Observe that

$$Z_n = \min\{X_1, \ldots, X_n\} = -\max\{-X_1, \ldots, -X_n\}$$

with $-X_n \sim Unif[-b, -a]$ for each $n \in \mathbb{N}^*$. We apply the above result for the sequence $(-X_n)_{n \in \mathbb{N}^*}$ and we obtain $-Z_n \xrightarrow{P} -a$. By Theorem 4.6 it follows that $Z_n \xrightarrow{P} a$.

The sequences $(Y_n - b)_{n \in \mathbb{N}^*}$ and $(a - Z_n)_{n \in \mathbb{N}^*}$ are increasing sequences of random variables which converge in probability to zero. By Problem 4.1.3 it follows that they converge a.s. to zero. Hence, $Y_n \xrightarrow{a.s.} b$ and $Z_n \xrightarrow{a.s.} a$.

Problem 4.1.5. Let $(p_n)_{n \in \mathbb{N}^*}$ be a sequence of numbers in $(0, 1)$ such that $\lim\limits_{n \to \infty} np_n = \lambda > 0$. Consider $(X_n)_{n \in \mathbb{N}^*}$ to be a sequence of random variables such that $X_n \sim Bino(n, p_n)$ for each $n \in \mathbb{N}^*$. Then $(X_n)_{n \in \mathbb{N}^*}$ converges in distribution to a random variable which is $Poiss(\lambda)$ distributed.

Solution 4.1.5: For each $n \in \mathbb{N}^*$ let M_{X_n} denote the moment generating function of X_n. We have by Example 3.16-(1) for each $t \in \mathbb{R}$ and $n \in \mathbb{N}^*$

$$M_{X_n}(t) = (1 + (e^t - 1)p_n)^n.$$

Then

$$\lim_{n \to \infty} M_{X_n}(t) = \lim_{n \to \infty} \left(1 + \frac{(e^t - 1)p_n n}{n}\right)^n = \exp\{\lambda(e^t - 1)\} \text{ for } t \in \mathbb{R},$$

where we used the following property

$$\lim_{n \to \infty} a_n = 0 \implies \lim_{n \to \infty} (1 + a_n)^{\frac{1}{a_n}} = e$$

and the assumption that $\lim\limits_{n \to \infty} np_n = \lambda$. Hence, the sequence $(M_{X_n})_{n \in \mathbb{N}^*}$ converges pointwise, with respect to $t \in \mathbb{R}$, to the moment generating function of the $Poiss(\lambda)$ distribution (see Problem 3.5.2). By Theorem 4.9, it follows that $(X_n)_{n \in \mathbb{N}^*}$ converges in distribution to a random variable which is $Poiss(\lambda)$ distributed.

4.2 Weak Law and Strong Law of Large Numbers

Definition 4.6. A sequence $(X_n)_{n\in\mathbb{N}^*}$ of integrable random variables obeys the **weak law of large numbers (WLLN)**, if

$$\frac{1}{n}\sum_{k=1}^{n}\left(X_k - E(X_k)\right) \xrightarrow{P} 0. \qquad \blacklozenge$$

Theorem 4.15. *Let $(X_n)_{n\in\mathbb{N}^*}$ be a sequence of random variables such that $E(|X_n|^2) < \infty$ for each $n \in \mathbb{N}^*$ and the following condition holds*

$$\lim_{n\to\infty}\frac{1}{n^2}V\left(X_1 + \cdots + X_n\right) = 0.$$

Then $(X_n)_{n\in\mathbb{N}^*}$ obeys the WLLN.

Example 4.1. (1) Let $(X_n)_{n\in\mathbb{N}^*}$ be a sequence of pairwise independent random variables satisfying the condition

$$V(X_n) \le L < \infty \text{ for all } n \in \mathbb{N}^*,$$

where $L > 0$ is a constant. Then, by Theorem 4.15 and Example 3.9, $(X_n)_{n\in\mathbb{N}^*}$ obeys the WLLN.

(2) Let $(X_n)_{n\in\mathbb{N}^*}$ be a sequence of pairwise independent random variables, where for every $n \in \mathbb{N}^*$ we have

$$P(X_n = 0) = 1 - p_n, \ P(X_n = 1) = p_n \text{ with } p_n \in [0, 1].$$

Then, by Theorem 4.15 and Example 3.9, $(X_n)_{n\in\mathbb{N}^*}$ obeys the WLLN.

Note that in both examples the pairwise independence condition may be replaced by the condition that the random variables from the sequence $(X_n)_{n\in\mathbb{N}^*}$ are pairwise uncorrelated. ▲

Theorem 4.16. **(WLLN for uniformly integrable and independent random variables)** [Knill (2009), Theorem 2.6.4, p. 54] *Let $(X_n)_{n\in\mathbb{N}^*}$ be a sequence of uniformly integrable and independent random variables. Then*

$$\frac{1}{n}\sum_{k=1}^{n}(X_k - E(X_k)) \xrightarrow{L^1} 0.$$

Definition 4.7. A sequence $(X_n)_{n\in\mathbb{N}^*}$ of integrable random variables obeys the **strong law of large numbers (SLLN)**, if

$$\frac{1}{n}\sum_{k=1}^{n}\left(X_k - E(X_k)\right) \xrightarrow{a.s.} 0. \qquad \blacklozenge$$

Theorem 4.17. [Gut (2005), Theorem 5.2, p. 286] *If $(X_n)_{n\in\mathbb{N}^*}$ is a sequence of independent random variables with $\sum_{n=1}^{\infty} V(X_n) < \infty$, then*

$$P\left(\sum_{k=1}^{\infty}(X_k - E(X_k)) < \infty\right) = 1,$$

i.e., the series $\sum_{k=1}^{\infty}(X_k - E(X_k))$ is a.s. convergent.

Theorem 4.18. (Kolmogorov) *If $(X_n)_{n\in\mathbb{N}^*}$ is a sequence of independent random variables with $\sum_{n=1}^{\infty} \frac{1}{n^2} V(X_n) < \infty$, then*

$$\frac{1}{n}\sum_{k=1}^{n}\left(X_k - E(X_k)\right) \xrightarrow{a.s.} 0,$$

i.e., $(X_n)_{n\in\mathbb{N}^*}$ obeys the SLLN.

Example 4.2. Let $r \geq 1$ and let $(X_n)_{n\in\mathbb{N}^*}$ be a sequence of independent random variables such that

$$P(X_n = n) = P(X_n = -n) = \frac{1}{2n^r \log n}, \quad P(X_n = 0) = 1 - \frac{1}{n^r \log n}$$

for $n \in \mathbb{N}$ with $n \geq 2$ and $X_1 = 0$. We prove that:
(1) If $r > 1$, then $(X_n)_{n\in\mathbb{N}^*}$ obeys the SLLN.
(2) If $r = 1$, then $(X_n)_{n\in\mathbb{N}^*}$ obeys the WLLN, but not the SLLN.
Proof:
(1) Since $r > 1$, we have

$$\sum_{n=1}^{\infty}\frac{1}{n^2}V(X_n) = \sum_{n=2}^{\infty}\frac{1}{n^r \log n} \leq \sum_{n=1}^{\infty}\frac{1}{n^r} < \infty.$$

Hence $(X_n)_{n\in\mathbb{N}^*}$ obeys the SLLN, by Theorem 4.18.
(2) For every $n \in \mathbb{N}$, $n \geq 2$, we have

$$\frac{1}{n^2}V\left(\sum_{k=1}^{n}X_k\right) = \frac{1}{n^2}\sum_{k=2}^{n}\frac{k}{\log k}.$$

By the Stolz–Cesàro Theorem (see [Choudary and Niculescu (2014), Theorem 2.7.1, p. 59]), we deduce that

$$\lim_{n\to\infty}\frac{1}{n^2}\sum_{k=2}^{n}\frac{k}{\log k} = \lim_{n\to\infty}\frac{n+1}{(2n+1)\log(n+1)} = 0$$

and thus, by Theorem 4.15, $(X_n)_{n\in\mathbb{N}^*}$ obeys the WLLN.

Next, suppose that $(X_n)_{n\in\mathbb{N}^*}$ obeys the SLLN, i.e.,

$$S_n = \frac{1}{n}\sum_{k=1}^{n} X_k \xrightarrow{a.s.} 0.$$

Then (see Theorem 4.6)

$$\frac{1}{n}X_n = S_n - \frac{n-1}{n}S_{n-1} \xrightarrow{a.s.} 0.$$

Let $\varepsilon \in (0,1)$ and for every $n \in \mathbb{N}^*$ denote $A_n = \left\{\omega \in \Omega : \frac{1}{n}|X_n(\omega)| > \varepsilon\right\}$. In view of Theorem 4.1, we have

$$P(\limsup_{n\to\infty} A_n) = 0$$

and thus, by the Borel–Cantelli Lemma (see Theorem 1.10), $\sum_{n=1}^{\infty} P(A_n) <$

∞, which is in contradiction with $\sum_{n=2}^{\infty} P(|X_n| > n\varepsilon) = \sum_{n=2}^{\infty} \frac{1}{n\log n} = \infty$, where we use the Cauchy (condensation) test (see [Schinazi (2012), p. 51]).

▲

Theorem 4.19. [Bauer (1996), Corollary 12.2, p. 81] *Let* $(X_n)_{n\in\mathbb{N}^*}$ *be a sequence of integrable, independent, identically distributed random variables. Then*

$$\frac{1}{n}\sum_{k=1}^{n} X_k \xrightarrow{a.s.} E(X_1),$$

i.e., $(X_n)_{n\in\mathbb{N}^*}$ *obeys the SLLN.*

Example 4.3. (1) Let $(X_n)_{n\in\mathbb{N}^*}$ be a sequence of independent random variables, where

$$P(X_n = 0) = 1 - p, P(X_n = 1) = p, \ n \in \mathbb{N}^*,$$

with $p \in [0,1]$. Then, by Theorem 4.19, $(X_n)_{n\in\mathbb{N}^*}$ obeys the SLLN, i.e.,

$$\frac{1}{n}\sum_{k=1}^{n} X_k \xrightarrow{a.s.} p.$$

(2) In view of (1), if in a sequence of independent trials of an experiment the probability that an event A occurs is $p \in [0,1]$ for each trial, then

$$P\left(\lim_{n\to\infty} |r_n(A) - p| = 0\right) = 1,$$

where $r_n(A)$ is the relative frequency of the occurrence of the event A after repeating the experiment n times. This means that the sequence of relative frequencies of the occurrence of an event tends a.s. to the probability of that event, i.e., $r_n(A) \xrightarrow{a.s.} P(A)$.

In the following, we exemplify the above observation, by simulating in Matlab 100 tosses of a fair coin: 1 for "heads" and 0 for "tails" (in the figure below, we have the plot of the corresponding relative frequencies of the "heads").

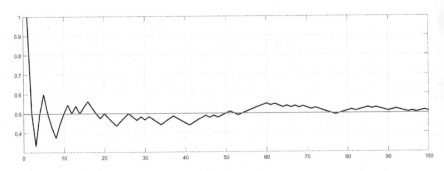

Fig. 4.1: Example 4.3, Strong Law of Large Numbers

```
>>plot(1:100,cumsum(randsample([0 1],100,'true'))./(1:100),'-k','LineWidth',1.8);
>>grid on;
```

▲

Example 4.4. Let $X_1, \ldots, X_n$ be a random sample corresponding to the random variable X having the distribution function F. Recall Definition 2.16 concerning the empirical distribution function corresponding to the random sample, which is defined as $\hat{F}_n : \mathbb{R} \to [0, 1]$

$$\hat{F}_n(x) = \frac{1}{n} \sum_{i=1}^{n} \mathbb{I}_{\{X_i \leq x\}}.$$

Observe that $E\big(\mathbb{I}_{\{X_i \leq x\}}\big) = E\big(\mathbb{I}^2_{\{X_i \leq x\}}\big) = P(X_i \leq x) = F(x)$ for each $x \in \mathbb{R}$, $i \in \{1, \ldots, n\}$. Hence, $\hat{F}_n$ has the following properties:

(1) $E\big(\hat{F}_n(x)\big) = F(x)$ and $V\big(\hat{F}_n(x)\big) = \frac{1}{n} F(x)\big(1 - F(x)\big)$ for each $x \in \mathbb{R}$;

(2) $\hat{F}_n(x) \xrightarrow{a.s.} F(x)$ for each $x \in \mathbb{R}$ (by Theorem 4.19).

Note that the above result is less strong than the result of Glivenko–Cantelli (see Remark 2.9). ▲

4.2.1 Solved Problems

Problem 4.2.1. Let $(X_n)_{n \in \mathbb{N}^*}$ be a sequence of independent random variables such that $X_n \sim Unif[0, 1]$ for each $n \in \mathbb{N}^*$. Let $Y_n = \prod_{i=1}^{n} X_i$, $n \in \mathbb{N}^*$. Prove that $Y_n \xrightarrow{a.s.} 0$.

Solution 4.2.1: Method 1: Note that for each $n \in \mathbb{N}^*$ we have $Y_n(\omega) > 0$ for a.e. $\omega \in \Omega$. By the SLLN (see Theorem 4.19) we can write

$$\frac{1}{n}\log(Y_n) = \frac{1}{n}\sum_{k=1}^{n} \log(X_k) \xrightarrow{a.s.} E\big(\log(X_1)\big),$$

where

$$E\big(\log(X_1)\big) = \int_0^1 \log t \, dt = -1.$$

Then $\log(Y_n) \xrightarrow{a.s.} -\infty$, which implies that $(Y_n)_{n \in \mathbb{N}^*}$ converges a.s. to 0.
Method 2: Observe that $(Y_n)_{n \in \mathbb{N}^*}$ is a.s. a decreasing sequence of strictly positive random variables. We use the Markov inequality (see Theorem 3.6) to write for each $\varepsilon > 0$

$$P(|Y_n| > \varepsilon) = P(Y_n > \varepsilon) \leq \frac{1}{\varepsilon}E(Y_n) = \frac{1}{\varepsilon \cdot 2^n} \to 0, \text{ as } n \to \infty.$$

Hence $(Y_n)_{n \in \mathbb{N}^*}$ converges in probability to 0 and, by Problem 4.1.3, it follows that it converges also a.s. to 0.

Problem 4.2.2. Let $(X_n)_{n \in \mathbb{N}^*}$ be a sequence of independent random variables such that $X_n \sim Poiss(\lambda_n)$, $\lambda_n > 0$, $n \in \mathbb{N}^*$. Let $Y_n = \frac{1}{n}\sum_{i=1}^{n} X_i$ for $n \in \mathbb{N}^*$. If $\lim_{n \to \infty} \lambda_n = \lambda$, show that $(Y_n)_{n \in \mathbb{N}^*}$ converges a.s. to λ.

Solution 4.2.2: We have $V(X_n) = \lambda_n$ for each $n \in \mathbb{N}^*$. By the convergence assumption, we have that the sequence $(\lambda_n)_{n \in \mathbb{N}^*}$ is bounded, hence $\sum_{n=1}^{\infty} \frac{1}{n^2}V(X_n) < \infty$. The convergence assumption of this problem together with the Stolz–Cesàro Theorem imply

$$\frac{1}{n}\sum_{k=1}^{n} \lambda_k \to \lambda. \tag{4.2}$$

We use Theorem 4.18 to conclude that

$$\frac{1}{n}\sum_{k=1}^{n} \big(X_k - E(X_k)\big) \xrightarrow{a.s.} 0,$$

which is equivalent to

$$\frac{1}{n}\sum_{k=1}^{n} X_k - \frac{1}{n}\sum_{k=1}^{n}\lambda_k \xrightarrow{a.s.} 0.$$

By Theorem 4.6 and (4.2), it follows that $Y_n \xrightarrow{a.s.} \lambda$.

Problem 4.2.3 (Monte Carlo Integration I). Let $g : [a,b] \to [0,M]$ be an integrable function, where $a < b$ and $M > 0$. Approximate the value of the integral $\int_a^b g(t)\,dt$ by the following method: Let $X_1, X_2, \ldots, Y_1, Y_2, \ldots$ be independent random variables such that

$$X_n \sim Unif[a,b], \ Y_n \sim Unif[0,M] \text{ for each } n \in \mathbb{N}^*.$$

Prove that

$$(b-a)M \cdot \frac{\#\{k \in \{1,\ldots,n\} : Y_k \leq g(X_k)\}}{n} \xrightarrow{a.s.} \int_a^b g(t)dt\,.$$

Note that for large n we have the approximation

$$\int_a^b g(t)dt \approx (b-a)M \cdot \frac{\#\{k \in \{1,\ldots,n\} : Y_k \leq g(X_k)\}}{n}\,.$$

As an application of the above result, estimate, by corresponding simulations in Matlab, the integral $\int_{-1}^{3} |\sin(e^t)|dt.$

Solution 4.2.3: Let $n \in \mathbb{N}^*$. Denote the random variable

$$N_n = \#\{k \in \{1, ..., n\} : Y_k \leq g(X_k)\} = \sum_{k=1}^{n} \mathbb{I}_{\{Y_k \leq g(X_k)\}}\,.$$

N_n is the random variable that shows how many pairs (X_k, Y_k), $k \in \{1, ..., n\}$, are under the graph of the function g and above the Ox-axis. We know that $\int_a^b g(t)dt$ represents the area (i.e., the $\lambda_{\mathbb{R}^2}$ measure) of the set $A = \{(x,y) \in [a,b] \times [0,M] : y \leq g(x)\}$.

Consider the event B that a uniformly generated random point from $[a,b] \times [0,M]$ is in A. By the SLLN (see Example 4.3 applied for the event B)

$$\frac{1}{n}N_n \xrightarrow{a.s.} P(B). \tag{4.3}$$

By geometric probability (see (1.1)) we have

$$P(B) = \frac{\lambda_{\mathbb{R}^2}(A)}{\lambda_{\mathbb{R}^2}([a,b] \times [0,M])} = \frac{1}{(b-a)M}\int_a^b g(t)dt\,.$$

At this moment we can better understand the reason behind the formula (1.1) of geometric probability, since we can write for the independent random variables $X_k \sim Unif[a,b]$ and $Y_k \sim Unif[0,M]$, $k \in \mathbb{N}^*$,

$$P(B) = P(Y_k \le g(X_k)) = \underbrace{\iint}_{\{(x,y)\in\mathbb{R}^2 : y\le g(x)\}} f_{X_k,Y_k}(x,y)dxdy$$

$$= \underbrace{\iint}_{\{(x,y)\in[a,b]\times[0,M] : y\le g(x)\}} \frac{1}{(b-a)M}dxdy = \frac{\lambda_{\mathbb{R}^2}(A)}{(b-a)M}.$$

Finally, we obtain by (4.3)

$$(b-a)M \cdot \frac{\#\{k \in \{1,...,n\} : Y_k \le g(X_k)\}}{n} \xrightarrow{a.s.} \int_a^b g(t)dt,$$

which leads for large n to the approximation

$$\int_a^b g(t)dt \approx (b-a)M \cdot \frac{\#\{k \in \{1,...,n\} : Y_k \le g(X_k)\}}{n}.$$

For the estimation of the integral we can use the following code in Matlab.

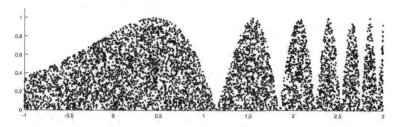

Fig. 4.2: Problem 4.2.3

```
g=@(t) abs(sin(exp(t)));
a=-1; b=3;
n=10000;
X=unifrnd(a,b,1,n);
M=1;
Y=unifrnd(0,M,1,n);
clf; axis equal; hold on; axis([a,b, 0, M+0.1]);
plot(X(Y<=g(X)),Y(Y<=g(X)),'.k');
int1=(b-a)*M*mean(Y<=g(X)); %Monte Carlo integration
fprintf('Estimated value by Monte Carlo integration: %5.4f.\n',int1);
int2=integral(g,a,b); %numerical integration using global adaptive quadrature
fprintf('Estimated value by numerical integration: %5.4f.\n',int2);
```

Estimated value by Monte Carlo integration: 2.6560.
Estimated value by numerical integration: 2.6517.

Problem 4.2.4 (Monte Carlo Integration II). Let $(U_n)_{n\in\mathbb{N}^*}$ be a sequence of independent random variables which are $Unif[a, b]$ distributed. Assume that $g : [a, b] \to \mathbb{R}$ is an integrable function. Denote

$$I_n = \frac{g(U_1) + \cdots + g(U_n)}{n}, n \in \mathbb{N}^*, \quad \text{and} \quad I = \frac{1}{b-a} \int_a^b g(t)dt.$$

Prove that $I_n \xrightarrow{a.s.} I$ and, if $\int_a^b g^2(t)dt < \infty$, then $I_n \xrightarrow{L^2} I$.

Note that for large n we have the approximation

$$\int_a^b g(t)dt \approx \frac{b-a}{n}\Big(g(U_1) + \cdots + g(U_n)\Big).$$

As an application of the above result, estimate, by some corresponding simulations in Matlab, the integral $\int_0^{2\pi} \exp\{\sin(t)\}dt$.

Solution 4.2.4: We have

$$E(g(U_1)) = \frac{1}{b-a} \int_a^b g(t)dt = I.$$

By the SLLN (see Theorem 4.19), it follows that $I_n \xrightarrow{a.s.} I$.
We compute for $n \in \mathbb{N}^*$

$$V(I_n) = \frac{1}{n^2}\Big(V(g(U_1)) + \cdots + V(g(U_n))\Big) = \frac{V(g(U_1))}{n}$$

$$= \frac{1}{n(b-a)} \int_a^b g^2(t)dt - \frac{1}{n}\left(\frac{1}{b-a} \int_a^b g(t)dt\right)^2,$$

since $U_1, \ldots, U_n \sim Unif[a, b]$ are independent random variables. Hence $\lim_{n\to\infty} V(I_n) = \lim_{n\to\infty} E\Big((I_n - I)^2\Big) = 0$. So, $I_n \xrightarrow{L^2} I$.
For the estimation of the integral we can use the following code in Matlab.

```
g=@(t) exp(sin(t));
a=0; b=2*pi;
n=10000;
int1=(b-a)*mean(g(unifrnd(a,b,1,n))); %Monte Carlo integration
fprintf('Estimated value by Monte Carlo integration: %5.4f.\n',int1);
int2=integral(g,a,b); %numerical integration using global adaptive quadrature
fprintf('Estimated value by numerical integration: %5.4f.\n',int2);
```

Estimated value by Monte Carlo integration: 7.9439.
Estimated value by numerical integration: 7.9549.

Problem 4.2.5. Estimate in Matlab the area of the interior of the plane curve that satisfies the following equation

$$(x^2 + y^2 - 1)^3 = x^2 y^3,$$

using two different Monte Carlo methods.

Solution 4.2.5:

```
function heart_sim(n)
%first method
clf; axis equal; hold on;
count=0;
for i=1:n
    x=unifrnd(-1.5,1.5); y=unifrnd(-1,1.5);
    if (x^2 + y^2 -1)^3<x^2*y^3
        plot(x,y,'pk');
        count=count+1;
    end
end
fprintf('First method: estimated area=%3.2f\n',count/n*7.5);
%second method
b=@(x) -(x.^2).^(1/3); c=@(x) x.^2-1;
delta=@(x) b(x).^2-4*c(x);
x0=abs(fzero(delta,0));
f1=@(x) (-b(x)+sqrt(delta(x)))./2;
f2=@(x) (-b(x)-sqrt(delta(x)))./2;
U=unifrnd(0,x0,1,n);
fprintf('Second method: estimated area=%3.2f\n',2*x0*mean(f1(U)-f2(U)));
end
```

```
>> heart_sim(5000)
First method: estimated area=3.67
Second method: estimated area=3.68
```

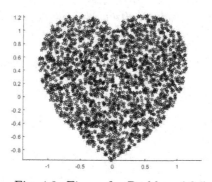

Fig. 4.3: Figure for Problem 4.2.5

4.3   Central Limit Theorems

Let $(X_n)_{n\in\mathbb{N}^*}$ be a sequence of independent random variables. Throughout this section we denote

$$\mu_k = E(X_k) \text{ for each } k \in \mathbb{N}^*$$

and

$$D_n = \Big(\sum_{k=1}^{n} V(X_k) \Big)^{\frac{1}{2}} \text{ for each } n \in \mathbb{N}^*$$

and we assume that $D_n \neq 0$ for all $n \in \mathbb{N}^*$. The following condition is called **Lindeberg's condition:**

$$\lim_{n\to\infty} \frac{1}{D_n^2} \sum_{k=1}^{n} E\Big((X_k - \mu_k)^2 \cdot \mathbb{I}_{\{|X_k - \mu_k| > \varepsilon D_n\}} \Big) = 0 \quad \text{for all} \quad \varepsilon > 0. \quad (4.4)$$

For details regarding central limit theorems, see [Billingsley (1986), Section 27] and [Skorokhod (2005), Section 5.3].

Theorem 4.20. (Lindeberg) *If $(X_n)_{n\in\mathbb{N}^*}$ is a sequence of independent random variables satisfying Lindeberg's condition (4.4), then*

$$\lim_{n\to\infty} P\left(\frac{1}{D_n} \sum_{k=1}^{n} (X_k - \mu_k) \leq x \right) = \frac{1}{\sqrt{2\pi}} \int_{-\infty}^{x} \exp\left\{ -\frac{t^2}{2} \right\} dt$$

for all $x \in \mathbb{R}$, i.e., the sequence

$$\left(\frac{1}{D_n} \sum_{k=1}^{n} (X_k - \mu_k) \right)_{n\in\mathbb{N}^*}$$

converges in distribution to a random variable having standard normal distribution $N(0,1)$.

Theorem 4.21. (Lindeberg–Lévy) *Let $(X_n)_{n\in\mathbb{N}^*}$ be a sequence of independent, identically distributed random variables with $\mu = E(X_n)$ and $\sigma^2 = V(X_n) > 0$ for all $n \in \mathbb{N}^*$. Then*

$$\lim_{n\to\infty} P\left(\frac{1}{\sqrt{n}\sigma} \sum_{k=1}^{n} (X_k - \mu) \leq x \right) = \frac{1}{\sqrt{2\pi}} \int_{-\infty}^{x} \exp\left\{ -\frac{t^2}{2} \right\} dt$$

for all $x \in \mathbb{R}$. In particular, if

$$S_n = X_1 + \cdots + X_n \ \text{ for all } \ n \in \mathbb{N}^*,$$

then

$$\lim_{n\to\infty} P\left(a < \frac{S_n - n\mu}{\sqrt{n}\sigma} \leq b \right) = \frac{1}{\sqrt{2\pi}} \int_{a}^{b} \exp\left\{ -\frac{t^2}{2} \right\} dt,$$

for all $a, b \in \mathbb{R}$ with $a < b$.

Theorem 4.22. (Moivre–Laplace) *We consider a sequence of independent trials such that the probability that an event A occurs in a trial of the experiment is p, where $p \in (0,1)$ is the same for each trial. Denote by A_j the event that A occurred in the jth trial and let $X_j = \mathbb{I}_{A_j}$ for $j \in \mathbb{N}^*$. Then $S_n = X_1 + \cdots + X_n$, which represents the number of occurrences of A in $n \in \mathbb{N}^*$ trials, has binomial distribution $Bino(n,p)$. Moreover,*

$$\lim_{n \to \infty} P\left(a < \frac{S_n - np}{\sqrt{np(1-p)}} \leq b \right) = \frac{1}{\sqrt{2\pi}} \int_a^b \exp\left\{ -\frac{t^2}{2} \right\} dt,$$

for all $a, b \in \mathbb{R}$ with $a < b$.

Example 4.5. Let $X \sim Ber(p)$, where $p \in (0,1)$, be a random variable used by a statistician to study the proportion of people of some population that prefer dark chocolate to milk chocolate: if a randomly chosen person from the population prefers dark chocolate, then X equals 1 and if the person prefers milk chocolate, then X equals 0. Consider a random sample $(X_n)_{n \in \mathbb{N}^*}$ corresponding to X. The statistician does not know p, but applies Theorem 4.22 to estimate p by using

$$\bar{X}_n = \frac{1}{n}\left(X_1 + \ldots + X_n\right), \quad n \in \mathbb{N}^*,$$

and estimates the least number n of persons to be included in the sample such that the following inequality holds

$$P\left(|\bar{X}_n - p| \leq 0.01 \right) \geq 0.99.$$

Using **norminv** from Matlab, the statistician can find a value $c > 0$ such that

$$P\left(|\bar{X}_n - p| \leq c\sqrt{\frac{p(1-p)}{n}} \right) \approx 0.99,$$

for sufficiently large $n \in \mathbb{N}^*$. Indeed, by Theorem 4.22, it holds

$$P\left(|\bar{X}_n - p| \leq c\sqrt{\frac{p(1-p)}{n}} \right) \approx 2F(c) - 1,$$

for sufficiently large $n \in \mathbb{N}^*$, where F is the distribution function for $N(0,1)$. Hence,

```
>> c=norminv(199/200)
c = 2.5758
```

In order to estimate the least number n of persons to be included in the sample the statistician writes

$$P\left(|\bar{X}_n - p| \leq \frac{c}{2\sqrt{n}}\right) \geq P\left(|\bar{X}_n - p| \leq c\sqrt{\frac{p(1-p)}{n}}\right) \approx 0.99$$

and chooses

```
>> n=floor((50*c)^2)
n = 16587
```

to obtain the (estimated) inequality

$$P\left(|\bar{X}_{16587} - p| \leq 0.01\right) \geq 0.99. \qquad \blacktriangle$$

4.3.1   Solved Problems

Problem 4.3.1. Let $(X_n)_{n\in\mathbb{N}^*}$ be a sequence of random variables such that $X_n \sim \chi^2(n,1)$ for $n \in \mathbb{N}^*$. Then the sequence $(Y_n)_{n\in\mathbb{N}^*}$ defined by

$$Y_n = \frac{X_n - E(X_n)}{\sqrt{V(X_n)}} \quad \text{for } n \in \mathbb{N}^*$$

converges in distribution to a random variable having standard normal distribution $N(0,1)$.

Solution 4.3.1: Let $Z_k \sim N(0,1)$ with $k \in \mathbb{N}^*$ be independent random variables. We have by Example 3.12

$$E(Z_k^2) = 1 \text{ and } V(Z_k^2) = 2 \text{ for each } k \in \mathbb{N}^*.$$

By Theorem 4.21 it follows that for all $x \in \mathbb{R}$

$$\lim_{n\to\infty} P\left(\frac{1}{\sqrt{2n}} \sum_{k=1}^{n}(Z_k^2 - 1) \leq x\right) = \frac{1}{\sqrt{2\pi}} \int_{-\infty}^{x} \exp\left\{-\frac{t^2}{2}\right\} dt.$$

By Problem 3.7.12, we deduce that $U_n = Z_1^2 + \cdots + Z_n^2 \sim \chi^2(n,1)$ and then $E(U_n) = n$, $V(U_n) = 2n$ for $n \in \mathbb{N}^*$. So, for $n \in \mathbb{N}^*$, U_n and X_n have the same distribution and we write for all $x \in \mathbb{R}$

$$P\left(\frac{1}{\sqrt{2n}} \sum_{k=1}^{n}(Z_k^2 - 1) \leq x\right) = P\left(\frac{U_n - n}{\sqrt{2n}} \leq x\right) = P\left(\frac{X_n - n}{\sqrt{2n}} \leq x\right).$$

Therefore,

$$\lim_{n\to\infty} P\left(Y_n \leq x\right) = \frac{1}{\sqrt{2\pi}} \int_{-\infty}^{x} \exp\left\{-\frac{t^2}{2}\right\} dt \quad \text{for all } x \in \mathbb{R}.$$

So, $(Y_n)_{n \in \mathbb{N}^*}$ converges in distribution to a random variable having standard normal distribution $N(0, 1)$.

Problem 4.3.2. Let $(X_n)_{n \in \mathbb{N}^*}$ be a sequence of independent random variables such that $X_n \sim Unif[\sqrt{n} - n, \sqrt{n} + n]$ for all $n \in \mathbb{N}^*$. Let

$$\bar{X}_n = \frac{1}{n\sqrt{n}} (X_1 + \ldots + X_n), \quad n \in \mathbb{N}^*.$$

1) Simulate in Matlab some values of $\bar{X}_n$, then estimate the mean and the variance, for some given $n \in \mathbb{N}^*$. Also, plot the corresponding histogram and the probability density function of the $N(\frac{2}{3}, \frac{1}{9})$ normal distribution.
2) Prove that $E(\bar{X}_n) \to \frac{2}{3}$ and $V(\bar{X}_n) \to \frac{1}{9}$.
3) Prove that $(\bar{X}_n)_{n \in \mathbb{N}^*}$ converges in distribution to a random variable that follows the $N(\frac{2}{3}, \frac{1}{9})$ normal distribution.

Solution 4.3.2: 1)

```
function [m,v]=unifconv(n,N_sim)
clf;hold on;
X=zeros(1,N_sim);
for i=1:N_sim
    for k=1:n
        X(i)=X(i)+unifrnd(sqrt(k)-k,sqrt(k)+k);
    end
    X(i)=X(i)/(n*sqrt(n));
end
h=histogram(X,'Normalization','pdf');
fplot(@(x) normpdf(x,2/3,1/3),h.BinLimits);
m=mean(X); v=var(X,1);
end

>>[m,v]=unifconv(20,1000)
m = 0.6778
v = 0.1177
```

2) We note that $E(X_n) = \sqrt{n}$ and $V(X_n) = \dfrac{n^2}{3}$ for $n \in \mathbb{N}^*$. We have

$$E(\bar{X}_n) = \frac{1}{n\sqrt{n}} \sum_{k=1}^{n} \sqrt{k} = \frac{1}{n} \sum_{k=1}^{n} \sqrt{\frac{k}{n}} \longrightarrow \int_0^1 \sqrt{t}\, dt = \frac{2}{3}.$$

Similarly, we have

$$V(\bar{X}_n) = \frac{1}{n^3} \sum_{k=1}^{n} \frac{k^2}{3} = \frac{1}{n} \sum_{k=1}^{n} \frac{1}{3} \left(\frac{k}{n} \right)^2 \longrightarrow \int_0^1 \frac{t^2}{3}\, dt = \frac{1}{9}.$$

3) We have

$$D_n^2 = \sum_{k=1}^{n} V(X_k) = \sum_{k=1}^{n} \frac{k^2}{3} = \frac{n(n+1)(2n+1)}{18} \quad \text{for } n \in \mathbb{N}^*.$$

Also, we have $|X_k - E(X_k)| = |X_k - \sqrt{k}| \le k \le n$ a.s., for all $k \in \{1, \dots, n\}$. But $\dfrac{n}{D_n} \to 0$, then for every $\varepsilon > 0$ and for sufficiently large $n \in \mathbb{N}^*$ we have

$$\max_{k \in \{1,\dots,n\}} |X_k - E(X_k)| \le n \le \varepsilon D_n \quad \text{a.s.},$$

which implies $\mathbb{I}_{\{|X_k - E(X_k)| > \varepsilon D_n\}} = 0$ a.s. for all $k \in \{1, \dots, n\}$. So, $(X_n)_{n \in \mathbb{N}^*}$ satisfies the Lindeberg condition (4.4), hence by Theorem 4.20 the sequence

$$\left(\frac{1}{D_n} \sum_{k=1}^{n} (X_k - \sqrt{k}) \right)_{n \in \mathbb{N}^*}$$

converges in distribution to a random variable having $N(0, 1)$ distribution. But

$$\frac{1}{D_n} \sum_{k=1}^{n} (X_k - \sqrt{k}) = \frac{n\sqrt{n}}{D_n} \left(\bar{X}_n - \frac{1}{n} \sum_{k=1}^{n} \sqrt{\frac{k}{n}} \right),$$

then in view of the proof given for 2), we can deduce by Theorem 4.7 and Example 2.24-(3) that $(\bar{X}_n)_{n \in \mathbb{N}^*}$ converges in distribution to a random variable having the $N(\frac{2}{3}, \frac{1}{9})$ normal distribution.

4.4 Problems for Chapter 4

Problem 4.4.1. Let $(X_n)_{n \in \mathbb{N}^*}$ be a sequence of independent random variables such that for each $n \in \mathbb{N}^*$

$$P(X_n = n) = \frac{1}{n}, \quad P(X_n = 1) = \frac{n-1}{n}.$$

Let $Y_n = \frac{1}{n} X_n$ for $n \in \mathbb{N}^*$. Verify if the following sequences: $(X_n)_{n \in \mathbb{N}^*}$, $(Y_n)_{n \in \mathbb{N}^*}$, $(Y_1 \cdot Y_2 \cdot \dots \cdot Y_n)_{n \in \mathbb{N}^*}$ converge: 1) a.s.; 2) in probability.

Solution 4.4.1: Observe that for each $\varepsilon > 0$ we have

$$P(|X_n - 1| > \varepsilon) = P(X_n > \varepsilon + 1) = \frac{1}{n} \to 0, \quad \text{as } n \to \infty.$$

Let $n \in \mathbb{N}^*$ be large enough such that $n\varepsilon > 1$. Then

$$P(|Y_n| > \varepsilon) = P(X_n > n\varepsilon) = \frac{1}{n} \to 0, \quad \text{as } n \to \infty.$$

Hence, $X_n \xrightarrow{P} 1$ and $Y_n \xrightarrow{P} 0$.

For $n \in \mathbb{N}^*$ consider $A_n = \{\omega \in \Omega : |X_n(\omega) - 1| > 1\}$, which are independent events and satisfy

$$\sum_{n=1}^{\infty} P(A_n) = \sum_{n=1}^{\infty} \frac{1}{n} = \infty.$$

By the Borel–Cantelli Lemma (see Theorem 1.10), it follows that

$$P(\limsup_{n \to \infty} A_n) = 1.$$

By Theorem 4.1, it follows that $(X_n)_{n \in \mathbb{N}^*}$ does not converge a.s. to 1, hence $(X_n)_{n \in \mathbb{N}^*}$ does not converge a.s. to a random variable.

Similarly, for each $n \in \mathbb{N}^*$ let $B_n = \left\{\omega \in \Omega : |Y_n(\omega)| > \frac{1}{2}\right\} = \left\{\omega \in \Omega : |X_n(\omega)| > \frac{n}{2}\right\}$, which are independent events and

$$\sum_{n=1}^{\infty} P(B_n) = \sum_{n=1}^{\infty} \frac{1}{n} = \infty.$$

Then

$$P(\limsup_{n \to \infty} B_n) = 1.$$

By Theorem 4.1, it follows that $(Y_n)_{n \in \mathbb{N}^*}$ does not converge a.s. to 0, hence $(Y_n)_{n \in \mathbb{N}^*}$ does not converge a.s. to a random variable.

Denote $Z_n = Y_1 \cdot Y_2 \cdot \ldots \cdot Y_n$ for $n \in \mathbb{N}^*$. Note that we have a.s.

$$0 \leq Z_{n+1} \leq Z_n \leq \frac{1}{n} X_n \leq 1.$$

Hence, $(Z_n)_{n \in \mathbb{N}^*}$ is a.s. a bounded and decreasing sequence, which implies that

$$(Z_n)_{n \in \mathbb{N}^*} \text{ is a.s. convergent.} \tag{4.5}$$

For each $\varepsilon > 0$ we write

$$\lim_{n \to \infty} P(|Z_n| > \varepsilon) \leq \lim_{n \to \infty} P(|X_n| > n\varepsilon) = 0.$$

Hence, $Z_n \xrightarrow{P} 0$. Combining this result with (4.5), we obtain that $Z_n \xrightarrow{a.s.} 0$ (we use here the property of uniqueness of the limit).

Problem 4.4.2. Give examples of sequences of independent two-valued random variables $(X_n)_{n \in \mathbb{N}^*}$ such that:

1) $X_n \xrightarrow{a.s.} 0$ and $X_n \xrightarrow{L^r} 0$ for all $r > 0$.

2) $X_n \xrightarrow{a.s.} 0$ and $X_n \xrightarrow{L^r} 0$ for all $r > 0$.

3) $X_n \xrightarrow{a.s.} 0$, $X_n \xrightarrow{P} 0$ and $X_n \xrightarrow{L^r} 0$ for a fixed $r > 0$.

Solution 4.4.2: For every $n \in \mathbb{N}^*$ let

$$P(X_n = 0) = p_n \text{ and } P(X_n = x_n) = 1 - p_n,$$

where $x_n \geq 1$ and $p_n \in (0,1)$, and for every $\varepsilon > 0$ consider the event $A_{n,\varepsilon} = \{\omega \in \Omega : X_n(\omega) \leq \varepsilon\}$. Also, we assume that $(X_n)_{n \in \mathbb{N}^*}$ is a sequence of independent random variables.

We have

$$X_n \overset{P}{\longrightarrow} 0 \iff \lim_{n \to \infty} P(A_{n,\varepsilon}) = 1 \text{ for all } \varepsilon > 0 \iff \lim_{n \to \infty} p_n = 1.$$

To justify the last equivalence, we note that for $\varepsilon \in (0,1)$

$$P(A_{n,\varepsilon}) = P(X_n = 0) = p_n, \quad n \in \mathbb{N}^*,$$

and thus

$$\lim_{n \to \infty} P(A_{n,\varepsilon}) = 1 \text{ for all } \varepsilon \in (0,1) \iff \lim_{n \to \infty} p_n = 1.$$

Since $P(A_{n,\varepsilon_1}) \leq P(A_{n,\varepsilon_2})$ (because $A_{n,\varepsilon_1} \subseteq A_{n,\varepsilon_2}$) for $0 < \varepsilon_1 \leq \varepsilon_2$ and $n \in \mathbb{N}^*$, we obtain

$$\lim_{n \to \infty} P(A_{n,\varepsilon}) = 1 \text{ for all } \varepsilon > 0 \iff \lim_{n \to \infty} P(A_{n,\varepsilon}) = 1 \text{ for all } \varepsilon \in (0,1)$$

and thus the desired equivalence holds.

In view of Theorem 4.1, we deduce

$$X_n \overset{a.s.}{\longrightarrow} 0 \iff \lim_{N \to \infty} P\left(\bigcap_{n=N}^{\infty} A_{n,\varepsilon}\right) = 1 \text{ for all } \varepsilon > 0 \iff \lim_{N \to \infty} \prod_{n=N}^{\infty} p_n = 1.$$

To justify the last equivalence, we note that for $\varepsilon \in (0,1)$ we have

$$P\left(\bigcap_{n=N}^{\infty} A_{n,\varepsilon}\right) = P\left(\bigcap_{n=N}^{\infty} \{X_n = 0\}\right) = \prod_{n=N}^{\infty} p_n, \quad N \in \mathbb{N}^*,$$

where we used the independence of the random variables, and thus

$$\lim_{N \to \infty} P\left(\bigcap_{n=N}^{\infty} A_{n,\varepsilon}\right) = 1 \text{ for all } \varepsilon \in (0,1) \iff \lim_{N \to \infty} \prod_{n=N}^{\infty} p_n = 1.$$

Since $P\left(\bigcap_{n=N}^{\infty} A_{n,\varepsilon_1}\right) \leq P\left(\bigcap_{n=N}^{\infty} A_{n,\varepsilon_2}\right)$ for $0 < \varepsilon_1 \leq \varepsilon_2$ and $N \in \mathbb{N}^*$, we obtain

$$\lim_{N \to \infty} P\left(\bigcap_{n=N}^{\infty} A_{n,\varepsilon}\right) = 1 \text{ for all } \varepsilon > 0 \iff \lim_{N \to \infty} P\left(\bigcap_{n=N}^{\infty} A_{n,\varepsilon}\right) = 1 \text{ for all } \varepsilon \in (0,1)$$

and thus the desired equivalence holds.

For every $r > 0$ we have

$$X_n \xrightarrow{L^r} 0 \iff \lim_{n \to \infty} x_n^r (1 - p_n) = 0.$$

Now, we can easily provide the desired examples.

For 1) let $p_n = \dfrac{n}{n+1}$ and $x_n = 1$, $n \in \mathbb{N}^*$.

For 2) let $p_n = \exp\left\{\dfrac{1}{n+1} - \dfrac{1}{n}\right\}$ and $x_n = \exp\left\{\dfrac{1}{1 - p_n}\right\}$, $n \in \mathbb{N}^*$.

For 3) let $p_n = \dfrac{n}{n+1}$ and $x_n = (1 - p_n)^{-\frac{1}{r}}$, $n \in \mathbb{N}^*$.

Problem 4.4.3. Let $(X_n)_{n \in \mathbb{N}}$, $(Y_n)_{n \in \mathbb{N}}$ be sequences of random variables and let X, Y be random variables such that $X_n \xrightarrow{d} X$ and $Y_n \xrightarrow{d} Y$.

1) Give an example such that $X_n + Y_n \xrightarrow{d} \not\to X + Y$.

2) Give an example such that $X_n \cdot Y_n \xrightarrow{d} \not\to X \cdot Y$.

Solution 4.4.3: 1) Let $X_n \sim Ber(0.5)$ and $Y_n = 1 - X_n$, for $n \in \mathbb{N}$, and let $X, Y \sim Ber(0.5)$ be independent. Then $X_n \xrightarrow{d} X$, $Y_n \xrightarrow{d} Y$ and $X_n + Y_n \xrightarrow{d} 1$, but, clearly the cumulative distribution function of $X + Y$ is different from the one of a constant random variable, because $P(X + Y = 0) = 0.25$.

2) Let $P(X_n = 2) = P(X_n = 0.5) = 0.5$ and $Y_n = \dfrac{1}{X_n}$, for $n \in \mathbb{N}$, and let X, Y be independent discrete random variables such that

$$P(X = 2) = P(X = 0.5) = P(Y = 2) = P(Y = 0.5) = 0.5.$$

Then $X_n \xrightarrow{d} X$ and $Y_n \xrightarrow{d} Y$, but $X_n \cdot Y_n = 1$ a.s. and $P(X \cdot Y = 4) = P(X \cdot Y = 0.25) = 0.25$ and thus $X_n \cdot Y_n \xrightarrow{d} \not\to X \cdot Y$.

Problem 4.4.4. Let $x \in \mathbb{R}$ and $(X_n)_{n \in \mathbb{N}}$ be a sequence of random variables. Prove that: $X_n \xrightarrow{P} x \iff X_n \xrightarrow{d} x$.

Solution 4.4.4: By Theorem 4.7 the statement of the problem can be reformulated as

$$X_n - x \xrightarrow{P} 0 \iff X_n - x \xrightarrow{d} 0.$$

One implication is clear. Let $\varepsilon > 0$. Then

$$P(|X_n - x| > \varepsilon) = P\Big(\{X_n - x < -\varepsilon\} \cup \{X_n - x > \varepsilon\}\Big)$$

$$\leq P(X_n - x \leq -\varepsilon) + 1 - P(X_n - x \leq \varepsilon), \quad n \in \mathbb{N}.$$

Since $X_n - x \xrightarrow{d} 0$, we deduce that $X_n - x \xrightarrow{P} 0$.

Problem 4.4.5. Let $(U_n)_{n \in \mathbb{N}^*}$ be a sequence of independent random variables such that $U_n \sim Unif[0,1]$ and define $X_n = U_n^n$ for each $n \in \mathbb{N}^*$. Does $(X_n)_{n \in \mathbb{N}^*}$ converge in probability, a.s., in mean of order r $(r > 0)$, in distribution?

Solution 4.4.5: We compute for $r > 0$

$$E(|X_n|^r) = E\big((U_n^n)^r\big) = \int_0^1 x^{nr} dx = \frac{1}{nr+1} \to 0 \text{ as } n \to \infty.$$

Hence, $(X_n)_{n \in \mathbb{N}^*}$ converges in mean of order r, in distribution, in probability (by using Theorem 4.3) to zero. Note that for these results the independence assumption is not needed.

For each $n \in \mathbb{N}^*$ we consider the events $A_n = \{\omega \in \Omega : |X_n(\omega)| > \varepsilon\}$, which are independent and satisfy

$$P(A_n) = 0, \text{ if } \varepsilon \geq 1, \text{ and } P(A_n) = 1 - \varepsilon^{\frac{1}{n}}, \text{ if } \varepsilon \in (0,1).$$

Let $\varepsilon \in (0,1)$. The series

$$\sum_{n=1}^{\infty} P(A_n) = \sum_{n=1}^{\infty} (1 - \varepsilon^{\frac{1}{n}}) = \infty,$$

because

$$\lim_{n \to \infty} \frac{1 - \varepsilon^{\frac{1}{n}}}{\frac{1}{n}} = -\log \varepsilon > 0 \text{ for } \varepsilon \in (0,1),$$

and then, by applying the limit comparison test (see [Schinazi (2012), p. 50]), the series $\sum_{n=1}^{\infty} (1 - \varepsilon^{\frac{1}{n}})$ diverges just as the series $\sum_{n=1}^{\infty} \frac{1}{n}$ does.

By the Borel–Cantelli Lemma (see Theorem 1.10), it follows that

$$P(\limsup_{n \to \infty} A_n) = 1.$$

By using Theorem 4.1, we obtain that $(X_n)_{n \in \mathbb{N}^*}$ does not converge a.s. to zero, hence it does not converge a.s. to a random variable.

Problem 4.4.6. Consider the experiment of rolling a die infinitely many times. For $n \in \mathbb{N}^*$ let X_n be the smallest number obtained in the first n rollings of a die. Does $(X_n)_{n \in \mathbb{N}^*}$ converge in probability? Does $(X_n)_{n \in \mathbb{N}^*}$ converge a.s.?

Solution 4.4.6: Let N_i be the random variable that shows the number obtained in the ith rolling, $i \in \mathbb{N}^*$. Then for all $\varepsilon > 0$ and $n \in \mathbb{N}^*$

$$P(|X_n - 1| > \varepsilon) = P(X_n > \varepsilon + 1) = P(\min\{N_1, \ldots, N_n\} > \varepsilon + 1)$$

$$= P\Big(\bigcap_{i=1}^{n} \{N_i > \varepsilon + 1\}\Big) = \prod_{i=1}^{n} \big(1 - P(N_i \leq \varepsilon + 1)\big)$$

$$= \big(1 - P(N_1 \leq \varepsilon + 1)\big)^n.$$

We get for all $\varepsilon > 0$

$$\lim_{n \to \infty} P(|X_n - 1| > \varepsilon) = 0.$$

Hence, $(X_n)_{n \in \mathbb{N}^*}$ converges in probability to 1. But it is a decreasing sequence and thus, by Problem 4.1.3, $(X_n)_{n \in \mathbb{N}^*}$ converges a.s. to 1.

Problem 4.4.7. Let $a, b > 0$. Consider $(X_n)_{n \in \mathbb{N}^*}$ to be a sequence of independent random variables such that $X_n \sim Gamma(a, b)$ for each $n \in \mathbb{N}^*$. Denote $S_n = \frac{1}{n}(X_1 + \cdots + X_n)$, $n \in \mathbb{N}^*$. Prove that $(S_n)_{n \in \mathbb{N}^*}$ converges in mean square to ab.

Solution 4.4.7: Observe that $S_n \sim Gamma\big(na, \frac{b}{n}\big)$ (see Problem 3.7.31-1)). Moreover, $E(S_n) = ab$ and $V(S_n) = \dfrac{ab^2}{n}$ (by Problem 3.4.3-1)), then

$$E\big(|S_n - ab|^2\big) = V(S_n) = \frac{ab^2}{n}.$$

So, $E(|S_n - ab|^2) \to 0$.

Problem 4.4.8. Let $(X_n)_{n \in \mathbb{N}^*}$ be a sequence of independent random variables such that $X_n \sim Poiss(\frac{1}{n})$ for $n \in \mathbb{N}^*$. Prove that $X_n \xrightarrow{P} 0$. Does $(X_n)_{n \in \mathbb{N}^*}$ converge a.s. to zero?

Solution 4.4.8: We have $E(X_n) = V(X_n) = \dfrac{1}{n}$ for $n \in \mathbb{N}^*$. Therefore,

$$E(X_n^2) = V(X_n) + E^2(X_n) = \frac{1}{n} + \frac{1}{n^2} \to 0 \text{ as } n \to \infty.$$

This implies that $(X_n)_{n \in \mathbb{N}^*}$ converges in L^2 to zero and then, by Theorem 4.3, it follows that $(X_n)_{n \in \mathbb{N}^*}$ converges in probability to zero.

For $\varepsilon \in (0, 1)$ and $n \in \mathbb{N}^*$ we consider $A_n = \{\omega \in \Omega : |X_n(\omega)| > \varepsilon\}$, which are independent events and satisfy

$$P(A_n) = P(X_n \geq 1) = 1 - P(X_n = 0) = 1 - e^{-\frac{1}{n}}.$$

The following series diverges

$$\sum_{n=1}^{\infty} P(A_n) = \sum_{n=1}^{\infty} (1 - e^{-\frac{1}{n}}) = \infty,$$

because

$$\lim_{n \to \infty} \frac{1 - e^{-\frac{1}{n}}}{\frac{1}{n}} = 1$$

and then, by applying the comparison test, the series $\sum_{n=1}^{\infty} (1 - e^{-\frac{1}{n}})$ diverges

just as the series $\sum_{n=1}^{\infty} \frac{1}{n}$ does.

By the Borel–Cantelli Lemma (see Theorem 1.10), it follows that

$$P(\limsup_{n \to \infty} A_n) = 1.$$

By using Theorem 4.1, we obtain that $(X_n)_{n \in \mathbb{N}^*}$ does not converge a.s. to zero.

Problem 4.4.9. In the below cases: 1), 2), 3), we consider $(U_n)_{n \in \mathbb{N}^*}$ to be a sequence of pairwise independent random variables. Does the sequence converge a.s., in L^2, in probability, in distribution? If yes, to which random variable?

1) $P\left(U_n = -\frac{n+2}{n}\right) = 0.3, P\left(U_n = \frac{n+1}{n}\right) = 0.7$ for each $n \in \mathbb{N}^*$;

2) $P\left(U_n = \frac{n-1}{n^2}\right) = \frac{n-1}{2n}, P\left(U_n = \frac{n}{2n-1}\right) = \frac{n+1}{2n}$ for each $n \in \mathbb{N}^*$;

3) $P\left(U_n = 1 - \frac{2}{n}\right) = P\left(U_n = 1 + \frac{1}{n}\right) = 0.5$ for each $n \in \mathbb{N}^*$.

Solution 4.4.9: We will formulate the above convergence problems in a more general context: Consider the sequence of pairwise independent random variables $(X_n)_{n \in \mathbb{N}^*}$ such that

$$P(X_n = a_n) = p_n \text{ and } P(X_n = b_n) = 1 - p_n \text{ for each } n \in \mathbb{N}^*,$$

where $(a_n)_{n \in \mathbb{N}^*}, (b_n)_{n \in \mathbb{N}^*}$ are sequences of real numbers such that $\lim_{n \to \infty} a_n = a \in \mathbb{R}$ and $\lim_{n \to \infty} b_n = b \in \mathbb{R}$ with $a_n < b_n$ for $n \in \mathbb{N}^*$, while $(p_n)_{n \in \mathbb{N}^*}$ is a sequence of probabilities from $[0,1]$ converging to some value $p \in (0,1)$.

We will use Theorem 4.10-(2), the Cauchy criterion for the convergence in L^2. Since X_m and X_n are independent random variables for $m \neq n$, $m, n \in \mathbb{N}^*$ we have

$$E\left((X_m - X_n)^2\right) = E(X_m^2) - 2E(X_m)E(X_n) + E(X_n^2)$$

$$= a_m^2 p_m + b_m^2(1 - p_m) - 2(a_m p_m + b_m(1 - p_m))(a_n p_n + b_n(1 - p_n))$$
$$+ a_n^2 p_n + b_n^2(1 - p_n).$$

Then

$$\lim_{\substack{m,n \to \infty \\ m \neq n}} E\big((X_m - X_n)^2\big) = 2(a - b)^2(1 - p)p.$$

First case: If $a \neq b$, then $(X_n)_{n \in \mathbb{N}^*}$ is not a Cauchy sequence in the L^2 sense, hence it does not converge in L^2 to a random variable. Observe that there exists a bound $M > 0$ such that $P(\sup_{n \in \mathbb{N}^*} |X_n| \leq M) = 1$, since $(a_n)_{n \in \mathbb{N}^*}$, $(b_n)_{n \in \mathbb{N}^*}$ are convergent sequences, hence bounded. We have

$$\lim_{c \to \infty} \sup_{n \in \mathbb{N}^*} E\big(|X_n|^2 \mathbb{I}_{\{|X_n|^2 > c\}}\big) \leq \lim_{c \to \infty} M^2 \cdot \mathbb{I}_{\{M^2 > c\}} = 0.$$

This implies that $\big(|X_n|^2\big)_{n \in \mathbb{N}^*}$ is uniformly integrable (see Definition 4.5). By Theorem 4.12 and the above reasoning it follows that $(X_n)_{n \in \mathbb{N}^*}$ does not converge in probability to a random variable, hence $(X_n)_{n \in \mathbb{N}^*}$ does not converge a.s. to a random variable (by Theorem 4.3).

For $n \in \mathbb{N}^*$ we write the cumulative distribution function for X_n

$$F_{X_n}(x) = \begin{cases} 0, & \text{if } x < a_n \\ p_n, & \text{if } a_n \leq x < b_n \\ 1, & \text{if } b_n \leq x. \end{cases}$$

Observe that $F_{X_n}(x) \to F(x)$ for each $x \in \mathbb{R} \setminus \{a, b\}$, where

$$F(x) = \begin{cases} 0, & \text{if } x < a \\ p, & \text{if } a \leq x < b \\ 1, & \text{if } b \leq x. \end{cases}$$

Hence, considering the random variable X with the distribution

$$P(X = a) = p \text{ and } P(X = b) = 1 - p,$$

we get $X_n \xrightarrow{d} X$.

Second case: If $a = b$, then $(X_n)_{n \in \mathbb{N}^*}$ is a Cauchy sequence in the L^2 sense, i.e.,

$$\lim_{m,n \to \infty} E\big((X_m - X_n)^2\big) = 0,$$

hence it converges in L^2. We write

$$X_n - a = (a_n - a)\mathbb{I}_{\{X_n = a_n\}} + (b_n - a)\mathbb{I}_{\{X_n = b_n\}} \text{ for each } n \in \mathbb{N}^*.$$

The indicator function is bounded, so $(X_n)_{n \in \mathbb{N}^*}$ converges a.s. to the value a. By Theorem 4.3 and Theorem 4.5 it follows that $(X_n)_{n \in \mathbb{N}^*}$ converges a.s., in L^2, in probability and in distribution to the value a.

Solution for the problem: 1) We consider $a_n = -\frac{n+2}{n}$, $b_n = \frac{n+1}{n}$ for each $n \in \mathbb{N}^*$. Then $\lim\limits_{n\to\infty} a_n \neq \lim\limits_{n\to\infty} b_n$, hence $(U_n)_{n\in\mathbb{N}^*}$ does not converge a.s. or in L^2 or in probability to a random variable, but it converges in distribution to a random variable U with $P(U = -1) = 0.3$, $P(U = 1) = 0.7$.

2) We consider $a_n = \frac{n-1}{n^2}$, $b_n = \frac{n}{2n-1}$ for each $n \in \mathbb{N}^*$. Then we have $\lim\limits_{n\to\infty} a_n \neq \lim\limits_{n\to\infty} b_n$, hence $(U_n)_{n\in\mathbb{N}^*}$ does not converge a.s. or in L^2 or in probability to a random variable, but converges in distribution to a random variable U with

$$P(U = 0) = P(U = 0.5) = 0.5.$$

3) We consider $a_n = \frac{n-2}{n}$, $b_n = \frac{n+1}{n}$ for each $n \in \mathbb{N}^*$. Then, $\lim\limits_{n\to\infty} a_n = \lim\limits_{n\to\infty} b_n = 1$, hence $(U_n)_{n\in\mathbb{N}^*}$ converges a.s., in L^2, in probability and in distribution to 1.

Problem 4.4.10. Consider a sequence $(X_n)_{n\in\mathbb{N}^*}$ of random variables such that $X_n \sim Unif[1, n^2 + 1]$ for each $n \in \mathbb{N}^*$. Prove that $(X_n)_{n\in\mathbb{N}^*}$ does not converge in distribution to a random variable and that $X_n \xrightarrow{a.s.} \infty$.

Solution 4.4.10: For $n \in \mathbb{N}^*$ the distribution function F_{X_n} of X_n is

$$F_{X_n}(x) = \begin{cases} 0, & \text{if } x < 1 \\ \dfrac{x-1}{n^2}, & \text{if } 1 \leq x < n^2 + 1 \\ 1, & \text{if } n^2 + 1 \leq x. \end{cases}$$

We see that

$$\lim_{n\to\infty} F_{X_n}(x) = 0 \quad \text{for all } x \in \mathbb{R}.$$

But the limit function is not a distribution function. Hence, $(X_n)_{n\in\mathbb{N}^*}$ does not converge in distribution to a random variable.

Let $\varepsilon > 0$. For $n \in \mathbb{N}^*$, consider $A_n = \left\{ \omega \in \Omega : \frac{1}{X_n(\omega)} > \varepsilon \right\}$, which satisfy

$$\sum_{n=1}^{\infty} P(A_n) = \sum_{n=1}^{\infty} P\left(X_n < \frac{1}{\varepsilon}\right) = N_\varepsilon - 1 + \left(\frac{1}{\varepsilon} - 1\right) \sum_{n=N_\varepsilon}^{\infty} \frac{1}{n^2} < \infty,$$

where $N_\varepsilon \in \mathbb{N}^*$ is the smallest number such that $N_\varepsilon^2 + 1 > \frac{1}{\varepsilon}$, if $\varepsilon \in (0,1)$ while $P(A_n) = 0$, for all $n \in \mathbb{N}^*$, if $\varepsilon \geq 1$. By the Borel–Cantelli Lemma (see Theorem 1.10), it follows that

$$P(\limsup_{n\to\infty} A_n) = 0.$$

By Theorem 4.1, it follows that $\left(\dfrac{1}{X_n}\right)_{n\in\mathbb{N}^*}$ converges a.s. to zero, which implies that $(X_n)_{n\in\mathbb{N}^*}$ converges a.s. to ∞.

Problem 4.4.11. We consider a sequence of pairwise independent random variables $(X_n)_{n\in\mathbb{N}^*}$ where $X_n \sim Exp(\lambda_n)$, $\lambda_n > 0$, $n \in \mathbb{N}^*$ and $\lim\limits_{n\to\infty} \lambda_n = \lambda > 0$. Prove that $X_n \xrightarrow{d} X$, where $X \sim Exp(\lambda)$, but $(X_n)_{n\in\mathbb{N}^*}$ does not converge: 1) a.s.; 2) in L^2; 3) in probability to a random variable.

Solution 4.4.11: For $n \in \mathbb{N}$ and $x \in \mathbb{R}$ we write the cumulative distribution function for $X_n \sim Exp(\lambda_n)$ and also for $X \sim Exp(\lambda)$

$$F_{X_n}(x) = \begin{cases} 0, & \text{if } x \leq 0 \\ 1 - e^{-\lambda_n x}, & \text{if } x > 0 \end{cases}$$

and

$$F_X(x) = \begin{cases} 0, & \text{if } x \leq 0 \\ 1 - e^{-\lambda x}, & \text{if } x > 0. \end{cases}$$

The limit $\lim\limits_{n\to\infty} \lambda_n = \lambda$ implies $X_n \xrightarrow{d} X$.

Let $\varepsilon > 0$ and $m, n \in \mathbb{N}$, $m \neq n$. We compute

$$P(|X_m - X_n| > \varepsilon) = 1 - \int_{-\varepsilon}^{\varepsilon} f_{X_m - X_n}(z)\,dz,$$

where $f_{X_m - X_n}$ is a density function of $X_m - X_n$ which can be given, in view of Example 2.15, by

$$f_{X_m - X_n}(z) = \int_{\mathbb{R}} f_{X_m}(u) f_{-X_n}(z - u)\,du \quad \text{for } z \in \mathbb{R},$$

where f_{-X_n} is a density function of $-X_n$ which for every $z \in \mathbb{R}$ satisfies

$$f_{-X_n}(z - u) = \begin{cases} 0, & \text{if } z \geq u \\ \lambda_n \exp\{\lambda_n(z - u)\}, & \text{if } z < u. \end{cases}$$

We compute for $z > 0$

$$f_{X_m - X_n}(z) = \int_z^\infty f_{X_m}(u) f_{-X_n}(z - u)\,du = \frac{\lambda_m \lambda_n e^{-\lambda_m z}}{\lambda_m + \lambda_n}.$$

For $z \leq 0$ we have

$$f_{X_m - X_n}(z) = \int_0^\infty f_{X_m}(u) f_{-X_n}(z - u)\,du = \frac{\lambda_m \lambda_n e^{\lambda_n z}}{\lambda_m + \lambda_n}.$$

Therefore,

$$P(|X_m - X_n| > \varepsilon) = 1 - \int_{-\varepsilon}^{\varepsilon} f_{X_m - X_n}(z)dz = \frac{\lambda_m e^{-\lambda_n \varepsilon} + \lambda_n e^{-\lambda_m \varepsilon}}{\lambda_m + \lambda_n}$$

and thus

$$\lim_{\substack{m,n \to \infty \\ m \neq n}} P(|X_m - X_n| > \varepsilon) = e^{-\lambda \varepsilon}.$$

By the Cauchy criterion from Theorem 4.10, we obtain that $(X_n)_{n \in \mathbb{N}^*}$ does not converge in probability to a random variable, hence it does not converge neither a.s., nor in L^2 (by Theorem 4.3).

Problem 4.4.12. Let $g : [0,1] \to \mathbb{R}$ be a continuous function. For every $x \in [0,1]$ let $(X_{x,n})_{n \in \mathbb{N}^*}$ be a sequence of independent random variables such that $P(X_{x,n} = 1) = 1 - P(X_{x,n} = 0) = x$ for $n \in \mathbb{N}^*$.
1) Prove the following uniform WLLN:

$$\frac{1}{n} \sum_{k=1}^{n} X_{x,k} \xrightarrow{P} x \text{ uniformly with respect to } x \in [0,1].$$

2) Prove that

$$g\left(\frac{1}{n} \sum_{k=1}^{n} X_{x,k}\right) \xrightarrow{L^1} g(x) \text{ uniformly with respect to } x \in [0,1].$$

3) Prove Weierstrass' theorem: there exists a sequence of polynomial functions that converges uniformly on $[0,1]$ to g.

Solution 4.4.12: 1) Let $\bar{X}_{x,n} = \frac{1}{n} \sum_{k=1}^{n} X_{x,k}$, $n \in \mathbb{N}^*$. Using the independence of the random variables, we get that $E(\bar{X}_{x,n}) = x$ and $V(\bar{X}_{x,n}) = \frac{x(1-x)}{n}$, $x \in [0,1]$, $n \in \mathbb{N}^*$. By Chebyshev's inequality (see Theorem 3.6-(2)), we have for every $\varepsilon > 0$

$$P\left(|\bar{X}_{x,n} - x| \geq \varepsilon\right) \leq \frac{x(1-x)}{\varepsilon^2 n} \leq \frac{1}{4\varepsilon^2 n}$$

for all $x \in [0,1]$ and $n \in \mathbb{N}^*$. Hence, for every $\varepsilon > 0$

$$\lim_{n \to \infty} P\left(|\bar{X}_{x,n} - x| > \varepsilon\right) = 0 \text{ uniformly with respect to } x \in [0,1].$$

2) g is uniformly continuous, because $[0,1]$ is a compact set, and thus, for every $\varepsilon > 0$ there exists $\delta_\varepsilon > 0$ such that $|g(x) - g(y)| \leq \varepsilon$ for all $x, y \in [0,1]$ such that $|x - y| \leq \delta_\varepsilon$. Thus, for every $\varepsilon > 0$

$$P\left(|\bar{X}_{x,n} - x| \leq \delta_\varepsilon\right) \leq P\left(|g(\bar{X}_{x,n}) - g(x)| \leq \varepsilon\right).$$

This implies

$$P\left(\left|g(\bar{X}_{x,n}) - g(x)\right| > \varepsilon\right) \leq P\left(\left|\bar{X}_{x,n} - x\right| > \delta_\varepsilon\right)$$

for all $x \in [0,1]$ and $n \in \mathbb{N}^*$. Using 1), we deduce that for every $\varepsilon > 0$

$$\lim_{n\to\infty} P\left(\left|g(\bar{X}_{x,n}) - g(x)\right| > \varepsilon\right) = 0 \text{ uniformly with respect to } x \in [0,1].$$

Let $M = \max_{x\in[0,1]} |g(x)|$. For every $\varepsilon > 0$ we have

$$E\left(\left|g(\bar{X}_{x,n}) - g(x)\right|\right)$$
$$\leq \varepsilon \cdot P\left(\left|g(\bar{X}_{x,n}) - g(x)\right| \leq \varepsilon\right) + 2M \cdot P\left(\left|g(\bar{X}_{x,n}) - g(x)\right| > \varepsilon\right),$$

for all $x \in [0,1]$ and $n \in \mathbb{N}^*$. Hence, for every $\varepsilon > 0$

$$\limsup_{n\to\infty} E\left(\left|g(\bar{X}_{x,n}) - g(x)\right|\right) \leq \varepsilon \text{ uniformly for all } x \in [0,1]$$

and thus

$$\lim_{n\to\infty} E\left(\left|g(\bar{X}_{x,n}) - g(x)\right|\right) = 0 \text{ uniformly with respect to } x \in [0,1].$$

3) Observe that

$$P\left(\bar{X}_{x,n} = \frac{k}{n}\right) = C(n,k)x^k(1-x)^{n-k} \text{ for each } k \in \{0,\dots,n\}.$$

For every $n \in \mathbb{N}^*$ denote

$$B_n(x) = \sum_{k=0}^n C(n,k)x^k(1-x)^{n-k}g\left(\frac{k}{n}\right), \quad x \in [0,1].$$

Then

$$E\left(\left|g(\bar{X}_{x,n}) - g(x)\right|\right) = \sum_{k=0}^n C(n,k)x^k(1-x)^{n-k}\left|g\left(\frac{k}{n}\right) - g(x)\right|$$
$$\geq |B_n(x) - g(x)| \text{ for all } x \in [0,1].$$

In view of 2), we get that the sequence of polynomial functions $(B_n)_{n\in\mathbb{N}^*}$ converges uniformly on $[0,1]$ to g.

The above results are connected to Bernstein's proof of Weierstrass' theorem (see, e.g., [Gut (2005), p. 277]).

Problem 4.4.13. Let $(a_n)_{n\in\mathbb{N}^*}$ be a sequence of strictly positive real numbers. Consider $(X_n)_{n\in\mathbb{N}^*}$ to be a sequence of independent random variables such that $X_n \sim Unif[0,a_n]$ for each $n \in \mathbb{N}^*$. Denote $Y_n = \frac{1}{n}\sum_{i=1}^n X_i$, $n \in \mathbb{N}^*$. Prove that, if $\lim_{n\to\infty} a_n = a \in \mathbb{R}$, then $Y_n \xrightarrow{a.s.} \frac{a}{2}$. Moreover, study the a.s. convergence of $(Y_n)_{n\in\mathbb{N}^*}$, if $a_n = n^{\frac{1}{4}}$, respectively if $a_n = n$, for $n \in \mathbb{N}^*$.

Solution 4.4.13: Assume that $\lim\limits_{n\to\infty} a_n = a$. We have $V(X_n) = \dfrac{a_n^2}{12}$ for each $n \in \mathbb{N}^*$. By the assumption, it follows that the sequence $(a_n)_{n\in\mathbb{N}^*}$ is bounded, hence we have $\sum\limits_{n=1}^{\infty} \dfrac{1}{n^2}V(X_n) < \infty$. By the convergence assumption of this problem together with the Stolz–Cesàro Theorem, we have

$$\frac{1}{n}\sum_{k=1}^{n} a_k \to a. \qquad (4.6)$$

We use Theorem 4.18 to conclude that

$$\frac{1}{n}\sum_{k=1}^{n}\left(X_k - E(X_k)\right) \xrightarrow{a.s.} 0,$$

which is equivalent to

$$\frac{1}{n}\sum_{k=1}^{n} X_k - \frac{1}{n}\sum_{k=1}^{n}\frac{a_k}{2} \xrightarrow{a.s.} 0.$$

By Theorem 4.6 and (4.6) it follows that $Y_n \xrightarrow{a.s.} \frac{a}{2}$.

If $a_n = n^{\frac{1}{4}}, n \in \mathbb{N}^*$, then

$$\sum_{n=1}^{\infty} \frac{1}{n^2}V(X_n) = \frac{1}{12}\sum_{n=1}^{\infty} \frac{1}{n^{\frac{3}{2}}} < \infty.$$

Since $\lim\limits_{n\to\infty} a_n = \infty$, using the Stolz–Cesàro Theorem, we have

$$\frac{1}{n}\sum_{k=1}^{n} a_k \to \infty. \qquad (4.7)$$

By Theorem 4.18, it follows

$$\frac{1}{n}\sum_{k=1}^{n}\left(X_k - E(X_k)\right) \xrightarrow{a.s.} 0,$$

which is equivalent to

$$\frac{1}{n}\sum_{k=1}^{n} X_k - \frac{1}{n}\sum_{k=1}^{n}\frac{a_k}{2} \xrightarrow{a.s.} 0.$$

By (4.7), it follows that $Y_n \xrightarrow{a.s.} \infty$.

Let $a_n = n$ for $n \in \mathbb{N}^*$. The Cauchy–Schwarz inequality (for real numbers) implies

$$(X_1 + \ldots + X_n)\cdot\left(1 + \ldots + \frac{1}{n}\right) \geq \left(\sqrt{\frac{X_1}{1}} + \ldots + \sqrt{\frac{X_n}{n}}\right)^2 \quad \text{a.s.,}$$

hence

$$\frac{1}{n}(X_1 + \ldots + X_n) \geq \left(\frac{\sqrt{\frac{X_1}{1}} + \ldots + \sqrt{\frac{X_n}{n}}}{n}\right)^2 \cdot \frac{n}{1 + \ldots + \frac{1}{n}} \quad \text{a.s.}$$

The first factor on the right side of the above inequality converges a.s. to $E^2(\sqrt{U}) = \frac{4}{9}$, where $U \sim Unif[0,1]$, in view of the SLLN (see Theorem 4.19), and for the second factor we have

$$\lim_{n \to \infty} \frac{n}{1 + \ldots + \frac{1}{n}} = \infty,$$

by using the Stolz–Cesàro Theorem. It follows that $Y_n \xrightarrow{a.s.} \infty$.

Problem 4.4.14. Let $(X_n)_{n \in \mathbb{N}^*}$ be a sequence of independent random variables such that $X_n \sim Unif[0,n]$ for all $n \in \mathbb{N}^*$ and let $0 < r < 2$. Prove that

$$\lim_{n \to \infty} \frac{1}{n^r}(X_1 + \ldots + X_n) = \infty \quad \text{a.s.}$$

Solution 4.4.14: The Cauchy–Schwarz inequality (for real numbers) implies

$$(X_1 + \ldots + X_n) \cdot \left(1 + \ldots + \frac{1}{n}\right) \geq \left(\sqrt{\frac{X_1}{1}} + \ldots + \sqrt{\frac{X_n}{n}}\right)^2$$

and thus

$$\frac{1}{n^r}(X_1 + \ldots + X_n) \geq \left(\frac{\sqrt{\frac{X_1}{1}} + \ldots + \sqrt{\frac{X_n}{n}}}{n}\right)^2 \cdot \frac{n^{2-r}}{1 + \ldots + \frac{1}{n}}.$$

The first factor on the right side of the above inequality converges a.s. to $E^2(\sqrt{Y}) = \frac{4}{9}$, where $Y \sim Unif[0,1]$, in view of the SLLN (see Theorem 4.19), and for the second factor we have

$$\lim_{n \to \infty} \frac{n^{2-r}}{1 + \ldots + \frac{1}{n}} = \infty,$$

by using the Stolz–Cesàro Theorem and L'Hôpital's rule.

Problem 4.4.15. Let $(X_n)_{n \in \mathbb{N}^*}$ be a sequence of independent random variables such that $X_n \sim Unif[0,n]$ for all $n \in \mathbb{N}^*$. Prove that

$$\frac{1}{n^2}(X_1 + \ldots + X_n) \xrightarrow{L^2} \frac{1}{4}.$$

Solution 4.4.15: For every $n \in \mathbb{N}^*$, let

$$\bar{X}_n = \frac{1}{n^2} \sum_{k=1}^{n} \left(X_k - \frac{k}{2} \right).$$

Then, $E(\bar{X}_n) = 0$ and $V(\bar{X}_n) = E(\bar{X}_n^2) = \frac{1}{36n} \left(1 + \frac{1}{n} \right) \left(1 + \frac{1}{2n} \right)$.

Note that by the above computations we obtain $\bar{X}_n \xrightarrow{L^2} 0$. But

$$\bar{X}_n + \frac{1}{4n} = \frac{1}{n^2} (X_1 + \ldots + X_n) - \frac{1}{4}$$

and thus we have

$$\frac{1}{n^2} (X_1 + \ldots + X_n) \xrightarrow{L^2} \frac{1}{4},$$

since $E(\bar{X}_n) = 0$ for every $n \in \mathbb{N}^*$ and $E\left(\left(\bar{X}_n + \frac{1}{4n} \right)^2 \right) \to 0$.

Problem 4.4.16. Let $(X_n)_{n \in \mathbb{N}^*}$ be a sequence of independent random variables such that $X_n \sim Unif[0, n]$ for each $n \in \mathbb{N}^*$. Prove that

$$\limsup_{n \to \infty} X_n = \infty \quad \text{and} \quad \liminf_{n \to \infty} X_n = 0 \quad \text{a.s.}$$

Solution 4.4.16: Let $M \in \mathbb{N}^*$. We have

$$\lim_{n \to \infty} P(X_n > M) = \lim_{n \to \infty} \left(1 - \frac{M}{n} \right) = 1.$$

We consider the events $A_n = \{ \omega \in \Omega : X_n(\omega) > M \}$, $n \in \mathbb{N}^*$, which are independent. Also, consider $B_n = \left\{ \omega \in \Omega : X_n(\omega) \leq \frac{1}{M} \right\}$, $n \in \mathbb{N}^*$, which are independent events. It holds

$$\sum_{n=1}^{\infty} P(A_n) = \sum_{n=1}^{\infty} P(B_n) = \infty.$$

By the Borel–Cantelli Lemma (see Theorem 1.10), it follows that

$$P(\limsup_{n \to \infty} A_n) = P(\limsup_{n \to \infty} B_n) = 1.$$

This implies

$$1 = \lim_{n \to \infty} P\left(\bigcup_{k \geq n} A_k \right) = \lim_{n \to \infty} P(\sup_{k \geq n} X_k > M)$$

and

$$1 = \lim_{n \to \infty} P\left(\bigcup_{k \geq n} B_k \right) = \lim_{n \to \infty} P\left(\inf_{k \geq n} X_k \leq \frac{1}{M} \right).$$

Hence, for each $M \in \mathbb{N}^*$

$$P\left(\bigcap_{n=1}^{\infty} \{ \sup_{k \geq n} X_k > M \} \right) = 1$$

and

$$P\left(\bigcap_{n=1}^{\infty} \{ \inf_{k \geq n} X_k \leq \frac{1}{M} \} \right) = 1.$$

We obtain

$$P\left(\bigcap_{M=1}^{\infty} \bigcap_{n=1}^{\infty} \{ \sup_{k \geq n} X_k > M \} \right) = P\left(\bigcap_{M=1}^{\infty} \bigcap_{n=1}^{\infty} \{ \inf_{k \geq n} X_k \leq \frac{1}{M} \} \right) = 1.$$

So, we conclude

$$\limsup_{n \to \infty} X_n = \lim_{n \to \infty} \sup_{k \geq n} X_k = \infty \quad \text{a.s.}$$

and

$$\liminf_{n \to \infty} X_n = \lim_{n \to \infty} \inf_{k \geq n} X_k = 0 \quad \text{a.s.}$$

Problem 4.4.17. Let $(X_n)_{n \in \mathbb{N}}$ be a sequence of random variables that converges in probability to a random variable X. Let Y be a random variable such that X and Y are independent and identically distributed. Prove that:

1) If $P(X \leq x) \in \{0, 1\}$ for every $x \in \mathbb{R}$, then $X_n \xrightarrow{P} Y$.

2) If there exists $x \in \mathbb{R}$ such that $P(X \leq x) \in (0, 1)$, then $X_n \xrightarrow{P} \!\!\!\!/ \; Y$, even though $X_n \xrightarrow{d} Y$.

Solution 4.4.17: 1) Since the distribution function $F_X : \mathbb{R} \to \{0, 1\}$ of X (and Y) is increasing and right-continuous, there exists $x_0 \in \mathbb{R}$ such that

$$F_X(x) = \begin{cases} 0, & x < x_0 \\ 1, & x \geq x_0. \end{cases}$$

Since

$$P(X \leq x_0) - P(X < x_0) = F(x_0) - \lim_{x \nearrow x_0} F(x) = P(Y \leq x_0) - P(Y < x_0),$$

we deduce that $P(X = x_0) = P(Y = x_0) = 1$ and thus $P(X = Y) = 1$. Hence, $X_n \xrightarrow{P} Y$.

2) Suppose that $X_n \xrightarrow{P} Y$. Then, by Theorem 4.5, we have $P(X = Y) = 1$.

Let $p = P(X \le x)$. We have also $p = P(Y \le x)$, because Y has the same distribution as X. But X and Y are independent,

$$P(\{X \le x\} \cap \{Y > x\}) = P(X \le x) \cdot P(Y > x) = p(1 - p) > 0,$$

then we deduce that $P(X \ne Y) > 0$, because

$$P(\{X \le x\} \cap \{Y > x\}) \le P(X \ne Y),$$

which gives a contradiction.

$X_n \xrightarrow{d} Y$, since X and Y have the same distribution function and convergence in probability implies convergence in distribution (see Theorem 4.3).

Problem 4.4.18. We consider a sequence of pairwise independent random variables $(X_n)_{n \in \mathbb{N}}$, where $X_n \sim Unif[a_n, b_n]$ with $a_n < b_n$ for each $n \in \mathbb{N}$. $\lim\limits_{n \to \infty} a_n = a \in \mathbb{R}$ and $\lim\limits_{n \to \infty} b_n = b \in \mathbb{R}$. Study the following types of convergence for $(X_n)_{n \in \mathbb{N}}$: a.s., in L^2, in probability, in distribution.

Solution 4.4.18:

First case $a \ne b$: For $n \in \mathbb{N}$ and $x \in \mathbb{R}$ we write the distribution function for $X_n \sim Unif[a_n, b_n]$ and also for a random variable $X \sim Unif[a, b]$:

$$F_{X_n}(x) = \begin{cases} 0, & \text{if } x \le a_n \\ \dfrac{x - a_n}{b_n - a_n}, & \text{if } a_n \le x < b_n \\ 1, & \text{if } b_n \le x \end{cases}$$

and

$$F_X(x) = \begin{cases} 0, & \text{if } x \le a \\ \dfrac{x - a}{b - a}, & \text{if } a \le x < b \\ 1, & \text{if } b \le x. \end{cases}$$

The limits $\lim\limits_{n \to \infty} a_n = a$ and $\lim\limits_{n \to \infty} b_n = b$ imply $X_n \xrightarrow{d} X$. We compute

$$\lim_{\substack{m,n \to \infty \\ m \ne n}} E(|X_m - X_n|^2) = \lim_{\substack{m,n \to \infty \\ m \ne n}} \left(E(X_m^2) + E(X_n^2) - 2E(X_m)E(X_n) \right)$$

$$= \frac{(b - a)^2}{6} \ne 0.$$

By the Cauchy criterion from Theorem 4.10, we obtain that $(X_n)_{n \in \mathbb{N}}$ does not converge in L^2 to a random variable.

But $X_n \sim Unif[a_n, b_n]$ for each $n \in \mathbb{N}$, hence there exists a bound $M > 0$ such that $P(\sup_{n \in \mathbb{N}} |X_n| \leq M) = 1$, since $(a_n)_{n \in \mathbb{N}}$, $(b_n)_{n \in \mathbb{N}}$ are convergent sequences, hence bounded. We have

$$\lim_{c \to \infty} \sup_{n \in \mathbb{N}} E\big(|X_n|^2 \mathbb{I}_{\{|X_n|^2 > c\}}\big) \leq \lim_{c \to \infty} M^2 \cdot \mathbb{I}_{\{M^2 > c\}} = 0.$$

This implies that $\big(|X_n|^2\big)_{n \in \mathbb{N}}$ is uniformly integrable (see Definition 4.5). By Theorem 4.12 and the above reasoning it follows that $(X_n)_{n \in \mathbb{N}}$ does not converge in probability, hence $(X_n)_{n \in \mathbb{N}}$ does not converge a.s. to a random variable (by Theorem 4.3).

Second case $a = b$: In this case

$$\lim_{m, n \to \infty} E\big(|X_m - X_n|^2\big) = \frac{(b-a)^2}{6} = 0.$$

By the Cauchy criterion Theorem 4.10-(2), we obtain that $(X_n)_{n \in \mathbb{N}}$ converges in L^2. Hence $(X_n)_{n \in \mathbb{N}}$ converges also in probability, respectively in distribution, to a.

For $n \in \mathbb{N}$ let $\Omega_n = \{\omega \in \Omega : a_n \leq X_n(\omega) \leq b_n\}$. Note that $P(\Omega_n) = 1$, since $X_n \sim Unif[a_n, b_n]$. Consider $\Omega^* = \bigcap_{n=1}^{\infty} \Omega_n$. We have $P(\Omega^*) = 1$. We obtain for each $n \in \mathbb{N}$ and $\omega \in \Omega^*$

$$a_n - a \leq X_n(\omega) - a \leq b_n - a.$$

Then $X_n(\omega) \to a$ for each $\omega \in \Omega^*$, hence $X_n \xrightarrow{a.s.} a$.

Problem 4.4.19. Let $r \geq 1$, $p \in (0,1)$, $x \in \mathbb{R}$ and let $(X_n)_{n \in \mathbb{N}^*}$ be a sequence of independent random variables such that

$$P(X_n = -n^r) = P(X_n = n^r) = \frac{p}{2} \quad \text{and} \quad P(X_n = x) = 1 - p$$

for all $n \in \mathbb{N}^*$. Prove that $(X_n)_{n \in \mathbb{N}^*}$ does not obey the WLLN.

Solution 4.4.19: Let $\bar{X}_n = \frac{1}{n}(X_1 + \ldots + X_n)$ for $n \in \mathbb{N}^*$. Suppose $(X_n)_{n \in \mathbb{N}^*}$ obeys the WLLN. Then $\bar{X}_n \xrightarrow{P} x(1 - p)$. By Theorem 4.6, we have

$$\frac{1}{n} X_n = \bar{X}_n - \frac{n-1}{n} \bar{X}_{n-1} \xrightarrow{P} 0.$$

Since $P\left(\frac{1}{n}|X_n| = n^{r-1}\right) = p$, $n \in \mathbb{N}^*$ and $r \geq 1$, we deduce that

$$P\left(\frac{1}{n}|X_n| > \varepsilon\right) \not\to 0, \text{ as } n \to \infty, \text{ for } \varepsilon \in (0,1)$$

and thus $\dfrac{1}{n} X_n \overset{P}{\not\to} 0$, which gives a contradiction.

Problem 4.4.20. Let $\lambda > 0$ and $(Z_n)_{n \in \mathbb{N}^*}$ be a sequence of independent random variables with the $Exp(\lambda)$ distribution. For every $n \in \mathbb{N}^*$ let

$$X_n = \frac{1}{n}(Z_1 + \ldots + Z_n)$$

and

$$Y_n = \max\{Z_1, \ldots, Z_n\} - \frac{\log n}{\lambda}.$$

1) Write a function in Matlab that, for given $n, N \in \mathbb{N}^*$ and $\lambda > 0$, generates a vector of N random numbers that follow the distribution of $X_n + Y_n$ and plots the corresponding empirical cumulative distribution function. Test your function for $n \in \{50, 100, 200\}$, $N \in \{100, 500, 1000\}$ and $\lambda \in \{1, 2, 3\}$.
2) Prove that $(X_n + Y_n)_{n \in \mathbb{N}^*}$ converges in distribution to a random variable Z having the following cumulative distribution function

$$F_Z(x) = \exp\left\{ -\exp\{1 - \lambda x\} \right\}, \quad x \in \mathbb{R}.$$

Add the corresponding plot of this function to the plot given for 1).
3) Prove that

$$\frac{\max\{Z_1, \ldots, Z_n\}}{\log n} \overset{P}{\longrightarrow} \frac{1}{\lambda}.$$

Solution 4.4.20: 1)

```
function exp_conv_dist(lambda,n,N)
z=exprnd(1/lambda,n,N);
x=mean(z,1);
y=max(z,[ ],1)-log(n)/lambda;
clf; hold on;
ecdf(x+y);
fplot(@(x)exp(-exp(1-lambda*x)),[min(x+y) max(x+y)]);
legend('Empirical CDF','CDF');
end
```

>> exp_conv_dist(2,100,500)

2) Since $(Z_n)_{n \in \mathbb{N}^*}$ obeys the SLLN (see Theorem 4.19), $X_n \overset{a.s.}{\longrightarrow} \frac{1}{\lambda}$ and thus $X_n \overset{P}{\longrightarrow} \frac{1}{\lambda}$. For every $n \in \mathbb{N}^*$ the cumulative distribution function F_{Y_n} of Y_n is given by

$$F_{Y_n}(x) = P\left(\bigcap_{k=1}^{n} \left\{ Z_k \leq x + \frac{\log n}{\lambda} \right\} \right) = \prod_{k=1}^{n} P\left(Z_k \leq x + \frac{\log n}{\lambda} \right)$$

$$= \begin{cases} 0, & \text{if } x \leq -\dfrac{\log n}{\lambda} \\ \left(1 - \exp\{-\log n - \lambda x\}\right)^n, & \text{if } x > -\dfrac{\log n}{\lambda}. \end{cases}$$

We deduce that $Y_n \xrightarrow{d} Y$, where Y is a random variable having the following cumulative distribution function:

$$F_Y(x) = \exp\{-e^{-\lambda x}\}, \quad x \in \mathbb{R}.$$

Since $Y + \frac{1}{\lambda}$ has the same cumulative distribution function as Z, we deduce, by Theorem 4.7, that $X_n + Y_n \xrightarrow{d} Z$.

3) Using Theorem 4.7, we deduce that $\dfrac{1}{\log n} Y_n \xrightarrow{d} 0$ and thus

$$\frac{\max\{Z_1, \ldots, Z_n\}}{\log n} \xrightarrow{d} \frac{1}{\lambda}.$$

Problem 4.4.4 implies that

$$\frac{\max\{Z_1, \ldots, Z_n\}}{\log n} \xrightarrow{P} \frac{1}{\lambda}.$$

Problem 4.4.21. Let $(X_n)_{n \in \mathbb{N}^*}$ be a sequence of strictly positive, independent and identically distributed random variables with mean $\mu > 0$ and variance $\sigma^2 > 0$. For every $n \in \mathbb{N}^*$ let $\bar{X}_n = \dfrac{1}{n} \sum_{i=1}^{n} X_n$.

1) Suppose that $X_n \sim Gamma(a, b)$ for all $n \in \mathbb{N}^*$, where $a, b > 0$. Write a function in Matlab that, for given $n, k \in \mathbb{N}^*$ and $a, b > 0$, generates a vector of k random numbers that follow the distribution of $\dfrac{1}{\bar{X}_n}$ and plots the corresponding histogram. Test your function for $n \in \{50, 100, 200\}$, $k \in \{100, 500, 1000\}$, $a \in \{1, 2, 3\}$, $b \in \{1, 2, 3\}$. What can we say about the shape of the plot?

2) Prove that $\left(\dfrac{\mu^2 \sqrt{n}}{\sigma} \left(\dfrac{1}{\bar{X}_n} - \dfrac{1}{\mu}\right)\right)_{n \in \mathbb{N}^*}$ converges in distribution to a random variable having standard normal distribution $N(0, 1)$.

Solution 4.4.21: 1) The following code is related to the *delta-method* (see [Billingsley (1986), Example 27.2, p. 368]).

```
function delta_method_gam(a,b,n,k)
x=gamrnd(a,b,n,k);
xx=1./mean(x,1);
clf; hold on;
histogram(xx,'Normalization','probability');
end
```

```
>>delta_method_gam(1,2,100,1000)
```

We observe that the histogram takes the shape of the graphic of a density function of a normal distribution.

2) By the SLLN (see Theorem 4.18) we have $\bar{X}_n \xrightarrow{a.s.} \mu$, hence

$$\frac{\mu}{\bar{X}_n} \xrightarrow{a.s.} 1 \text{ and } \frac{\mu}{\bar{X}_n} \xrightarrow{P} 1.$$

By Theorem 4.21, we have that $\left(\dfrac{\sqrt{n}}{\sigma}(\bar{X}_n - \mu)\right)_{n \in \mathbb{N}^*}$ converges in distribution to a random variable $Z \sim N(0,1)$. Then, by using Theorem 4.7, we obtain

$$\frac{\mu^2 \sqrt{n}}{\sigma}\left(\frac{1}{\bar{X}_n} - \frac{1}{\mu}\right) = \frac{\mu}{\bar{X}_n} \cdot \frac{\sqrt{n}}{\sigma}(\mu - \bar{X}_n) \xrightarrow{d} -Z, \text{ where } -Z \sim N(0,1).$$

Problem 4.4.22. Let $(X_n)_{n \in \mathbb{N}^*}$ be a sequence of independent random variables such that $P(X_n = n) = \frac{1}{n}$ and $P(X_n = 0) = 1 - \frac{1}{n}$ for all $n \in \mathbb{N}^*$. Prove that $(X_n)_{n \in \mathbb{N}^*}$ does not obey the WLLN.

Solution 4.4.22: Suppose that $(X_n)_{n \in \mathbb{N}^*}$ obeys the WLLN. We denote $\bar{X}_n = \frac{1}{n}(X_1 + \ldots + X_n)$ for $n \in \mathbb{N}^*$. Then $\bar{X}_n \xrightarrow{P} 1$, because $E(X_n) = 1$. Let $n \in \mathbb{N}$ with $n \geq 3$. In view of $E(\bar{X}_n) = 1$ and the independence, we have

$$E((\bar{X}_n - 1)^2) = V(\bar{X}_n) = \frac{1}{n^2}\sum_{k=1}^{n} V(X_k) = \frac{1}{n^2}\sum_{k=1}^{n}(k-1) = \frac{n-1}{2n}.$$

Also, we have

$$E(|\bar{X}_n|^3) = E(\bar{X}_n^3) = \frac{1}{n^3}\left(\sum_{k=1}^{n} E(X_k^3) + \sum_{k=1}^{n}\sum_{\substack{l=1 \\ l \neq k}}^{n} 3E(X_k^2)E(X_l)\right.$$

$$\left. + \sum_{1 \leq k < l < m \leq n} 6E(X_k)E(X_l)E(X_m)\right)$$

$$= \frac{1}{n^3}\left(\sum_{k=1}^{n} k^2 + \sum_{k=1}^{n}\sum_{\substack{l=1 \\ l \neq k}}^{n} 3k + \sum_{1 \leq k < l < m \leq n} 6\right)$$

$$= \frac{1}{n^3}\left(\frac{1}{6}n(n+1)(2n+1) + \frac{3}{2}(n-1)n(n+1) + 6C(n,3)\right)$$

$$= \frac{1}{6}\left(1 + \frac{1}{n}\right)\left(2 + \frac{1}{n}\right) + \frac{3}{2}\left(1 - \frac{1}{n}\right)\left(1 + \frac{1}{n}\right) + \left(1 - \frac{1}{n}\right)\left(1 - \frac{2}{n}\right).$$

By the above computations there exists $M > 0$ such that
$$\sup_{n \in \mathbb{N}^*} E(|\bar{X}_n|^3) \leq M.$$
Then $\left(|\bar{X}_n|^2\right)_{n \in \mathbb{N}^*}$ is uniformly integrable (see Definition 4.5), because
$$\lim_{c \to \infty} \sup_{n \in \mathbb{N}^*} E\left(|\bar{X}_n|^2 \mathbb{I}_{\{|\bar{X}_n|^2 > c\}}\right) \leq \lim_{c \to \infty} \sup_{n \in \mathbb{N}^*} c^{-\frac{1}{2}} E(|\bar{X}_n|^3)$$
$$\leq M \cdot \lim_{c \to \infty} c^{-\frac{1}{2}} = 0.$$
Since $\bar{X}_n \xrightarrow{P} 1$, in view of Theorem 4.12, we have $\bar{X}_n \xrightarrow{L^2} 1$, which is in contradiction with $E\left((\bar{X}_n - 1)^2\right) \to \frac{1}{2}$.

Problem 4.4.23. Let $r \in \mathbb{R}$, $p \in (0,1)$ and let $(X_n)_{n \in \mathbb{N}^*}$ be a sequence of independent random variables such that $P(X_n = -n^r) = P(X_n = n^r) = \frac{p}{2}$ and $P(X_n = 0) = 1 - p$ for all $n \in \mathbb{N}^*$. Prove that:
1) If $r < \frac{1}{2}$, then $(X_n)_{n \in \mathbb{N}^*}$ obeys the SLLN.
2) If $r \geq \frac{1}{2}$, then $(X_n)_{n \in \mathbb{N}^*}$ does not obey the WLLN.

Solution 4.4.23: We note that $E(X_n) = 0$ and $V(X_n) = E(X_n^2) = pn^{2r}$ for $n \in \mathbb{N}^*$.
1) Since
$$\sum_{n=1}^\infty \frac{1}{n^2} V(X_n) = p \sum_{n=1}^\infty \frac{1}{n^\alpha} < \infty,$$
where $\alpha = 2 - 2r > 1$, we deduce by Theorem 4.18, that $(X_n)_{n \in \mathbb{N}^*}$ obeys the SLLN.
2) Suppose that $(X_n)_{n \in \mathbb{N}^*}$ obeys the WLLN. Then $\bar{X}_n \xrightarrow{P} 0$, where $\bar{X}_n = \frac{1}{n}(X_1 + \ldots + X_n)$, $n \in \mathbb{N}^*$. In particular, we have $\bar{X}_n \xrightarrow{d} 0$.
We have for $n \in \mathbb{N}^*$
$$\frac{D_n^2}{n^{2r+1}} = \frac{1}{n^{2r+1}} \sum_{k=1}^n V(X_k) = \frac{p}{n} \sum_{k=1}^n \left(\frac{k}{n}\right)^{2r} \to p \int_0^1 t^{2r} dt = \frac{p}{2r+1}. \quad (4.8)$$
Hence $\frac{n^r}{D_n} \to 0$. Let $\varepsilon > 0$. Then for every n sufficiently large we have $n^r < \varepsilon D_n$ and therefore $\mathbb{I}_{\{|X_k| \geq \varepsilon D_n\}} = 0$ a.s. for all $k \in \{1, \ldots, n\}$. So, the Lindeberg condition (4.4) is satisfied. Thus $\left(\frac{1}{D_n} \sum_{k=1}^n X_k\right)_{n \in \mathbb{N}^*}$ converges in distribution to a random variable having normal distribution $N(0,1)$. Since
$$\frac{1}{D_n} \sum_{k=1}^n X_k = \frac{1}{n^{r-\frac{1}{2}}} \cdot \sqrt{\frac{n^{2r+1}}{D_n^2}} \cdot \bar{X}_n, n \in \mathbb{N}^*,$$

and in view of $r \geq \frac{1}{2}$, (4.8) and our supposition (see also Theorem 4.7), we deduce that

$$\frac{1}{D_n} \sum_{k=1}^{n} X_k \xrightarrow{d} 0,$$

which produces a contradiction.

Problem 4.4.24. Let $(\lambda_n)_{n \in \mathbb{N}}$ be a sequence of strictly positive real numbers with $\lim_{n \to \infty} \lambda_n = \infty$ and let $(X_n)_{n \in \mathbb{N}}$ be a sequence of random variables such that $X_n \sim Poiss(\lambda_n)$ for each $n \in \mathbb{N}$. Prove that the sequence $(Y_n)_{n \in \mathbb{N}}$ defined by

$$Y_n = \frac{X_n - E(X_n)}{\sqrt{V(X_n)}} \quad \text{for each } n \in \mathbb{N}$$

converges in distribution to a random variable having $N(0, 1)$ distribution.

Solution 4.4.24: For each $n \in \mathbb{N}$ let M_{Y_n} denote the moment generating function of Y_n. First, we prove that the sequence $(M_{Y_n})_{n \in \mathbb{N}}$ converges to the moment generating function of the standard normal distribution. Since $E(X_n) = \lambda_n$ and $V(X_n) = \lambda_n$, we have

$$Y_n = \frac{X_n - E(X_n)}{\sqrt{V(X_n)}} = \frac{1}{\sqrt{\lambda_n}} X_n - \sqrt{\lambda_n} \quad \text{for all } n \in \mathbb{N}.$$

By the properties of the moment generating function (see Theorem 3.9-(2)), it follows that

$$M_{Y_n}(t) = \exp\left\{ -\sqrt{\lambda_n} t \right\} M_{X_n}\left(\frac{t}{\sqrt{\lambda_n}} \right) \quad \text{for all } n \in \mathbb{N}, \ t \in \mathbb{R}.$$

But the moment generating function of the Poisson distributed random variable X_n is

$$M_{X_n}(t) = \exp\left\{ \lambda_n(e^t - 1) \right\} \quad \text{for each } n \in \mathbb{N}, \ t \in \mathbb{R},$$

see Problem 3.5.2. Therefore,

$$\log(M_{Y_n}(t)) = \lambda_n \left(\exp\left\{ \frac{t}{\sqrt{\lambda_n}} \right\} - 1 \right) - \sqrt{\lambda_n} t \quad \text{for each } n \in \mathbb{N}.$$

Using the Taylor expansion of the exponential function, it follows that

$$\log(M_{Y_n}(t)) = \frac{t^2}{2} + O\left(\frac{t^3}{\sqrt{\lambda_n}} \right) \quad \text{as } n \to \infty \text{ for each } t \in \mathbb{R}.$$

Here, we used the "Big-O notation"

$$g(n) = O(h(n)) \quad n \to \infty$$

which refers to

$$\limsup_{n \to \infty} \left| \frac{g(n)}{h(n)} \right| < \infty.$$

Hence,

$$\lim_{n \to \infty} M_{Y_n}(t) = \exp\left\{ \frac{t^2}{2} \right\} \quad \text{for each } t \in \mathbb{R}.$$

Then, by Theorem 4.9 (see also the solution of Problem 3.5.4), it follows that $(Y_n)_{n \in \mathbb{N}}$ converges in distribution to a random variable having normal distribution $N(0,1)$.

Problem 4.4.25. Let $Y_1, Y_2, \ldots, U_1, U_2, \ldots$ be independent random variables such that $Y_n \sim Beta(1,2)$ and $U_n \sim Unif[0,1]$ for each $n \in \mathbb{N}^*$. Denote $X_n = Y_n(2U_n - 1)$ and $\bar{X}_n = \frac{1}{\sqrt{n}} \sum_{k=1}^{n} X_k$ for each $n \in \mathbb{N}^*$. Prove that $E\big(\sin(\bar{X}_n) \big) \to 0$.

Solution 4.4.25: We note that $(X_n)_{n \in \mathbb{N}^*}$ is a sequence of independent identically distributed random variables with mean value 0 and finite strictly positive standard deviation σ. Theorem 4.21 implies that $(\bar{X}_n)_{n \in \mathbb{N}^*}$ converges in distribution to a random variable $X \sim N(0, \sigma^2)$. By Theorem 4.8, we have $\sin(\bar{X}_n) \xrightarrow{d} \sin(X)$. Since $\big(\sin(\bar{X}_n) \big)_{n \in \mathbb{N}^*}$ is uniformly integrable (because it is bounded), we have, by Theorem 4.13, that $E\big(\sin(\bar{X}_n) \big) \to E\big(\sin(X) \big)$. Since the sine function is odd and bounded, and $X \sim N(0, \sigma^2)$, we deduce that $E\big(\sin(X) \big) = 0$.

Problem 4.4.26. Let $m \in \mathbb{N}^*$ and let $(X_n)_{n \in \mathbb{N}^*}$ be a sequence of independent random variables such that $\frac{1}{n} X_n \sim Beta(m, n)$ for all $n \in \mathbb{N}^*$.
1) For given (large) $n, N \in \mathbb{N}^*$ simulate N values of X_n and plot, on the same figure, the corresponding histogram and a probability density function of the $Gamma(m, 1)$ distribution.
2) Prove that

$$\frac{1}{n} \sum_{j=1}^{n} X_j^k \xrightarrow{L^1} \frac{(m+k-1)!}{(m-1)!} \quad \text{as } n \to \infty, \quad \text{for all } k \in \mathbb{N}^*.$$

Solution 4.4.26: 1)

```
function betagam(m,n,N)
clf; hold on;
X=n*betarnd(m,n,1,N);
h=histogram(X,'Normalization','pdf');
fplot(@(x) gampdf(x,m,1),h.BinLimits);
end
```

>>betagam(3,100,1000)

2) Let $k, m, n \in \mathbb{N}^*$. If $X \sim Gamma(m, 1)$, then, from Problem 3.4.3-1), we have

$$E(X^k) = \prod_{r=0}^{k-1} (m + r).$$

If $X \sim Beta(m, n)$, then

$$E(X^k) = \frac{B(m + k, n)}{B(m, n)} = \prod_{r=0}^{k-1} \frac{m + r}{n + m + r},$$

where we use Proposition A.2.

Since the function

$$f_{X_n}(x) = \frac{(m + n - 1)!}{(m - 1)!(n - 1)!} \left(\frac{x}{n}\right)^{m-1} \left(1 - \frac{x}{n}\right)^{n-1} \frac{1}{n}$$

$$= \frac{1}{(m - 1)!} x^{m-1} \left(\left(1 - \frac{x}{n}\right)^{-\frac{n}{x}}\right)^{-\frac{x(n-1)}{n}} \prod_{r=0}^{m-1} \left(1 + \frac{r}{n}\right)$$

for $x \in [0, n]$ and $f_{X_n}(x) = 0$ for $x \notin [0, n]$ is a density function of X_n, we deduce by Theorem 4.14 that $X_n \xrightarrow{d} X$, where X is a random variable that follows the $Gamma(m, 1)$ distribution.

Also, we have

$$E(X_n^k) = n^k \prod_{r=0}^{k-1} \frac{m + r}{n + m + r} = \prod_{r=0}^{k-1} \frac{m + r}{1 + \frac{m+r}{n}} \longrightarrow \prod_{r=0}^{k-1} (m + r) = E(X^k).$$

By Theorem 4.13, we deduce that $(X_n^k)_{n \in \mathbb{N}^*}$ is uniformly integrable and thus, by Theorem 4.16, we have

$$\frac{1}{n} \sum_{j=1}^{n} (X_j^k - E(X_j^k)) \xrightarrow{L^1} 0.$$

Since, using the Stolz–Cesàro Theorem, we have

$$\lim_{n\to\infty} \frac{1}{n} \sum_{j=1}^{n} E(X_j^k) = \lim_{n\to\infty} \frac{1}{n} \sum_{j=1}^{n} \prod_{r=0}^{k-1} \frac{m+r}{1+\frac{m+r}{j}} = \lim_{n\to\infty} \prod_{r=0}^{k-1} \frac{m+r}{1+\frac{m+r}{n+1}}$$
$$= \frac{(m+k-1)!}{(m-1)!},$$

we obtain the desired relation.

Problem 4.4.27. Using Theorem 4.21, prove that

$$\lim_{n\to\infty} e^{-n} \sum_{k=0}^{n} \frac{n^k}{k!} = \frac{1}{2}.$$

Solution 4.4.27: Let $(X_n)_{n\in\mathbb{N}^*}$ be a sequence of independent random variables, where $X_n \sim Poiss(1)$ for each $n \in \mathbb{N}^*$. Then, by Problem 3.5.2, it follows that $X_1 + \cdots + X_n \sim Poiss(n)$ and $E(X_n) = V(X_n) = 1$ for each $n \in \mathbb{N}^*$. Then, by Theorem 4.21, we get

$$\lim_{n\to\infty} P\left(\frac{1}{\sqrt{n}}(X_1 + \cdots + X_n - n) \le 0\right) = \frac{1}{\sqrt{2\pi}} \int_{-\infty}^{0} \exp\left\{-\frac{t^2}{2}\right\} dt = \frac{1}{2}.$$

Using the properties of the Poisson distribution, we have

$$P\left(\frac{1}{\sqrt{n}}(X_1 + \cdots + X_n - n) \le 0\right) = P(X_1 + \cdots + X_n \le n)$$
$$= \sum_{k=0}^{n} P(X_1 + \cdots + X_n = k) = e^{-n} \sum_{k=0}^{n} \frac{n^k}{k!}.$$

Chapter 5

Examples of Stochastic Processes

5.1 Some General Remarks

Throughout this chapter $(\Omega, \mathcal{F}, P)$ denotes a probability space. Let $\mathcal{T} \subseteq \mathbb{R}$ be a nonempty set. A family $(X_t)_{t \in \mathcal{T}}$ of random variables X_t is called **stochastic process**, this means X_t is a random variable for every $t \in \mathcal{T}$. We can also say that a stochastic process is a function $X : \Omega \times \mathcal{T} \to \mathbb{R}$ such that $\omega \in \Omega \mapsto X(\omega, t)$ is a random variable for every $t \in \mathcal{T}$. For arbitrary fixed $\omega \in \Omega$ the function $t \in \mathcal{T} \to X(\omega, t)$ is called a **trajectory** (**sample path**) of the stochastic process. Usually t has the interpretation of time. In this sense we choose either discrete time, e.g., $\mathcal{T} = \{0, 1, \ldots\}$, or continuous time, e.g., $\mathcal{T} = [0, \infty)$. The range of values $\mathcal{S} = X(\Omega \times \mathcal{T})$ of a stochastic process is called **state space**. $(X_t)_{t \in \mathcal{T}}$ has **discrete state space** $\mathcal{S}$, if $\mathcal{S}$ is a countable set such that for every $i \in \mathcal{S}$ there exists $s \in \mathcal{T}$ such that $P(X_s = i) > 0$. Obviously, sequences of random variables (see Chapter 4) are special stochastic processes.

Definition 5.1. Let $(X_t)_{t \in \mathcal{T}}$ and $(Y_t)_{t \in \mathcal{T}}$ be two stochastic processes. $(Y_t)_{t \in \mathcal{T}}$ is a **modification** (**version**) of $(X_t)_{t \in \mathcal{T}}$, if $P(X_t = Y_t) = 1$ for each $t \in \mathcal{T}$. The two processes are **indistinguishable** if

$$P(X_t = Y_t \text{ for all } t \in \mathcal{T}) = 1. \qquad \blacklozenge$$

Definition 5.2. A stochastic process $(X_t)_{t \in \mathcal{T}}$ is said to be **càdlàg**, if a.s. it has sample paths which are right-continuous, with left limits (the acronym càdlàg comes from French for "continue à droite, limites à gauche"). $\blacklozenge$

Definition 5.3. Let $(X_t)_{t \in \mathcal{T}}$ be a stochastic process. If

$$P(\lim_{t \to s} X_t = X_s \text{ for each } s \in \mathcal{T}) = 1,$$

then $(X_t)_{t \in \mathcal{T}}$ is said to be **continuous a.s.** (has **continuous trajectories, continuous sample paths**).

$(X_t)_{t \in \mathcal{T}}$ is said to be **continuous in probability** on $\mathcal{T}$, if for all $s \in \mathcal{T}$ and for all $\varepsilon > 0$

$$\lim_{t \to s} P\left(|X_t - X_s| \geq \varepsilon\right) = 0. \qquad \blacklozenge$$

Theorem 5.1. [Protter (2004), Theorem 2, p. 4] *Let $(X_t)_{t \in \mathcal{T}}$ and $(Y_t)_{t \in \mathcal{T}}$ be two stochastic processes such that $(Y_t)_{t \in \mathcal{T}}$ is a modification of $(X_t)_{t \in \mathcal{T}}$. If the two processes have a.s. right-continuous paths, then they are indistinguishable.*

There are a variety of books on stochastic processes, see for example [Gikhman and Skorokhod I (2004)], [Gikhman and Skorokhod II (2004)], [Girardin and Limnios (2018)], [Borodin (2017)], [Schilling and Partzsch (2014)] and the literature cited therein. Applications in finance, physics and biology can be found, for example in [Davis and Lleo (2015)], [Vitali, Motta and Galli (2018)] and [Holcman (2017)].

We know, from Remark 2.4, that a random variable is completely characterized by its distribution function. In the case of stochastic processes the role of the distribution is taken over by the family of finite dimensional distributions (see the Theorem of Kolmogorov, for example, in [Gikhman and Skorokhod I (2004)]).

Definition 5.4. The **family of finite dimensional distributions** of a stochastic process $(X_t)_{t \in \mathcal{T}}$ is the family of distributions of the random vectors $(X_{t_1}, X_{t_2}, \ldots, X_{t_n})$ with $t_1 < t_2 < \ldots < t_n \in \mathcal{T}$, $n \in \mathbb{N}^*$. $\qquad \blacklozenge$

5.2 Random Walks

A random walk is a stochastic process, describing a succession of random steps on some mathematical space, such as the axis of integer numbers, equidistant points placed on the circle, the points having integer coordinates in the 2 or 3-dimensional Euclidean space, the vertices of a graph etc. Random walks have applications in computer science, physics, chemistry, biology, economics, etc. For a detailed account of the theory and its applications we refer to the works [Feller (1968); Henze (2013); Weiss (1983)] (see also [Iancu and Lisei (2017)] for an elementary approach).

Definition 5.5. (Random walk on the integer line) Let $p \in (0, 1)$. A random walk, which starts at some $k \in \mathbb{Z}$, takes place along the axis o numbers, from one integer to another, in the following way: for each step, we go with probability p to the first larger integer and with probability $1-p$ to the first smaller integer (see Figure 5.1). Each step is independent of the previous ones. Let $(X_n)_{n\in\mathbb{N}^*}$ be the sequence of independent **random steps**, i.e., $X_n = -1$, if the nth step of the random walk is to the left, and $X_n = 1$, if the nth step of the random walk is to the right. The **random walk on the integer line** which starts at $k \in \mathbb{Z}$ is characterized by the sequence $(S_n)_{n\in\mathbb{N}}$, where $S_0 = k$ and $S_n = S_0 + X_1 + X_2 + \cdots + X_n$ is the position of the random walk after the nth step for $n \in \mathbb{N}^*$. If $p = \frac{1}{2}$, then the random walk is called **symmetric**. ◆

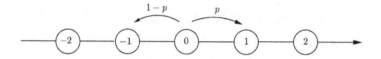

Fig. 5.1: Random walk on the integer line starting at 0

Let $p \in (0,1)$ and consider a random walk $(S_n)_{n\in\mathbb{N}}$ which starts at $k \in \mathbb{Z}$ and takes place along the integer line as described in Definition 5.5. Note that S_n is a discrete random variable taking values in the set $\{k - n, \ldots, k - 1, k, k + 1, \ldots, k + n\}$. Obviously, if $P(S_n = j) > 0$, then

$$P(S_{n+1} = j + 1 | S_n = j) = p \text{ and } P(S_{n+1} = j - 1 | S_n = j) = 1 - p$$

for $j \in \{k - n, \ldots, k - 1, k, k + 1, \ldots, k + n\}$.

Example 5.1. For $p \in (0,1)$, $n \in \mathbb{N}$ and $j \in \mathbb{Z}$ with $|j| \leq n$ given, the probability to end up in the integer number j after a random walk of n steps, which starts at 0, is

$$P(S_n = j) = \mathrm{C}\left(n, \frac{n+j}{2}\right) p^{\frac{n+j}{2}} (1 - p)^{\frac{n-j}{2}},$$

if j and n have the same parity, and $P(S_n = j) = 0$, if j and n have opposite parity.

Proof: Denote by n_R the number of steps taken by the random walk rightward and by n_L the number of steps taken leftward. Obviously, $n_R, n_L \in \{0, 1, \ldots, n\}$ and we have

$$n_R + n_L = n \text{ and } n_R - n_L = j. \tag{5.1}$$

If j and n have the same parity, we have

$$n_R = \frac{n+j}{2} \text{ and } n_L = \frac{n-j}{2}.$$

If j and n do not have the same parity, there are no integer numbers n_R and n_L satisfying (5.1), in fact there exists no random walk of n steps that ends up in j. Every random walk of n steps can be identified with n Bernoulli trials. Hence, the probability to make n_R steps rightward and $n_L = n - n_R$ steps leftward is $C(n, n_R)p^{n_R}(1-p)^{n_L}$. Therefore, if j and n have the same parity, then the probability to end up in j is $C\left(n, \frac{n+j}{2}\right) p^{\frac{n+j}{2}}(1-p)^{\frac{n-j}{2}}$; if j and n have opposite parity, then the probability to end up in j is 0.   ▲

Example 5.2. For a random walk $(S_n)_{n\in\mathbb{N}}$ on the integer line, which starts at 0, we compute:

(1) $P(S_6 = S_{10}) = P(X_7 + X_8 + X_9 + X_{10} = 0) = 6p^2(1-p)^2$;

(2) $P(S_5 = -1, S_{11} = 3) = P(S_5 = -1, S_{11} - S_5 = 4)$

$= P(X_1 + X_2 + X_3 + X_4 + X_5 = -1, X_6 + X_7 + X_8 + X_9 + X_{10} + X_{11} = 4)$

$= C(5, 2)p^2(1-p)^3 \cdot C(6, 5)p^5(1-p) = 60p^7(1-p)^4$;

(3) $P(S_6 = S_{11}) = P(X_7 + X_8 + X_9 + X_{10} + X_{11} = 0) = 0.$ ▲

Representation of a random walk on the integer line: We represent a random walk as a continuous path (polygonal line) in the Cartesian coordinate system, where the horizontal axis represents the discrete (time) steps and the vertical axis represents the position of the random walk, i.e. the point (j, S_j) $(j \in \mathbb{N})$ indicates that after j steps the random walk is at the position S_j on the integer line.

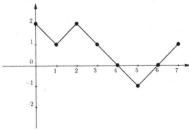

Fig. 5.2: A path of a random walk on the integer line

Let $n \in \mathbb{N}^*$. We plot the points $(0, S_0), (1, S_1), \ldots, (n, S_n)$; then for each $j \in \{0, \ldots, n-1\}$ we connect (j, S_j) and $(j+1, S_{j+1})$ with a straight line segment. We shall identify each random walk of n steps along the integer axis, which starts at $S_0 = k$, with the path that starts at $(0, k)$ and in each step connects the points (j, S_j) and $(j + 1, S_{j+1})$, $j \in \{0, \ldots, n - 1\}$ (see Figure 5.2 for $k = 2$ and $n = 7$). For $n = 0$ the path is just the point $(0, k)$. We observe that $S_n - S_0$ denotes the difference between the number of rightward steps and the number of leftward steps after n steps.

Example 5.3. The probability that after 6 steps a symmetric random walk on the integer line, which starts at 0, passes the number 1 exactly 2 times is $\dfrac{16}{2^6}$. To compute this we use the representation with paths and count the number of all possible paths: there are 8 paths connecting the points with coordinates $(1, 1)$ and $(3, 1)$ (see Figure 5.3), 4 paths connecting the points with coordinates $(1, 1)$ and $(5, 1)$ and 4 paths connecting the points with coordinates $(3, 1)$ and $(5, 1)$. ▲

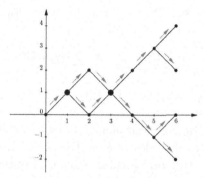

Fig. 5.3: Representing 6 steps for a symmetric random walk on the integer line, starting at 0: 8 paths connecting the points with coordinates $(1, 1)$ and $(3, 1)$

Definition 5.6. (Random walk on the circle) Let $p \in (0, 1)$, $m \in \mathbb{N}^*$. Consider the integers $0, 1, \ldots, m - 1$ to be placed equidistantly and anticlockwise on a circle. A random walk takes place along this circle, from one integer to another one, in the following way: the walk starts at 0 and, for each step, it moves anticlockwise with probability p and clockwise with probability $1 - p$ to the closest integer, see Figure 5.4. Each step is independent of the previous ones. The **random walk on the circle** is characterized by the sequence $(Z_n)_{n \in \mathbb{N}}$, where $Z_0 = 0$ and Z_n is the position on the circle after the nth step for $n \in \mathbb{N}^*$. If $p = \dfrac{1}{2}$, then the random walk on the circle is called **symmetric**. ◆

Note that Z_n is a discrete random variable taking values in the set $\{0, 1, \ldots, m - 1\}$ for each $n \in \mathbb{N}$.

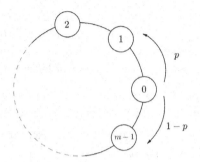

Fig. 5.4: Random walk on the circle

Example 5.4. For $p \in (0,1)$, $n \in \mathbb{N}$, $m \in \mathbb{N}^*$ and $j \in \{0, 1, \ldots, m-1\}$ given, the probability to end up in the position j after a random walk of $n \in \mathbb{N}$ steps on the circle is given by

$$P(Z_n = j) = \sum_{l \in I_j} C(n, l) p^l (1-p)^{n-l}, \tag{5.2}$$

where $I_j = \{l \in \{0, 1, \ldots, n\} : 2l - n \ (\mathrm{mod} \ m) = j\}$ and $P(Z_n = j) = 0$, if $I_j = \emptyset$.

Proof: The random walk along the circle can be identified with the random walk along the axis, described in Example 5.1, by replacing each integer of the axis with the corresponding remainder of the division by m. Hence, the random walk can end up only at one of the following integers:

$$2l - n \ (\mathrm{mod} \ m), \ l \in \{0, 1, \ldots, n\}.$$

Therefore, the probability to end up in $j \in \{0, 1, \ldots, m-1\}$ after $n \in \mathbb{N}$ steps is given as in (5.2).   ▲

Example 5.5. (1) The probabilities for a symmetric random walk on the circle with $m = 4$ points, after $0, 1, 2, 3$ and 4 steps are:

- $P(Z_0 = 0) = 1$;   • $P(Z_1 = 1) = P(Z_1 = 3) = \dfrac{1}{2}$;

- $P(Z_2 = 0) = P(Z_2 = 2) = \dfrac{2}{2^2}$;   • $P(Z_3 = 1) = P(Z_3 = 3) = \dfrac{4}{2^3}$;

- $P(Z_4 = 0) = P(Z_4 = 2) = \dfrac{8}{2^4}$.

(2) The probability that after 6 steps a symmetric random walk on the circle with $m = 4$, passes the number 1 exactly 2 times is $\dfrac{24}{2^6}$. The picture of all such paths starting from 0 and moving in the first step anticlockwise is given in Figure 5.5. There are 16 such paths. For example, such a random walk passes successively the points: 0, 1, 2, 3, 2, 1, 2. Similarly, one can count that the number of paths starting from 0, moving in the first step clockwise and passing the number 1 exactly 2 times is 8. ▲

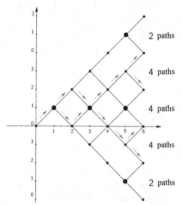

Fig. 5.5: A symmetric random walk on the circle ($m = 4$): 16 paths starting at 0 and moving anticlockwise for the first step, such that the number 1 is passed exactly 2 times

5.2.1 Solved Problems

Problem 5.2.1. Let $x_0 \in \mathbb{R}$, $p \in (0,1)$ and $n \in \mathbb{N}^*$. A particle moves along the real axis in the following way: it starts at x_0; for each step $i \in \mathbb{N}^*$ it chooses randomly a direction D_i, either to go left with probability p, or to go right with probability $1 - p$; the length of each step is a continuous random variable L_i uniformly distributed on $[0,1]$. Assume that $D_1, D_2, \ldots, L_1, L_2, \ldots$ are independent. Let X_n be the position of the particle after n steps. Find $E(X_n)$ and $V(X_n)$. Simulate the movement of the particle and estimate the asked values.

Solution 5.2.1: For every $i \in \mathbb{N}^*$ the random direction D_i is a discrete random variable that takes the value -1, if the particle goes left, and the value 1, if the particle goes right, and $L_i \sim Unif[0,1]$ gives the random length of the ith step. Then we have

$$E(X_n) = E\left(x_0 + \sum_{i=1}^{n} D_i \cdot L_i\right) = x_0 + \sum_{i=1}^{n} E(D_i) \cdot E(L_i)$$

$$= x_0 + \sum_{i=1}^{n}(-p + 1 - p) \cdot \frac{1}{2} = x_0 + n\left(\frac{1}{2} - p\right)$$

and by using Example 3.9 and Theorem 3.4 we obtain

$$V(X_n) = V\left(x_0 + \sum_{i=1}^{n} D_i \cdot L_i\right) = \sum_{i=1}^{n} V(D_i \cdot L_i)$$

$$= \sum_{i=1}^{n} V(D_i)V(L_i) + E^2(D_i)V(L_i) + E^2(L_i)V(D_i)$$

$$= n\left(\frac{p(1-p)}{3} + \frac{(1-2p)^2}{12} + p(1-p)\right) = n\left(\frac{1}{12} + p(1-p)\right).$$

```
%two graphical representations for a simulation
function rndunifwalk_simul
p=0.55; n=25; x0=0;
[~,step]=rndunifwalk(p,n,x0,1);
poz=cumsum([x0 step]);
clf; subplot(2,1,1); ax = subplot(2,1,1); hold on;
ax.XAxisLocation = 'origin'; ax.XLim=[0,n+1];
ax.YLim=[min(poz)-0.4,max(poz)+0.4];
subplot(2,1,2); ax = subplot(2,1,2); axis equal; hold on;
ax.XAxisLocation = 'origin'; ax.LineWidth=1.2;
ax.XLim=[min(poz)-0.2,max(poz)+0.2];
ax.YLim=[-0.4,0.4]; ax.YAxis.Visible = 'off';
for i=1:n
 subplot(2,1,1);
 plot([i-1,i],[poz(i),poz(i+1)],'k','LineWidth',1.2);
 plot([i,i],[0,poz(i+1)],':k');
 subplot(2,1,2);
 title(['Step ' num2str(i) ' = ' num2str(step(i))]);
 pic=plot(poz(i+1),0,'or','MarkerFaceColor','r');
 pause(0.5); set(pic,'Visible','off');
end
set(pic,'Visible','on');

%estimation of the expected value, by 10000 simulations
[x1,~]=rndunifwalk(p,n,x0,10000);
fprintf('The estimated expected value is %5.4f.\n',mean(x1));
fprintf('The expected value is %5.4f.\n',x0+n*(1/2-p));
fprintf('The estimated variance is %5.4f.\n',var(x1,1));
fprintf('The variance is %5.4f.\n',n*(1/12+p*(1-p)));
end


function [x1,step]=rndunifwalk(p,n,x0,N)
step=(-1).^binornd(1,p,N,n).*unifrnd(0,1,N,n);
x1=x0+sum(step,2);
end


>> rndunifwalk_simul
The estimated expected value is -1.2532.
The expected value is -1.2500.
The estimated variance is 8.2370.
The variance is 8.2708.
```

Problem 5.2.2. Let $(S_n)_{n\in\mathbb{N}}$ be a symmetric random walk starting at the origin with $(X_n)_{n\in\mathbb{N}^*}$ the sequence of independent random steps, i.e.,

$$P(X_n = -1) = P(X_n = 1) = \frac{1}{2} \text{ for each } n \in \mathbb{N}^*.$$

Compute $E(S_n)$, $E(S_n^2)$, $E(S_n^3)$, $E(S_n^4)$ for $n \in \mathbb{N}^*$.

Solution 5.2.2: Observe that $E(S_n) = E(X_1) + ... + E(X_n) = 0$, then

$$E(S_n^2) = V(S_n) = V(X_1 + ... + X_n) = V(X_1) + ... + V(X_n) = n$$

for $n \in \mathbb{N}^*$. On the other hand, $E(X_n) = E(X_n^3) = 0$ and $E(X_n^2) = E(X_n^4) = 1$ for each $n \in \mathbb{N}^*$. The steps X_n, $n \in \mathbb{N}^*$, are independent, therefore

$$E(S_n^3) = \sum_{i=1}^{n}\sum_{j=1}^{n}\sum_{k=1}^{n} E(X_i X_j X_k) = 0,$$

while

$$E(S_n^4) = \sum_{i=1}^{n}\sum_{j=1}^{n}\sum_{k=1}^{n}\sum_{l=1}^{n} E(X_i X_j X_k X_l) = n + 3n(n-1) = n(3n-2).$$

We used here that the nonzero terms are including
• n terms of the type $E(X_i^4)$, $i = \overline{1,n}$;
• $3n(n-1)$ terms of the type $E(X_i^2 X_j^2)$ corresponding to the indices $i \neq j$, $i,j = \overline{1,n}$.

Problem 5.2.3. Let $(S_n)_{n\in\mathbb{N}}$ be a symmetric random walk starting at the origin with $(X_n)_{n\in\mathbb{N}^*}$ the sequence of independent random steps, i.e.,

$$P(X_n = -1) = P(X_n = 1) = \frac{1}{2} \text{ for each } n \in \mathbb{N}^*.$$

Prove that:

1) $E\big(|S_{2n}|\big) = \dfrac{2n\mathrm{C}(2n, n)}{2^{2n}}$ and $E\big(|S_{2n+1}|\big) = \dfrac{(2n+1)\mathrm{C}(2n, n)}{2^{2n}}$, for $n \in \mathbb{N}^*$;

2) $\lim_{n\to\infty} \dfrac{1}{\sqrt{n}} E\big(|S_n|\big) = \sqrt{\dfrac{2}{\pi}}$;

3) $\lim_{n\to\infty} \sqrt{n} \cdot \dfrac{\mathrm{C}(2n, n)}{2^{2n}} = \dfrac{1}{\sqrt{\pi}}$;

4) $\dfrac{1}{n^q} S_n \xrightarrow{P} 0$, for all $q > \dfrac{1}{2}$;

5) the sequence $\left(\dfrac{1}{\sqrt{n}} S_n\right)_{n\in\mathbb{N}^*}$ does not converge in probability to a random variable.

Solution 5.2.3: 1) Let $n \in \mathbb{N}^*$. By using Example 5.1 we have

$$E\big(|S_{2n}|\big) = \sum_{k=-n}^{n} 2|k|C(2n, n+k)\frac{1}{2^{2n}} = \frac{1}{2^{2n-2}} \sum_{k=0}^{n} kC(2n, n+k)$$

$$= \frac{1}{2^{2n-2}} \left(\sum_{k=0}^{n}(n+k)C(2n, n+k) - n\sum_{k=0}^{n}C(2n, k) \right)$$

$$= \frac{1}{2^{2n-2}} \left(2n\sum_{k=0}^{n}C(2n-1, n+k-1) - n \cdot \frac{2^{2n} + C(2n, n)}{2} \right)$$

$$= \frac{1}{2^{2n-2}} \left(2n\sum_{k=n-1}^{2n-1}C(2n-1, k) - n \cdot \frac{2^{2n} + C(2n, n)}{2} \right)$$

$$= \frac{1}{2^{2n-2}} \left(2n \cdot \left(\frac{2^{2n-1}}{2} + C(2n-1, n-1)\right) \right) - n \cdot \frac{2^{2n} + C(2n, n)}{2}$$

$$= \frac{2nC(2n, n)}{2^{2n}}.$$

Similarly, we have

$$E\big(|S_{2n+1}|\big) = \sum_{k=-n-1}^{n} |2k+1|C(2n+1, n+k+1)\frac{1}{2^{2n+1}}$$

$$= \frac{1}{2^{2n}} \sum_{k=0}^{n}(2k+1)C(2n+1, n+k+1)$$

$$= \frac{1}{2^{2n-1}} \sum_{k=0}^{n}(n+k+1)C(2n+1, n+k+1)$$

$$- \frac{2n+1}{2^{2n}} \sum_{k=0}^{n}C(2n+1, n+k+1)$$

$$= \frac{2n+1}{2^{2n-1}} \sum_{k=0}^{n}C(2n, n+k) - \frac{2n+1}{2^{2n}} \cdot \frac{2^{2n+1}}{2}$$

$$= \frac{2n+1}{2^{2n-1}} \cdot \frac{2^{2n} + C(2n, n)}{2} - (2n+1) = \frac{(2n+1)C(2n, n)}{2^{2n}}.$$

2) Denote $Y_n = \frac{1}{\sqrt{n}}|S_n|$ for $n \in \mathbb{N}^*$. By Theorem 4.21 and Theorem 4.5 it follows that $Y_n \xrightarrow{d} |Z|$, where $Z \sim N(0, 1)$. But $E(Y_n^2) = 1$ for each $n \in \mathbb{N}^*$, therefore $(Y_n)_{n \in \mathbb{N}^*}$ is uniformly integrable, i.e.,

$$\lim_{c \to \infty} \sup_{n \in \mathbb{N}^*} E\big(Y_n \mathbb{I}_{\{|Y_n| > c\}}\big) \leq \lim_{c \to \infty} \sup_{n \in \mathbb{N}^*} \frac{1}{c}E(Y_n^2) = 0.$$

By Theorem 4.13 it follows

$$E(Y_n) \to E(|Z|) = \sqrt{\frac{2}{\pi}}.$$

3) It follows from 1) and 2):

$$\lim_{n\to\infty} \sqrt{n} \cdot \frac{C(2n,n)}{2^{2n}} = \lim_{n\to\infty} \frac{1}{2\sqrt{n}} E(|S_{2n}|) = \frac{1}{\sqrt{2}} \lim_{n\to\infty} \frac{1}{\sqrt{n}} E(|S_n|) = \frac{1}{\sqrt{\pi}}.$$

4) From 2) we deduce that $\dfrac{1}{n^q} S_n \xrightarrow{L^1} 0$, for all $q > \dfrac{1}{2}$. The desired result follows from Theorem 4.3.

5) Assume that $(U_n)_{n\in\mathbb{N}^*} = \left(\dfrac{1}{\sqrt{n}} S_n\right)_{n\in\mathbb{N}^*}$ converges in probability to a random variable which we denote by U. By using Problem 5.2.2 we compute

$$E(U_n^4) = \frac{1}{n^2} E(S_n^4) = 3 - \frac{2}{n} \text{ for each } n \in \mathbb{N}^*$$

and write

$$\lim_{c\to\infty} \sup_{n\in\mathbb{N}^*} E\left(U_n^2 \mathbb{I}_{\{U_n^2 > c\}}\right) \leq \lim_{c\to\infty} \sup_{n\in\mathbb{N}^*} \frac{1}{c^2} E(U_n^4) = 0.$$

Hence, $\left(U_n^2\right)_{n\in\mathbb{N}^*}$ is uniformly integrable. By Theorem 4.12 we obtain that $(U_n)_{n\in\mathbb{N}^*}$ converges in L^2.

Denote

$$\tilde{S}_n = X_{n+1} + \ldots + X_{2n} = S_{2n} - S_n.$$

By the properties of the random walk starting at 0, $\tilde{S}_n$ is independent of S_n and has the same distribution as S_n and thus $E(\tilde{S}_n) = E(S_n) = 0$, $E(\tilde{S}_n^2) = E(S_n^2) = n$ for all $n \in \mathbb{N}^*$. We write

$$E(U_n - U_{2n})^2 = E\left(\left(\frac{S_n}{\sqrt{n}} - \frac{S_{2n}}{\sqrt{2n}}\right)^2\right)$$

$$= \frac{1}{2n} E\left(S_n^2(\sqrt{2}-1)^2 - 2(\sqrt{2}-1)S_n\tilde{S}_n + \tilde{S}_n^2\right) = 2 - \sqrt{2}.$$

By the Cauchy criterion for random variables (see Theorem 4.10) it follows that the sequence $(U_n)_{n\in\mathbb{N}^*}$ does not converge in L^2 to a random variable, which contradicts our above reasoning. This implies that our assumption was wrong, hence $\left(\dfrac{1}{\sqrt{n}} S_n\right)_{n\in\mathbb{N}^*}$ does not converge in probability to a random variable.

5.3   Markov Processes

A Markov process is a stochastic process whose future probabilities are determined by its most recent values. Its definition is based on conditional probabilities, see Definition 3.11. For bibliography concerning Markov processes we refer the reader to [Breiman (1992)], [Doob (1953)] and [Kolokoltsov (2011)].

Definition 5.7. A stochastic process $(X_t)_{t \in \mathcal{T}}$ is called a **Markov process**, if for every $n \in \mathbb{N}$, $n \geq 2$, $t_1, \ldots, t_n \in \mathcal{T}$ with $t_1 < \ldots < t_n$ and $x \in \mathbb{R}$ it holds a.s.

$$P(X_{t_n} \leq x \,|\, X_{t_1}, \ldots, X_{t_{n-1}}) = P(X_{t_n} \leq x \,|\, X_{t_{n-1}}). \qquad \blacklozenge$$

Example 5.6. Let $(X_n)_{n \in \mathbb{N}}$ be a stochastic process with discrete state space $\mathcal{S}$. Then the following statements are equivalent:
1) $(X_n)_{n \in \mathbb{N}}$ is a Markov process;
2) for every $n \in \mathbb{N}$, $n \geq 2$, $x_1, \ldots, x_n \in \mathcal{S}$, $t_1, \ldots, t_n \in \mathbb{N}$ with $t_1 < \ldots < t_n$ and $P(X_{t_{n-1}} = x_{n-1}, \ldots, X_{t_1} = x_1) > 0$ we have (recall Definition 1.8)

$$P(X_{t_n} = x_n | X_{t_{n-1}} = x_{n-1}, \ldots, X_{t_1} = x_1) = P(X_{t_n} = x_n | X_{t_{n-1}} = x_{n-1}).$$

Proof: Suppose that $(X_n)_{n \in \mathbb{N}}$ is a Markov process. Let $n \in \mathbb{N}$, $n \geq 2$, $x \in \mathbb{R}$, $x_1, \ldots, x_{n-1} \in \mathcal{S}$, $t_1, \ldots, t_n \in \mathbb{N}$ with $t_1 < \ldots < t_n$ and $P(X_{t_{n-1}} = x_{n-1}, \ldots, X_{t_1} = x_1) > 0$. By using Example 3.8-(1) we deduce that for a.e. $\omega \in \{X_{t_{n-1}} = x_{n-1}, \ldots, X_{t_1} = x_1\}$

$$P(X_{t_n} \leq x | X_{t_{n-1}}, \ldots, X_{t_1})(\omega)$$
$$= P(X_{t_n} = x_n | X_{t_{n-1}} = x_{n-1}, \ldots, X_{t_1} = x_1)$$

and

$$P(X_{t_n} \leq x | X_{t_{n-1}})(\omega) = P(X_{t_n} = x_n | X_{t_{n-1}} = x_{n-1}).$$

Since $(X_n)_{n \in \mathbb{N}}$ is a Markov process, we get

$$P(X_{t_n} \leq x | X_{t_{n-1}} = x_{n-1}, \ldots, X_{t_1} = x_1) = P(X_{t_n} \leq x | X_{t_{n-1}} = x_{n-1})$$

for all $x \in \mathbb{R}$. Now, we can easily deduce the desired equality.

In view of Example 3.8-(1), it is not difficult to prove the converse implication.   ◂

Example 5.7. A sequence $(X_n)_{n \in \mathbb{N}}$ of independent random variables with discrete state space $\mathcal{S}$ is a Markov process.

Proof: Let $n \in \mathbb{N}, n \geq 2, x_1, \ldots, x_n \in \mathcal{S}$ and $t_1, \ldots, t_n \in \mathbb{N}$ with $t_1 < \ldots < t_n$ be such that $P(X_{t_{n-1}} = x_{n-1}, \ldots, X_{t_1} = x_1) > 0$. Then

$$
P(X_{t_n} = x_n \mid X_{t_{n-1}} = x_{n-1}, \ldots, X_{t_1} = x_1)
$$
$$
= \frac{P(X_{t_n} = x_n, X_{t_{n-1}} = x_{n-1}, \ldots, X_{t_1} = x_1 = x_{n-1})}{P(X_{t_{n-1}} = x_{n-1}, \ldots, X_{t_1} = x_1)} = P(X_{t_n} = x_n)
$$

and

$$
P(X_{t_n} = x_n \mid X_{t_{n-1}} = x_{n-1}) = P(X_{t_n} = x_n).
$$

Thus, by Example 5.6, we obtain that $(X_n)_{n \in \mathbb{N}}$ is a Markov process. ▲

Definition 5.8. Let $(X_t)_{t \in \mathcal{T}}$ be a stochastic process.
(1) $(X_t)_{t \in \mathcal{T}}$ has **independent increments**, if for every $n \in \mathbb{N}$, $n \geq 3$, and all $t_1, \ldots, t_n \in \mathcal{T}$ with $t_1 < \ldots < t_n$ the random variables $X_{t_2} - X_{t_1}, \ldots, X_{t_n} - X_{t_{n-1}}$ are independent.
(2) $(X_t)_{t \in \mathcal{T}}$ has **stationary increments**, if for every $s, t, u, v \in \mathcal{T}$ with $t - s = v - u > 0$ the random variables $X_t - X_s$ and $X_v - X_u$ have the same distribution. ◆

Example 5.8. Let $\mathcal{T} \subseteq [0, \infty)$ be such that $0 \in \mathcal{T}$ and let $c \in \mathbb{R}$. If $(X_t)_{t \in \mathcal{T}}$ is a stochastic process with independent increments and $X_0 = c$, then $(X_t)_{t \in \mathcal{T}}$ is a Markov process.
Proof: Let $n \in \mathbb{N}^*$, $n \geq 2$, $t_1, \ldots, t_n \in \mathcal{T}$ with $0 = t_0 \leq t_1 < t_2 < \ldots < t_n$ and $x \in \mathbb{R}$. Denote $Z_k = X_{t_k} - X_{t_{k-1}}$ for $k \in \{1, \ldots, n\}$. Let $\mathcal{G}$ be the σ-field generated by $X_{t_1}, \ldots, X_{t_{n-1}}$ and $\mathcal{K}$ be the σ-field generated by $Z_1, \ldots, Z_{n-1}$.

Since $Z_1, Z_2, \ldots, Z_{n-1}, Z_n$ are independent (see Definition 5.8-(1)), we have that Z_n is independent of $\mathcal{K}$ (see Remark 2.8-(2)). Consider the $\mathcal{B}^{n-1}/\mathcal{B}^{n-1}$ measurable function given by

$$
g(z_1, \ldots, z_{n-1}) = (c + z_1, c + z_1 + z_2, \ldots, c + z_1 + \ldots + z_{n-1})
$$

for $(z_1, \ldots, z_{n-1}) \in \mathbb{R}^{n-1}$. Since $g(Z_1, \ldots, Z_{n-1}) = (X_{t_1}, \ldots, X_{t_{n-1}})$, Z_n is independent of $\mathcal{G}$, see Remark 2.8-(1).

Let $f : \mathbb{R}^2 \to \mathbb{R}$ be given by

$$
f(u, v) = \begin{cases} 1, & \text{if } u + v \leq x \\ 0, & \text{otherwise.} \end{cases}
$$

By Theorem 3.3-(8), we deduce that

$$
E\big(f(Z_n, X_{t_{n-1}}) | \mathcal{G}\big)(\omega) = h\big(X_{t_{n-1}}(\omega)\big) \text{ for a.e. } \omega \in \Omega,
$$

where $h(y) = E\big(f(Z_n, y)\big)$ for $y \in \mathbb{R}$. Since h is $\mathcal{B}/\mathcal{B}$ measurable, we have that $h(X_{t_{n-1}})$ is $\mathcal{H}/\mathcal{B}$ measurable, where $\mathcal{H}$ is the σ-field generated by $X_{t_{n-1}}$. Clearly, $\mathcal{H} \subseteq \mathcal{G}$. In view of Theorem 3.3-(4)-(9), we get

$$E\big(f(Z_n, X_{t_{n-1}})|\mathcal{H}\big)(\omega) = h\big(X_{t_{n-1}}(\omega)\big) \text{ for a.e. } \omega \in \Omega,$$

and thus we have a.s.

$$E\big(f(Z_n, X_{t_{n-1}})|\mathcal{G}\big) = E\big(f(Z_n, X_{t_{n-1}})|\mathcal{H}\big).$$

Hence, we deduce by Definition 3.11 that

$$P(X_{t_n} \leq x \,|\, X_{t_1}, \ldots, X_{t_{n-1}}) = P(X_{t_n} \leq x \,|\, X_{t_{n-1}}). \qquad \blacktriangle$$

Example 5.9. The random walk $(S_n)_{n \in \mathbb{N}}$ from Definition 5.5 is a processes with independent and stationary increments. Moreover, for $j \in \mathbb{Z}$ and $n \in \mathbb{N}$, if $P(S_n = j) > 0$, it holds

$$P(S_{n+1} = j + 1|S_n = j) = p \text{ and } P(S_{n+1} = j - 1|S_n = j) = 1 - p.$$

Using Example 5.8, it follows that this process is a Markov process. $\qquad \blacktriangle$

Definition 5.9. Let $(X_t)_{t \in \mathcal{T}}$ be a Markov process having a discrete state space $\mathcal{S}$.
(1) We define for $s, t \in \mathcal{T}$ with $s < t$ and $i, j \in \mathcal{S}$ the **transition probabilities**

$$\tilde{p}_{ij}(s, t) = P(X_t = j \,|\, X_s = i), \text{ if } P(X_s = i) > 0.$$

(2) $(X_t)_{t \in \mathcal{T}}$ is said to be **homogeneous**, if for $i, j \in \mathcal{S}$ there exists a function $p_{ij} : (0, \infty) \to [0, 1]$ such that for $s, t \in \mathcal{T}$ with $s < t$ we have

$$\tilde{p}_{ij}(s, t) = p_{ij}(t - s), \text{ if } P(X_s = i) > 0. \qquad \blacklozenge$$

Next, we consider the **Chapman–Kolmogorov equations** for a Markov process with discrete state space.

Theorem 5.2. *Let $(X_t)_{t \in \mathcal{T}}$ be a Markov process with discrete state space $\mathcal{S}$. For $i, j \in \mathcal{S}$, $s, t, \tau \in \mathcal{T}$ with $s < \tau < t$ and $P(X_s = i) > 0$ we have*

$$\tilde{p}_{ij}(s, t) = \sum_{\substack{k \in \mathcal{S} \\ \tilde{p}_{ik}(s, \tau) > 0}} \tilde{p}_{ik}(s, \tau)\tilde{p}_{kj}(\tau, t).$$

Remark 5.1. Let $(X_t)_{t \geq 0}$ be a homogeneous Markov process with discrete state space $\mathcal{S}$ and let $i, j \in \mathcal{S}$.
(1) The function p_{ij} given in Definition 5.9-(2) for $\mathcal{T} = [0, \infty)$ can be

uniquely determined as follows: since $i \in \mathcal{S}$, there exists $s \geq 0$ such that $P(X_s = i) > 0$ and thus

$$p_{ij}(h) = P(X_{s+h} = j | X_s = i) \quad \text{for each } h > 0.$$

(2) We have the following Chapman–Kolmogorov equation for $u, v > 0$

$$p_{ij}(u + v) = \sum_{k \in \mathcal{S}} p_{ik}(u) p_{kj}(v). \qquad \blacktriangledown$$

In what follows we consider stochastic processes in **discrete time**.

Definition 5.10. Let $\mathcal{S} = \{0, 1, \ldots, m\}$, where $m \in \mathbb{N}^*$. A Markov process $(X_n)_{n \in \mathbb{N}}$ with state space $\mathcal{S}$ is called **Markov chain**, if there exists a matrix

$$\mathcal{P} = (p_{ij}^{(1)})_{i,j=\overline{0,m}} = \begin{pmatrix} p_{00}^{(1)} & p_{01}^{(1)} & \cdots & p_{0m}^{(1)} \\ p_{10}^{(1)} & p_{11}^{(1)} & \cdots & p_{1m}^{(1)} \\ \vdots & \vdots & \vdots & \vdots \\ p_{m0}^{(1)} & p_{m1}^{(1)} & \cdots & p_{mm}^{(1)} \end{pmatrix}$$

called **matrix of transition probabilities**, such that for every $i, j \in \{0, 1, \ldots, m\}$ and $n \in \mathbb{N}$, if $P(X_n = i) > 0$, then

$$P(X_{n+1} = j | X_n = i) = p_{ij}^{(1)}. \qquad \blacklozenge$$

Remark 5.2. Let $\mathcal{S} = \{0, 1, \ldots, m\}$, where $m \in \mathbb{N}^*$, and let $(X_n)_{n \in \mathbb{N}}$ be a Markov chain with state space $\mathcal{S}$. Then the matrix of transition probabilities $\mathcal{P} = (p_{ij}^{(1)})_{i,j=\overline{0,m}}$ given in Definition 5.10 can be uniquely determined as follows: let $i, j \in \mathcal{S}$; there exists $n \in \mathbb{N}$ such that $P(X_n = i) > 0$ and thus we have

$$p_{ij}^{(1)} = P(X_{n+1} = j | X_n = i). \qquad \blacktriangledown$$

Example 5.10. (1) Consider $(X_n)_{n \in \mathbb{N}}$ to be a Markov chain with state space $\mathcal{S} = \{0, 1, \ldots, m\}$, $m \in \mathbb{N}^*$, and matrix $\mathcal{P} = (p_{ij}^{(1)})_{i,j=\overline{0,m}}$ of transition probabilities. Denote by $\mathcal{P}^r = (p_{ij}^{(r)})_{i,j=\overline{0,m}}$ the rth power of the matrix $\mathcal{P}$, for $r \in \mathbb{N}^*$. Then we have for every $i, j \in \{0, 1, \ldots, m\}$, $n \in \mathbb{N}$ and $r \in \mathbb{N}^*$, if $P(X_n = i) > 0$,

$$P(X_{n+r} = j \mid X_n = i) = p_{ij}^{(r)}.$$

Proof: Let $i, j \in \{0, 1, \ldots, m\}$ and $n \in \mathbb{N}$ be such that $P(X_n = i) > 0$. We apply Theorem 5.2 with $s = n$, $\tau = n + 1$, $t = n + 2$ and we use Definition 5.10 to get

$$P(X_{n+2} = j | X_n = i) = \sum_{\substack{k=0 \\ p_{ik}^{(1)} > 0}}^{m} p_{ik}^{(1)} p_{kj}^{(1)} = \sum_{k=0}^{m} p_{ik}^{(1)} p_{kj}^{(1)} = p_{ij}^{(2)}.$$

By induction with respect to $r \in \mathbb{N}$, $r \geq 2$, using Theorem 5.2, we deduce the desired equalities.

(2) Note that each Markov chain is a homogeneous Markov process.   ▲

Example 5.11. Let $(X_n)_{n\in\mathbb{N}}$ be a Markov chain with $X_0 = 0$ and whose matrix of transition probabilities is

$$\mathcal{P} = (p_{ij}^{(1)})_{i,j=\overline{0,3}} = \begin{pmatrix} 0 & 1 & 0 & 0 \\ 0.5 & 0 & 0.5 & 0 \\ 0 & 0.5 & 0 & 0.5 \\ 0 & 0 & 1 & 0 \end{pmatrix}.$$

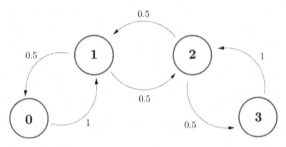

Fig. 5.6: Representation of the reflecting Markov chain from Example 5.11

We simulate values of $(X_n)_{n\in\mathbb{N}}$ as follows:

(1) Generate a value of the random variable that takes the values: $0, 1, 2, 3$ with the corresponding probabilities: $p_{00}^{(1)}, p_{01}^{(1)}, p_{02}^{(1)}, p_{03}^{(1)}$ (if a probability is 0, then the corresponding value is not generated). Let j_1 be the result. Obviously, j_1 is a value of X_1.

(2) Generate a value of the random variable that takes the values: $0, 1, 2, 3$ with the corresponding probabilities $p_{j_10}^{(1)}, p_{j_11}^{(1)}, p_{j_12}^{(1)}, p_{j_13}^{(1)}$. Let j_2 be the result. Obviously, j_2 is a value of X_2.

(3) We continue as before, until we get the desired number of values of the Markov chain.

We give the corresponding Matlab code:

```
function X=markov_chain(P,m,n)
X=zeros(1,n+1);
j=0;
for k=2:n+1
    j=randsample(0:m,1,'true',P(j+1,:));
    X(k)=j;
```

end
end

>>P=[0 1 0 0;0.5 0 0.5 0;0 0.5 0 0.5;0 0 1 0];
>>X=markov_chain(P,3,30)
X = 0 1 2 1 2 1 2 1 2 3 2 3 2 1 0 1 2 1 2 1 2 1 0 1 2 1 0 1 2 3 2

Next, we compute the probability mass function of X_3. We have

$$\mathcal{P}^3 = (p_{ij}^{(3)})_{i,j=\overline{0,3}} = \begin{pmatrix} 0 & 0.75 & 0 & 0.25 \\ 0.375 & 0 & 0.625 & 0 \\ 0 & 0.625 & 0 & 0.375 \\ 0.25 & 0 & 0.75 & 0 \end{pmatrix}.$$

By Example 5.10, we have

$$P(X_3 = j) = P(X_3 = j | X_0 = 0)P(X_0 = 0) = p_{0j}^{(3)} \text{ for } j \in \{0,1,2,3\}.$$

So, $P(X_3 = 0) = P(X_3 = 2) = 0$, $P(X_3 = 1) = 0.75$ and $P(X_3 = 3) = 0.25$.
▲

Example 5.12. Consider a random walk on the circle $(Z_n)_{n\in\mathbb{N}}$ (as presented in Definition 5.6) with $m = 4$ points, which starts at 0 and in each step, the random walk moves anticlockwise with probability $p \in (0,1)$ and clockwise with probability $1 - p$ to the closest point. This processes is a Markov chain having the following matrix of transition probabilities

$$\mathcal{P} = \begin{pmatrix} 0 & p & 0 & 0 & 1-p \\ 1-p & 0 & p & 0 & 0 \\ 0 & 1-p & 0 & p & 0 \\ 0 & 0 & 1-p & 0 & p \\ p & 0 & 0 & 1-p & 0 \end{pmatrix}.$$

▲

Definition 5.11. We say that a Markov chain $(Z_n)_{n\in\mathbb{N}}$ is a **branching process** , if it describes the evolution of a population as follows:
• Z_n is the number of members of the nth generation for $n \in \mathbb{N}$;
• Z_0 is a given number in $\mathbb{N}^*$;
• each member of the nth generation gives birth to a family, possibly empty, of members of the $(n + 1)$th generation for $n \in \mathbb{N}$;
• the family sizes of the members are independent and follow the same discrete distribution. ♦

Remark 5.3. Let $(Z_n)_{n\in\mathbb{N}}$ be a branching process.

(1) Then there exists a family $(X_{n,k})_{(n,k)\in\mathbb{N}\times\mathbb{N}^*}$ of independent and identically distributed random variables such that

$$Z_{n+1} = \sum_{k=1}^{\infty} X_{n,k} \cdot \mathbb{I}_{\{Z_n \geq k\}} \text{ for each } n \in \mathbb{N}.$$

Note that $X_{n,k}$ denotes the number of direct successors of member k of the nth generation for $n \in \mathbb{N}$ and $k \in \mathbb{N}^*$.

(2) Let $(X_{n,k})_{(n,k)\in\mathbb{N}\times\mathbb{N}^*}$ be as above and $h : \mathbb{R} \to \mathbb{R}$ be a $\mathcal{B}/\mathcal{B}$ measurable function. Let $n \in \mathbb{N}$. Since $Z_n, X_{n,1}, X_{n,2}, \ldots$ are independent, we have

$$E\big(h(Z_{n+1})\big) = \sum_{k=1}^{\infty} E\big(h(X_{n,1} + \ldots + X_{n,k})\big) P(Z_n = k),$$

provided the above expected values exist (see [Gut (2005), (15.4), p. 84]).

▼

Example 5.13. A certain species of mammal was infected with a harmful virus: each female can give birth, during its lifetime, to at most one female offspring. Suppose females are born independently with the same probability $p \in (0,1)$. Let Z_n be the number of females in the nth generation for $n \in \mathbb{N}$ and let $Z_0 = k$ for some fixed $k \in \mathbb{N}^*$. Prove that $(Z_n)_{n\in\mathbb{N}}$ is a branching process with $Z_n \sim Bino(k, p^n)$ for each $n \in \mathbb{N}^*$.

Proof: From the hypothesis we have, in view of Definition 5.11, that $(Z_n)_{n\in\mathbb{N}}$ is a branching process. Following Remark 5.3, we have

$$Z_{n+1} = \sum_{k=1}^{\infty} X_{n,k} \cdot \mathbb{I}_{\{Z_n \geq k\}} \text{ for each } n \in \mathbb{N},$$

where $(X_{n,k})_{(n,k)\in\mathbb{N}\times\mathbb{N}^*}$ is a family of independent random variables having distribution $Ber(p)$.

For $i \in \{1, \ldots, k\}$ let $Y_{i,0} = 1$ and for $n \in \mathbb{N}^*$ consider

$$Y_{i,n} = \begin{cases} 1, & \text{the } i\text{th female in the 0th generation has a female} \\ & \text{descendant in the } n\text{th generation} \\ 0, & \text{otherwise}. \end{cases}$$

We have: $Y_{i,n} = 1 \iff Y_{i,n} = Y_{i,n-1} = \ldots = Y_{i,0} = 1$, hence

$$P(Y_{i,n} = 1) = pP(Y_{i,n-1} = 1) = \ldots = p^n P(Y_{i,0} = 1) = p^n,$$

where we use the fact that the probability that a female gives birth to a female offspring is p for all $n \in \mathbb{N}$ and $i \in \{1, \ldots, k\}$. Since

$$Z_n = Y_{1,n} + \ldots + Y_{k,n},$$

we deduce that $Z_n \sim Bino(k, p^n)$ for each $n \in \mathbb{N}$.

For bibliography concerning branching processes we refer the reader to [Grimmet and Stirzaker (2001)], [Ghahramani (2005)] and [Harris (1963)]. Further we consider stochastic processes in **continuous time**.

Definition 5.12. A stochastic process $(N_t)_{t \geq 0}$ is called a **homogeneous Poisson process** (shortly Poisson process), if there exists a parameter $\lambda > 0$ and

(1) $P(N_0 = 0) = 1$;
(2) $(N_t)_{t \geq 0}$ has independent increments;
(3) For $0 \leq s < t$ the random variables $N_t - N_s \sim Poiss(\lambda(t - s))$ are Poisson distributed with parameter $\lambda(t - s)$ (see Section 2.4.7). ◆

Remark 5.4. Let $(N_t)_{t \geq 0}$ be a Poisson process. The random variable $(N_t)_{t \geq 0}$ is in fact a **counting process**, for example, $N_t - N_s$ counts the number of a certain event that occurs in the time interval $(s, t]$. Denote by S_n the time of the nth occurrence of the event (called also: nth arrival time or nth waiting time), $n \in \mathbb{N}^*$, $S_0 = 0$. The random variable $T_n = S_n - S_{n-1}, n \in \mathbb{N}^*$ is called nth **interarrival time**. Observe that for $t > 0$, $n \in \mathbb{N}^*$ we have the following relations: $\{S_n \leq t\} = \{N_t \geq n\}$, $\{S_n \leq t < S_{n+1}\} = \{N_t = n\}$, $S_n = \inf\{t \geq 0 : N_t = n\}$,

$$N_t = \max\{n \geq 0 : S_n \leq t\} = \#\{n \geq 1 : S_n \leq t\},$$

i.e., N_t counts the number events that occur in the interval $(0, t]$,

$$\{T_n > t\} = \{S_n - S_{n-1} > t\} = \{N_{S_{n-1}+t} - N_{S_{n-1}} = 0\}. \qquad \blacktriangledown$$

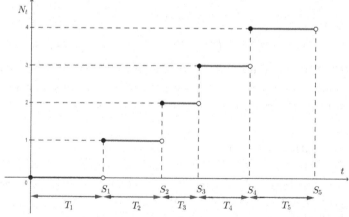

Fig. 5.7: A simulated sample path of a Poisson process

Remark 5.5. We list some properties of a Poisson process $(N_t)_{t\geq 0}$ with parameter $\lambda > 0$:

(1) $(N_t)_{t\geq 0}$ is continuous in probability (see [Protter (2004), Theorem 23, p. 13]). Moreover, there exists a unique modification of $(N_t)_{t\geq 0}$, which is càdlàg, i.e., a.s. the paths are right-continuous and have left limits (see [Protter (2004), Theorem 30, p. 20]).

(2) The interarrival times $T_1, \ldots, T_n, \ldots$ (see Remark 5.4) are independent random variables with the same exponential distribution $Exp(\lambda)$ (see [Karr (1993), Theorem 3.39, p. 93], [Gut (2009), Theorem 1.3, p. 228]). Conversely, if $(T_n)_{n\in\mathbb{N}^*}$ is a sequence of independent random variables with the same exponential distribution $Exp(\lambda)$ and $N_t = \#\{n \geq 1 : S_n \leq t\}$, for $t \geq 0$, where $S_n = T_1 + \ldots + T_n$, for $n \geq 1$, then $(N_t)_{t\geq 0}$ is a Poisson process with parameter λ (see [Gut (2009), Theorem 1.4, p. 231]). ▲

Notation: Recall the Landau **"little-o notation"**

$$g(h) = o(h), \ h \searrow 0$$

refers to

$$\lim_{h\searrow 0} \frac{g(h)}{h} = 0.$$

Remark 5.6. **(1)** By Example 5.8 and Definitions 5.8 and 5.12, it follows that $(N_t)_{t\geq 0}$ is a homogeneous Markov process with stationary increments.

(2) We see that $P(N_t = j) = \dfrac{(\lambda t)^j}{j!} e^{-\lambda t}, \ j \in \mathbb{N}$.

(3) For all $t \geq 0$ we have

$$P(N_{t+h} - N_t = 0) = P(N_h = 0) = e^{-\lambda h} = 1 - \lambda h + o(h), \ h \searrow 0$$

$$P(N_{t+h} - N_t = 1) = P(N_h = 1) = \lambda h e^{-\lambda h} = \lambda h + o(h), \ h \searrow 0$$

and

$$P(N_{t+h} - N_t \geq 2) = P(N_h \geq 2) = o(h), \ h \searrow 0.$$

(4) If we replace $t \mapsto \lambda t$ by a nonlinear positive monotone increasing function $t \mapsto \lambda(t)$ with $\lambda(0) = 0$, then the corresponding process is called **inhomogeneous Poisson process**. In this case, the increments are not necessarily stationary. ▼

Definition 5.13. Let $(X_t)_{t\geq 0}$ be a homogeneous Markov process with discrete state space $\mathcal{S} = \mathbb{N}$. Let $(\lambda_n)_{n\in\mathbb{N}}$ and $(\mu_n)_{n\in\mathbb{N}^*}$ be sequences of positive real numbers. $(X_t)_{t\geq 0}$ is called **birth and death process** with **birth rates** $(\lambda_n)_{n\in\mathbb{N}}$ and **death rates** $(\mu_n)_{n\in\mathbb{N}^*}$, if for every $i \in \mathbb{N}$ and $s \geq 0$ with $P(X_s = i) > 0$ the following conditions are fulfilled:

(1) $P(X_{s+h} = i+1|X_s = i) = \lambda_i h + o(h)$, $h \searrow 0$;

(2) $P(X_{s+h} = i-1|X_s = i) = \mu_i h + o(h)$, $h \searrow 0$, if $i \geq 1$;

(3) $P(|X_{s+h} - X_s| > 1|X_s = i) = o(h)$, $h \searrow 0$. ♦

Example 5.14. Let $(X_t)_{t \geq 0}$ be a birth and death process with birth rates $(\lambda_n)_{n \in \mathbb{N}}$ and death rates $(\mu_n)_{n \in \mathbb{N}^*}$. Set $\mu_0 = 0$.

(1) We have

$$P(X_{s+h} = i|X_s = i) = 1 - P(X_{s+h} = i-1|X_s = i)$$
$$- P(X_{s+h} = i+1|X_s = i) - P(|X_{s+h} - X_s| > 1|X_s = i)$$

for $i \in \mathbb{N}^*$, $s \geq 0$ with $P(X_s = i) > 0$ and $h > 0$. Also, we have

$$P(X_{s+h} = 0|X_s = 0) = 1 - P(X_{s+h} = 1|X_s = 0)$$
$$- P(|X_{s+h} - X_s| > 1|X_s = 0),$$

for $s \geq 0$ with $P(X_s = 0) > 0$ and $h > 0$. It follows from Definition 5.13 that for $i \in \mathbb{N}$ and $s \geq 0$ with $P(X_s = i) > 0$ we have

$$P(X_{s+h} = i|X_s = i) = 1 - (\lambda_i + \mu_i)h + o(h), \ h \searrow 0.$$

(2) In view of Remark 5.1, for every $i, j \in \mathbb{N}$ let $p_{ij} : (0, \infty) \to [0, 1]$ be such that for $s \geq 0$ with $P(X_s = i) > 0$ we have

$$p_{ij}(h) = P(X_{s+h} = j|X_s = i) \text{ for each } h > 0.$$

In view of Definition 5.13 and property (1) of this example, we define the **intensity** of $(X_t)_{t \geq 0}$ for $i, j \in \mathbb{N}$ by

$$q_{ij} = \begin{cases} \lim_{h \searrow 0} \dfrac{p_{ij}(h)}{h}, & \text{if } |i - j| \geq 1 \\ \lim_{h \searrow 0} \dfrac{1 - p_{ii}(h)}{h}, & \text{if } i = j \end{cases}$$

and we note that

$$q_{ij} = \begin{cases} \lambda_i, & \text{if } j = i+1 \\ \mu_i, & \text{if } j = i-1 \geq 0 \\ \lambda_i + \mu_i, & \text{if } j = i \\ 0, & \text{otherwise.} \end{cases}$$

(3) If $X_0 = 0$ a.s., $\mu_i = 0$ for all $i \in \mathbb{N}^*$ and there exists $\lambda > 0$ such that $\lambda_i = \lambda$ for all $i \in \mathbb{N}$, then $(X_t)_{t \geq 0}$ is related to a Poisson process with parameter λ (see [Gut (2009), Theorem 1.1, p. 223]). ▲

We mention that birth and death processes have applications in biology, ecology, epidemiology, genetics, queuing theory, etc. For bibliography concerning birth and death processes we refer the reader for example to [Grimmet and Stirzaker (2001)], [Ghahramani (2005)], [Sericola (2013)] and [Kirkwood (2015)].

Definition 5.14. A stochastic process $(X_t)_{t\in\mathcal{T}}$ is called a **Gaussian process**, if for all $n \in \mathbb{N}^*$ and all $t_1, \ldots, t_n \in \mathcal{T}$ with $t_1 < \ldots < t_n$ the random vectors $(X_{t_1}, \ldots, X_{t_n})$ have multivariate normal distribution. For a Gaussian process $(X_t)_{t\in\mathcal{T}}$ we define the **expectation function**

$$\tilde{m}(t) = E(X_t), \ t \in \mathcal{T},$$

and the **covariance function**

$$\tilde{\gamma}(s,t) = E\big((X_s - \tilde{m}(s))(X_t - \tilde{m}(t))\big), \ \ s,t \in \mathcal{T}. \qquad \blacklozenge$$

Remark 5.7. **(1)** Note that the multivariate normal distribution mentioned in Definition 5.14 refers mainly to the nondegenerate case, see Section 2.5.4, but in some special situations it can be the degenerate case, see Remark 3.7.

(2) $(X_t)_{t\in\mathcal{T}}$ is a Gaussian process if and only if for arbitrary $n \in \mathbb{N}^*$ $t_1, \ldots, t_n \in \mathcal{T}$, $t_1 < \ldots < t_n$, $(b_1, \ldots, b_n) \in \mathbb{R}^n \setminus \{0_n\}$ the linear combination $b_1 X_{t_1} + \ldots + b_n X_{t_n}$ is a normally distributed random variable (see Problem 3.6.1).

(3) It follows from the Definition 5.14 that for all $t_1, \ldots, t_n \in \mathcal{T}$ with $t_1 < \ldots < t_n$, $n \in \mathbb{N}^*$, the multivariate normal distribution of $\mathbb{X} = (X_{t_1}, \ldots, X_{t_n})$ is characterized by $\mathbb{m} = E(\mathbb{X}) = \big(\tilde{m}(t_1), \ldots, \tilde{m}(t_n)\big)$ and $A = V(\mathbb{X}) = \big(\tilde{\gamma}(t_i, t_j)\big)_{i,j=\overline{1,n}}.$

Definition 5.15. A stochastic process $(W_t)_{t\geq 0}$ is called **Wiener process** or **Brownian motion**, if
(1) $P(W_0 = 0) = 1$.
(2) $(W_t)_{t\geq 0}$ has independent increments.
(3) For all $t \geq 0$ and $h > 0$ it holds $W_{t+h} - W_t \sim N(0, h)$. $\qquad \blacklozenge$

Example 5.15. Let $(W_t)_{t\geq 0}$ be a Wiener process. Fix $0 < s < t$. Taking into account Definition 2.20, we prove that

$$F_{W_t|W_s}(x|y) = F_{W_{t-s}}(x - y) \text{ for } x, y \in \mathbb{R}.$$

Proof: The transformation

$$(W_t, W_s) = (W_t - W_s, W_s) \begin{pmatrix} 1 & 0 \\ 1 & 1 \end{pmatrix}$$

together with Example 2.14 and the fact that $(W_t - W_s, W_s)$ is a continuous random vector with independent components imply that

$$f_{W_t, W_s}(u, v) = f_{W_t - W_s}(u - v) f_{W_s}(v) \quad \text{for } u, v \in \mathbb{R}$$

is a joint density function of (W_t, W_s), where f_{W_s}, $f_{W_t - W_s}$ are density functions of W_s, respectively $W_t - W_s$.

We have for each $x, y \in \mathbb{R}$

$$F_{W_t | W_s}(x | y) = P(W_t \le x | W_s = y) = \int_{-\infty}^{x} f_{W_t | W_s}(u | y) du$$

$$= \frac{1}{\sqrt{2\pi(t - s)}} \int_{-\infty}^{x} \exp\left\{ -\frac{(u - y)^2}{2(t - s)} \right\} du = P(W_{t-s} \le x - y),$$

where we consider the computations

$$f_{W_t | W_s}(u | y) = \frac{f_{W_t, W_s}(u, y)}{f_{W_s}(y)} = f_{W_t - W_s}(u - y)$$

$$= \frac{1}{\sqrt{2\pi(t - s)}} \exp\left\{ -\frac{(u - y)^2}{2(t - s)} \right\}.$$

Hence for each $h, t > 0$ and $x, y \in \mathbb{R}$ we have

$$P(W_{t+h} \le x | W_t = y) = P(W_h \le x - y).$$

This property is a continuous analog of the homogeneity property of a Markov process having discrete state space given in Definition 5.9. ▲

Remark 5.8. (1) From Example 5.8 and Definitions 5.8 and 5.15 it follows that a Wiener process is a Markov process with stationary increments.
(2) A Wiener process has a.s. continuous trajectories: If $(W_t)_{t \ge 0}$ is a Wiener process, then there exists a modification of it $(\tilde{W}_t)_{t \ge 0}$ (i.e., for all $t \ge 0$ we have $W_t = \tilde{W}_t$ a.s.) which has a.s. continuous paths (for a.e. $\omega \in \Omega$ the mapping $t \mapsto \tilde{W}_t(\omega)$ is continuous). In what follows we identify W with $\tilde{W}$. So, we can say that the paths of a Wiener process are continuous with probability 1 (for details see [Baldi (2017), Example 2.4, p. 38]). Often, this property is included in the definition of the Wiener process. ▼

Example 5.16. Let $(W_t)_{t \ge 0}$ be a stochastic process with $P(W_0 = 0) = 1$. We prove that $(W_t)_{t \ge 0}$ is a Wiener process if and only if it is a Gaussian process with expectation function $\tilde{m}(t) = 0$ for each $t \ge 0$ and covariance function $\tilde{\gamma}(s, t) = \min\{s, t\}$ for each $s, t \ge 0$.
Proof: Let $0 < t_1 < \ldots < t_n$. Denote $t_0 = 0$ and

$$\mathbb{U}_1 = (W_{t_1}, W_{t_2}, \ldots, W_{t_n}),$$

$$\mathbb{U}_2 = \left(W_{t_1} - W_{t_0}, W_{t_2} - W_{t_1}, \ldots, W_{t_n} - W_{t_{n-1}} \right),$$

$$B = \begin{pmatrix} 1 & 1 & \ldots & 1 & 1 \\ 0 & 1 & \ldots & 1 & 1 \\ \vdots & \vdots & \ddots & \vdots & \vdots \\ 0 & 0 & \ldots & 0 & 1 \end{pmatrix}.$$

Note that B is an invertible matrix. We have $\mathbb{U}_1 = \mathbb{U}_2 B$ a.s.

"$\Longrightarrow$" By the definition of the Wiener process

$$W_{t_1} - W_{t_0}, W_{t_2} - W_{t_1}, \ldots, W_{t_n} - W_{t_{n-1}}$$

are independent random variables, each of them following a normal distribution with zero mean. We have $\mathbb{U}_2 \sim MVN(0_n, C)$, where C is a diagonal matrix, with $c_{ii} = V(W_{t_i} - W_{t_{i-1}}) = t_i - t_{i-1}$, $i \in \{1, \ldots, n\}$. By Example 2.24-(2) it follows that $\mathbb{U}_1 \sim MVN(0_n, B^T C B)$.

Observe that in the case when $t_1 = 0$ we have $W_{t_1} = 0$ a.s. and $(W_{t_1}, \ldots, W_{t_n}) = (0, \ldots, W_{t_n})$ follows a multivariate normal distribution included in the degenerate case (see Remark 5.7-(1)), since the corresponding covariance matrix has the first line and first column filled with zeros.

Hence $(W_t)_{t \geq 0}$ is a Gaussian process.

The property $W_t \sim N(0, t)$ implies $\tilde{m}(t) = 0$ for each $t \geq 0$. Further assume that $0 \leq s < t$. We write

$$\tilde{\gamma}(s, t) = \operatorname{cov}(W_s, W_t) = E(W_s W_t) = E(W_s - W_0)E(W_t - W_s) + E(W_s^2)$$
$$= 0 + s = \min\{s, t\},$$

where we used that $W_t - W_s$ and W_s are independent random variables.

"$\Longleftarrow$" We assume that $(W_t)_{t \geq 0}$ is a Gaussian process. Then $\mathbb{U}_1$ has non-degenerate multivariate normal distribution (see Definition 5.14). We have $\mathbb{U}_2 = \mathbb{U}_1 B^{-1}$ a.s., hence $\mathbb{U}_2$ has also multivariate normal distribution (by Example 2.24-(2)). By using the expectation function and covariance function of $(W_t)_{t \geq 0}$ we compute

$$E(\mathbb{U}_2) = \left(E(W_{t_1} - W_{t_0}), E(W_{t_2} - W_{t_1}), \ldots, E(W_{t_n} - W_{t_{n-1}}) \right) = 0_n$$

and

$$V(\mathbb{U}_2) = \left(\operatorname{cov}(W_{t_i} - W_{t_{i-1}}, W_{t_j} - W_{t_{j-1}}) \right)_{i,j=\overline{1,n}}.$$

Let $i, j \in \{1, \ldots, n\}$ be such that $t_{i-1} < t_i \leq t_{j-1} < t_j$. We have

$$\operatorname{cov}(W_{t_i} - W_{t_{i-1}}, W_{t_j} - W_{t_{j-1}})$$
$$= \tilde{\gamma}(t_i, t_j) - \tilde{\gamma}(t_{i-1}, t_j) - \tilde{\gamma}(t_i, t_{j-1}) + \tilde{\gamma}(t_{i-1}, t_{j-1}) = 0.$$

By Problem 3.4.5 it follows that the components of the random vector $\mathbb{U}_2$ are independent. Since $(W_t)_{t\geq 0}$ is a Gaussian process it follows that for $0 \leq s < t$ the random variable $W_t - W_s$ has normal distribution (see Remark 5.7-(2)) with expectation

$$E(W_t - W_s) = \tilde{m}(t) - \tilde{m}(s) = 0$$

and variance

$$V(W_t - W_s) = \operatorname{cov}(W_t - W_s, W_t - W_s)$$
$$= \tilde{\gamma}(t,t) - \tilde{\gamma}(s,t) - \tilde{\gamma}(t,s) + \tilde{\gamma}(s,s) = t - s.$$

Thus, $W_t - W_s \sim N(0, t-s)$. The statements of Definition 5.15 are fulfilled, hence $(W_t)_{t\geq 0}$ is a Wiener process. ▲

5.3.1 Solved Problems

Problem 5.3.1. Prove the Chapman–Kolmogorov equations (see Theorem 5.2): if $(X_t)_{t\in\mathcal{T}}$ is a Markov process with discrete state space $\mathcal{S}$, then for every $i, j \in \mathcal{S}$, $s, t, \tau \in \mathcal{T}$ with $s < \tau < t$ and $P(X_s = i) > 0$ we have

$$\tilde{p}_{ij}(s,t) = \sum_{\substack{k\in\mathcal{S} \\ \tilde{p}_{ik}(s,\tau)>0}} \tilde{p}_{ik}(s,\tau)\tilde{p}_{kj}(\tau,t).$$

Solution 5.3.1: Let $i, j \in \mathcal{S}$, $s, t, \tau \in \mathcal{T}$ be such that $s < \tau < t$ and $P(X_s = i) > 0$. First, we note that the terms in the sum are well defined. Indeed, in view of Definition 5.9, since $P(X_s = i) > 0$, we have that $\tilde{p}_{ik}(s,\tau) = P(X_\tau = k | X_s = i)$ is well defined for every $k \in \mathcal{S}$. If $\tilde{p}_{ik}(s,\tau) > 0$ for some $k \in \mathcal{S}$, then $P(X_\tau = k, X_s = i) > 0$ and thus $P(X_\tau = k) > 0$, which implies that $\tilde{p}_{kj}(\tau,t) = P(X_t = j | X_\tau = k)$ is well defined.

We have

$$P(X_t = j, X_s = i) = \sum_{k\in\mathcal{S}} P(X_s = i, X_\tau = k, X_t = j)$$

$$= \sum_{\substack{k\in\mathcal{S} \\ P(X_\tau=k,X_s=i)>0}} P(X_s = i)P(X_\tau = k | X_s = i)P(X_t = j | X_\tau = k, X_s = i)$$

$$= P(X_s = i) \sum_{\substack{k\in\mathcal{S} \\ P(X_\tau=k|X_s=i)>0}} P(X_\tau = k | X_s = i)P(X_t = j | X_\tau = k),$$

where we use the Markov property for the last equality. Now, it is not difficult to get the desired equality.

Problem 5.3.2. 1) Simulate a sample path for a homogeneous Poisson process with $\lambda = 0.8$ on the time interval $[0, 15]$ and plot the obtained result.

2) Prove that, if $(N_t)_{t \geq 0}$ is a Poisson process with parameter $\lambda > 0$, then $\frac{1}{t} N_t \xrightarrow{P} \lambda$, as $t \to \infty$, i.e.,

$$\lim_{t \to \infty} P\left(\left|\frac{1}{t} N_t - \lambda\right| \leq \varepsilon\right) = 1 \text{ for every } \varepsilon > 0.$$

3) Prove that, if $(N_t)_{t \geq 0}$ is a Poisson process with parameter $\lambda > 0$, then $\frac{1}{n} N_n \xrightarrow{a.s.} \lambda$. Using this result, estimate in Matlab the value of the parameter λ of a simulated sample path of a Poisson process.

Solution 5.3.2: 1) We follow Remark 5.5-(2) to write the following Matlab code.

```
function poiss_proc_sim(lambda,t0)
clf; hold on;
n=0; S=0;
while S<t0
    T=exprnd(1/lambda);
    plot([S S+T],[n n],'-k','LineWidth',2);
    plot([S S],[0 n],'--k');
    n=n+1; S=S+T;
end
plot([S t0],[n n],'-k','LineWidth',2);
plot([S S],[0 n],'--k');
grid on; axis([0 t0 0 n+1]); xticks(0:ceil(t0));
end

>>poiss_proc_sim(0.8,15)
```

2) By Chebyshev's inequality (see Theorem 3.6) and Problem 3.4.2, we have for every $\varepsilon > 0$

$$P\left(\left|\frac{1}{t} N_t - \lambda\right| > \varepsilon\right) \leq \frac{\lambda}{\varepsilon^2 t} \quad \text{for } t > 0$$

and thus

$$\lim_{t \to \infty} P\left(\left|\frac{1}{t} N_t - \lambda\right| \leq \varepsilon\right) = 1.$$

3) The increments $N_1 - N_0, N_2 - N_1, \ldots, N_n - N_{n-1}, \ldots$ are independent Poisson distributed random variables with parameter λ. By the SLLN (see Theorem 4.19), we have

$$\frac{1}{n} \sum_{k=1}^{n} (N_k - N_{k-1}) \xrightarrow{a.s.} \lambda$$

and thus $\frac{1}{n}N_n \xrightarrow{a.s.} \lambda$.

Using the above result, we estimate in Matlab the value of a parameter λ of a simulated sample path of a Poisson process.

```
function poiss_proc_sim2(n)
lambda=rand;
N=0; S=0;
while S<n
    T=exprnd(1/lambda);
    N=N+1; S=S+T;
end
fprintf('Lambda is %4.3f.\n',lambda);
fprintf('Estimated lambda is %4.3f.\n',N/n);
end
```

```
>> poiss_proc_sim2(1000)
```

```
Lambda is 0.909.
Estimated lambda is 0.897.
```

Problem 5.3.3. Let $(W_t)_{t \geq 0}$ be a Wiener process.

1) Prove that $\lim_{t \to \infty} \frac{1}{t} W_t = 0$ a.s.

Hint: Use Markov's inequality (Theorem 3.6) and for suitable $0 \leq r < s$ the following inequality

$$E\left(\sup_{t \in [r,s]} |W_t|^2 \right) \leq 4E\left(|W_s|^2\right),$$

which is Doob's maximal inequality applied for a Wiener process (see [Mörters and Peres (2010), Proposition 2.43, p. 54]; the proof of this inequality involves martingale theory).

2) Define

$$B_t = \begin{cases} tW_{1/t}, & \text{if } t > 0 \\ 0, & \text{if } t = 0. \end{cases}$$

Then $(B_t)_{t \geq 0}$ is a Wiener process.

Solution 5.3.3: 1) For $n \in \mathbb{N}^*$ and arbitrary $t \in [2^n, 2^{n+1})$ it holds

$$\left| \frac{1}{t}W_t - \frac{1}{2^n}W_{2^n} \right| \leq \left| \frac{1}{t} - \frac{1}{2^n} \right| \cdot |W_{2^n}| + \frac{1}{t}|W_t - W_{2^n}|$$

$$\leq \frac{2}{2^n}|W_{2^n}| + \frac{1}{2^n} \sup_{t \in [2^n, 2^{n+1}]} |W_t - W_{2^n}|. \quad (5.3)$$

It follows from SLLN (see Theorem 4.19) that

$$\frac{2}{n}|W_n| = 2\left|\frac{1}{n}\sum_{k=1}^{n}(W_k - W_{k-1})\right| \xrightarrow{a.s.} 0 \,.$$

Consequently, the subsequence $\left(\dfrac{2}{2^n}|W_{2^n}|\right)_{n\in\mathbb{N}^*}$ converges a.s. to 0. In what follows we consider the summand $\dfrac{1}{2^n}\sup\limits_{t\in[2^n,2^{n+1}]}|W_t - W_{2^n}|$. By using Markov's inequality, the Cauchy–Schwarz inequality for random variables and Doob's maximal inequality we write for arbitrary $\varepsilon > 0$

$$P\left(\frac{1}{2^n}\sup_{t\in[2^n,2^{n+1}]}|W_t - W_{2^n}| > \varepsilon\right) = P\left(\sup_{t\in[2^n,2^{n+1}]}|W_t - W_{2^n}| > \varepsilon 2^n\right)$$

$$\leq \frac{1}{\varepsilon 2^n}E\left(\sup_{t\in[2^n,2^{n+1}]}|W_t - W_{2^n}|\right) \leq \frac{1}{\varepsilon 2^n}\left(E\sup_{t\in[2^n,2^{n+1}]}|W_t - W_{2^n}|^2\right)^{\frac{1}{2}}$$

$$\leq \frac{1}{\varepsilon 2^{n-1}}\left(E|W_{2^{n+1}} - W_{2^n}|^2\right)^{\frac{1}{2}} = \frac{1}{\varepsilon 2^{n-1}}\left(2^{n+1} - 2^n\right)^{\frac{1}{2}} = \frac{2}{\varepsilon 2^{n/2}}\,.$$

Consequently,

$$\sum_{n=1}^{\infty}P\left(\frac{1}{2^n}\sup_{t\in[2^n,2^{n+1}]}|W_t - W_{2^n}| > \varepsilon\right) < \infty$$

and thus we get by Theorem 4.2

$$\frac{1}{2^n}\sup_{t\in[2^n,2^{n+1}]}|W_t - W_{2^n}| \xrightarrow{a.s.} 0\,.$$

Consequently, we can state by using inequality (5.3) that

$$\lim_{t\to\infty}\frac{W_t}{t} = 0 \text{ a.s.}$$

2) Observe that from 1) we have

$$\lim_{t\to 0}B_t = \lim_{r\to\infty}\frac{1}{r}W_r = 0 \text{ a.s.}$$

Let $0 < t_1 < \ldots < t_n$. Denote

$$\mathbb{U}_1 = (B_{t_1}, B_{t_2}, \ldots, B_{t_n}),$$

$$\mathbb{U}_2 = (W_{1/t_n} - W_0, W_{1/t_{n-1}} - W_{1/t_n}, \ldots, W_{1/t_1} - W_{1/t_2}),$$

$$A = \begin{pmatrix} t_1 \, t_2 \, \ldots \, t_{n-1} \, t_n \\ t_1 \, t_2 \, \ldots \, t_{n-1} \, 0 \\ \vdots \quad \vdots \quad \ddots \quad \vdots \quad \vdots \\ t_1 \, t_2 \, \ldots \quad 0 \quad 0 \\ t_1 \quad 0 \, \ldots \quad 0 \quad 0 \end{pmatrix}.$$

Note that A is an invertible matrix and $\mathbb{U}_1 = \mathbb{U}_2 A$ a.s. By the definition of the Wiener process

$$W_{1/t_n} - W_0, W_{1/t_{n-1}} - W_{1/t_n}, \ldots, W_{1/t_1} - W_{1/t_2}$$

are independent random variables, each of them following a normal distribution with zero mean. We have $\mathbb{U}_2 \sim MVN(0_n, C)$, where C is a diagonal matrix, with the values $c_{ii} = V(W_{1/t_{i-1}} - W_{1/t_i}) = 1/t_{i-1} - 1/t_i$, $i \in \{1, \ldots, n\}$ on the main diagonal. By Example 2.24-(2) it follows that $\mathbb{U}_1 \sim MVN(0_n, A^T C A)$. Observe that, in the case when $t_1 = 0$, $B_{t_1} = 0$ a.s. and $(B_{t_1}, \ldots, B_{t_n})$ follows a multivariate normal distribution included in the degenerate case (see Remark 5.7-(1)). We conclude that $(B_t)_{t \geq 0}$ is a Gaussian process. We have $E(B_t) = 0$ for each $t \geq 0$ and

$$\text{cov}(B_s, B_t) = st E(W_s W_t) = st \min\{1/t, 1/s\} = \min\{s, t\}$$

for each $s, t \geq 0$. Then it follows by Example 5.16 that $(B_t)_{t \geq 0}$ is a Wiener process.

Problem 5.3.4. Based on Definition 5.15, develop a simulation method for a Wiener process $(W_t)_{t \geq 0}$ on the time interval $[0, T]$, where $T > 0$, using only random numbers that follow the $N(0, 1)$ distribution. Write a corresponding Matlab code and plot some simulated sample paths.

Solution 5.3.4: Let $n \in \mathbb{N}^*$ be a large number (e.g., $n = 1000$) and let $\Delta_n = \dfrac{T}{n}$. Also, let $t_k^n = k\Delta_n$ for $k \in \{0, 1, \ldots, n\}$. Obviously,

$$W_{t_k^n} = \sum_{j=1}^k (W_{t_j^n} - W_{t_{j-1}^n})$$

holds for $k \in \{1, \ldots, n\}$. Since the increments are $N(0, \Delta_n)$ distributed and independent we can write

$$W_{t_k^n} = \sqrt{\Delta_n} \sum_{j=1}^k Z_j^n \quad \text{for } k \in \{1, \ldots, n\}$$

where $Z_1^n, \ldots, Z_n^n$ are independent $N(0, 1)$ distributed random variables.

The following Matlab function simulates K sample paths for a Wiener process on the time interval $[0, T]$.

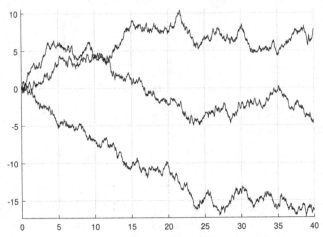

Fig. 5.8: Three simulated sample paths of a Wiener process

```
function WP(T,K,n)
clf; hold on; grid on;
for i=1:K
    Z=normrnd(0,sqrt(T/n),1,n);
    W=[0 cumsum(Z)];
    plot(0:T/n:T,W);
end
end
```

```
>>WP(40,3,2000)
```

5.4 Problems for Chapter 5

Problem 5.4.1. Let $(S_n)_{n\in\mathbb{N}}$ be a symmetric random walk starting at the origin with $(X_n)_{n\in\mathbb{N}^*}$ the sequence of independent random steps, i.e.,

$$P(X_n = -1) = P(X_n = 1) = \frac{1}{2} \text{ for each } n \in \mathbb{N}^*.$$

1) Prove that the sequence $\left(\dfrac{1}{n}S_n^2\right)_{n\in\mathbb{N}^*}$ converges in distribution to a random variable having $\chi^2(1,1)$ distribution.

2) Prove that the sequence $\left(\dfrac{1}{n}S_n^2\right)_{n\in\mathbb{N}^*}$ does not converge in L^2 to a random variable.

3) Prove that the sequence $\left(\dfrac{1}{n}S_n^2\right)_{n\in\mathbb{N}^*}$ does not converge in probability to a random variable.

Solution 5.4.1: 1) By Problem 5.2.2 we have for $n \in \mathbb{N}^*$: $E(S_n) = 0$, $E(S_n^2) = n$, $E(S_n^4) = n(3n - 2)$. For $x > 0$ we have by Theorem 4.21 that

$$\lim_{n \to \infty} P\left(\frac{1}{n} S_n^2 \leq x\right) = \lim_{n \to \infty} P\left(-\sqrt{x} \leq \frac{1}{\sqrt{n}} S_n \leq \sqrt{x}\right)$$

$$= P\left(-\sqrt{x} \leq Z \leq \sqrt{x}\right) = P(Z^2 \leq x),$$

where $Z \sim N(0,1)$. Then $\frac{1}{n} S_n^2 \xrightarrow{d} Z^2$, where $Z^2 \sim \chi^2(1,1)$, by Problem 3.7.12.

2) Denote

$$\tilde{S}_n = X_{n+1} + ... + X_{2n} = S_{2n} - S_n.$$

By the properties of the random walk, $\tilde{S}_n$ is independent of S_n and has the same distribution as S_n, in particular we have $E(\tilde{S}_n) = E(S_n) = 0$, $E(\tilde{S}_n^2) = E(S_n^2) = n$, $E(\tilde{S}_n^4) = E(S_n^4) = n(3n - 2)$ for all $n \in \mathbb{N}^*$. We write

$$E\left(\left(\frac{1}{n} S_n^2 - \frac{1}{2n} S_{2n}^2\right)^2\right) = \frac{1}{4n^2} E\left((S_n^2 - 2S_n \tilde{S}_n - \tilde{S}_n^2)^2\right)$$

$$= \frac{1}{4n^2} E\left(S_n^4 + 2S_n^2 \tilde{S}_n^2 + \tilde{S}_n^4 - 4S_n^3 \tilde{S}_n + 4S_n \tilde{S}_n^3\right)$$

$$= \frac{1}{4n^2}\left(E(S_n^4) + 2E(S_n^2)E(\tilde{S}_n^2) + E(\tilde{S}_n^4)\right)$$

$$= \frac{2n - 1}{n} \to 2, \text{ as } n \to \infty.$$

By the Cauchy criterion for random variables (see Theorem 4.10) it follows that the sequence $\left(\frac{1}{n} S_n^2\right)_{n \in \mathbb{N}^*}$ does not converge in L^2 to a random variable.

3) Suppose that $\left(\frac{1}{n} S_n^2\right)_{n \in \mathbb{N}^*}$ converges in probability to a random variable which we denote by U. By 1), Theorem 4.3 and Theorem 4.5 it follows that U has the same distribution as Z^2, namely $\chi^2(1,1)$. From $\frac{1}{n} S_n^2 \xrightarrow{P} U$ it follows by Theorem 4.6 that $\frac{1}{n^2} S_n^4 \xrightarrow{P} U^2$, as well as by Theorem 4.3 that $\frac{1}{n^2} S_n^4 \xrightarrow{d} U^2$. But $E(U^2) = E(Z^4) = 3$, since $Z \sim N(0,1)$, see Example 3.12. We have $E\left(\frac{1}{n^2} S_n^4\right) = 3 - \frac{2}{n}$ for each $n \in \mathbb{N}^*$ (by the computations given in the solution for 2)), thus we obtain

$$\lim_{n \to \infty} E\left(\frac{1}{n^2} S_n^4\right) = E(U^2) = 3.$$

Using the equivalences from Theorem 4.12 it follows that

$$\lim_{n\to\infty} E\left(\left(\frac{1}{n}S_n^2 - U\right)^2\right) = 0,$$

which gives a contradiction with 2). We obtain that, the sequence $\left(\frac{1}{n}S_n^2\right)_{n\in\mathbb{N}^*}$ does not converge in probability to a random variable.

Problem 5.4.2. Let $(S_n)_{n\in\mathbb{N}}$ be a random walk starting at the origin with $(X_n)_{n\in\mathbb{N}^*}$ the sequence of independent random steps such that

$$P(X_n = 1) = p, \ P(X_n = -1) = 1 - p \text{ for each } n \in \mathbb{N}^*,$$

where $p \in (0, 1)$ is given. Prove that:

1) If $p > \frac{1}{2}$, then $S_n \xrightarrow{a.s.} \infty$ and if $p < \frac{1}{2}$, then $S_n \xrightarrow{a.s.} -\infty$.

2) If $p = \frac{1}{2}$, then $\limsup_{n\to\infty} |S_n| = \infty$ a.s.

Solution 5.4.2: 1) If $p > \frac{1}{2}$, then $E(X_n) = 2p - 1 > 0$ for each $n \in \mathbb{N}^*$ and by the SLLN (see Theorem 4.19) we have

$$\frac{1}{n}S_n \xrightarrow{a.s.} 2p - 1 > 0.$$

This implies, $S_n \xrightarrow{a.s.} \infty$. Analogously the case $p < \frac{1}{2}$ is treated.

2) If $p = \frac{1}{2}$, then, by Theorem 4.21, $\frac{1}{\sqrt{n}}S_n \xrightarrow{d} Z$, where $Z \sim N(0, 1)$. Since the random walk is symmetric, we have that $S_{2^{n+1}} - S_{2^n}$ has the same distribution as S_{2^n} for $n \in \mathbb{N}$, hence

$$\lim_{n\to\infty} P(S_{2^{n+1}} - S_{2^n} > \sqrt{2^n}) = 1 - \lim_{n\to\infty} P(S_{2^n} \le \sqrt{2^n}) = 1 - F_Z(1) > 0,$$

where F_Z is the distribution function of Z. We consider the events $A_n = \{S_{2^{n+1}} - S_{2^n} > \sqrt{2^n}\}$, $n \in \mathbb{N}^*$, which are independent and, by the above reasoning, satisfy $\sum_{n=1}^{\infty} P(A_n) = \infty$. By the Borel–Cantelli Lemma (see Theorem 1.10), it follows that $P(\limsup_{n\to\infty} A_n) = 1$. We obtain that $(S_{2^n}(\omega))_{n\in\mathbb{N}^*}$ is an unbounded sequence for a.e. $\omega \in \Omega$, i.e.,

$$\limsup_{n\to\infty} |S_n| = \infty \quad \text{a.s.}$$

Remark 5.9. Let $(S_n)_{n\in\mathbb{N}}$ be a symmetric random walk starting at the origin as in Problem 5.4.2-2). The so-called law of iterated logarithm yields

$$\limsup_{n\to\infty} \frac{S_n}{\sqrt{2n \log(\log n)}} = 1 \quad \text{a.s.}$$

and

$$\liminf_{n\to\infty} \frac{S_n}{\sqrt{2n\log(\log n)}} = -1 \quad \text{a.s.}$$

see [Gut (2005), Theorem 1.2, p. 384]. These results imply

$$\limsup_{n\to\infty} S_n = \infty \text{ a.s.} \quad \text{and} \quad \liminf_{n\to\infty} S_n = -\infty \text{ a.s.}$$

which obviously are stronger properties than the one obtained in Problem 5.4.2-2). ▾

Problem 5.4.3. Let $(Y_n)_{n\in\mathbb{N}^*}$ be a sequence of independent random variables having the distribution

$$P(Y_n = -1) = P(Y_n = 1) = 0.5 \text{ for each } n \in \mathbb{N}^*.$$

If $Z_n = \sum_{k=1}^{n} \frac{1}{2^k} Y_k$ for each $n \in \mathbb{N}^*$, prove that $(Z_n)_{n\in\mathbb{N}^*}$ converges a.s., in L^2 and in distribution to a random variable which is $Unif[-1,1]$ distributed.

Solution 5.4.3:
One can prove by induction that Z_n, $n \in \mathbb{N}^*$, has the following distribution

$$P\left(Z_n = \frac{2k+1}{2^n} - 1\right) = \frac{1}{2^n} \text{ for } k \in \{0, 1, \dots, 2^n - 1\}.$$

The distribution function of Z_n is given by

$$F_{Z_n}(x) = \begin{cases} 0, & \text{if } x < -1 + \dfrac{1}{2^n} \\[2mm] \dfrac{k}{2^n}, & \text{if } \dfrac{2k-1}{2^n} - 1 \le x < \dfrac{2k+1}{2^n} - 1, k \in \{1, \dots, 2^n - 1\} \\[2mm] 1, & \text{if } 1 - \dfrac{1}{2^n} \le x. \end{cases}$$

The distribution function of a random variable $Z \sim Unif[-1,1]$ is for arbitrary $x \in \mathbb{R}$

$$F_Z(x) = \begin{cases} 0, & \text{if } x < -1 \\[2mm] \dfrac{x+1}{2}, & \text{if } -1 \le x < 1 \\[2mm] 1, & \text{if } 1 \le x. \end{cases}$$

Moreover, it is obvious that

$$\lim_{n\to\infty} F_{Z_n}(x) = F_Z(x) = 0 \quad \text{for every } x \in (-\infty, -1) \quad (5.4)$$

and

$$\lim_{n \to \infty} F_{Z_n}(x) = F_Z(x) = 1 \quad \text{for every } x \in [1, \infty). \tag{5.5}$$

Now, we prove

$$\lim_{n \to \infty} F_{Z_n}(x) = F_Z(x) \quad \text{for each } x \in [-1, 1).$$

Let $x \in [-1, 1)$ and $\varepsilon > 0$ be arbitrary. Then there exists $n_\varepsilon \in \mathbb{N}$ such that for every $n \geq n_\varepsilon$ we have $\frac{1}{2^{n+1}} < \varepsilon$. Let $n \geq n_\varepsilon$ be fixed. Then there exists a unique $k \in \{1, \ldots, 2^n - 1\}$ (depending on n and x) such that

$$\frac{2k-1}{2^n} - 1 < x < \frac{2k+1}{2^n} - 1.$$

This implies

$$\left| \frac{x+1}{2} - \frac{k}{2^n} \right| < \frac{1}{2^{n+1}} < \varepsilon.$$

Therefore

$$|F_Z(x) - F_{Z_n}(x)| < \varepsilon \quad \text{for any } n \geq n_\varepsilon. \tag{5.6}$$

Then, by (5.4), (5.5), (5.6), it follows that $Z_n \overset{d}{\longrightarrow} Z$.

We observe that

$$\sum_{k=1}^{\infty} V\left(\frac{Y_k}{2^k} \right) = \sum_{k=1}^{\infty} \frac{1}{2^{2k}} < \infty,$$

then, by Theorem 4.17, it follows that the series $\sum_{k=1}^{\infty} \frac{Y_k}{2^k}$ is a.s. convergent i.e., there exists $\lim_{n \to \infty} Z_n$ a.s. This limit must be a.s. a random variable which has the same distribution as Z, since $Z_n \overset{d}{\longrightarrow} Z$. For simplicity, we denote this random variable also by Z.

Since $|Y_n| = 1$ a.s. for each $n \in \mathbb{N}^*$, it follows that $(Z_n)_{n \in \mathbb{N}^*}$ is a.s. bounded by 1. By Theorem 4.12, it follows that $Z_n \overset{L^2}{\longrightarrow} Z$.

Problem 5.4.4. Give an example of a homogeneous Markov process $(X_n)_{n \in \mathbb{N}}$ that has no independent increments.

Solution 5.4.4: Let $(X_n)_{n \in \mathbb{N}}$ be a sequence of independent random variables such that $P(X_n = -1) = P(X_n = 1) = 0.5$ for $n \in \mathbb{N}$. Note that for $i, j \in \{-1, 1\}$ we have

$$P(X_{n+k} = j | X_n = i) = P(X_{n+k} = j) = 0.5 \text{ for each } n \in \mathbb{N}, k \in \mathbb{N}^*,$$

by the independence of X_{n+k} and X_n. We deduce, by Example 5.7 and Definition 5.9-(2), that $(X_n)_{n\in\mathbb{N}}$ is a homogeneous Markov process.

It holds $E(X_n) = 0$ for all $n \in \mathbb{N}$. We compute the covariance of the increments $X_n - X_{n-1}$ and $X_{n-1} - X_{n-2}$ for $n \in \mathbb{N}$ with $n \geq 2$:

$$\text{cov}(X_n - X_{n-1}, X_{n-1} - X_{n-2}) = E\big((X_n - X_{n-1})(X_{n-1} - X_{n-2})\big)$$
$$= -E\left(X_{n-1}^2\right) = -1 \neq 0.$$

Consequently, the increments are not independent.

Problem 5.4.5. Let $(X_t)_{t\geq 0}$ with $X_0 = 0$ be a stochastic process with independent increments such that the moment generating functions of the increments exist on $\mathbb{R}$. Prove that the finite dimensional distributions of this process are given by the distributions of the increments.

Solution 5.4.5: Recall the joint moment generating function for random vectors from Section 3.6. Let $n \in \mathbb{N}^*$, $0 \leq t_1 < \ldots < t_n$ and let $r_1, \ldots, r_n \in \mathbb{R}$ be chosen arbitrary. By the independent increments property we have

$$M_{X_{t_1},\ldots,X_{t_n}}(r_1,\ldots,r_n) = E\left(\exp\left\{\sum_{k=1}^{n} r_k X_{t_k}\right\}\right)$$

$$= M_{X_{t_1}-X_0, X_{t_2}-X_{t_1},\ldots,X_{t_n}-X_{t_{n-1}}}(r_1 + \ldots + r_n, r_2 + \ldots + r_n, \ldots, r_n)$$

$$= M_{X_{t_1}-X_0}(r_1 + \ldots + r_n) M_{X_{t_2}-X_{t_1}}(r_2 + \ldots + r_n) \cdot \ldots \cdot M_{X_{t_n}-X_{t_{n-1}}}(r_n).$$

The moment generating functions of the increments of the last line of the above relation are determined uniquely by the distributions of the increments, see Theorem 3.11. It follows from Theorem 3.13 that the distribution of $(X_{t_1}, \ldots, X_{t_n})$ is uniquely determined by $M_{X_{t_1},\ldots,X_{t_n}}$. Hence, the finite dimensional distributions of this process are given by the distributions of the increments.

Problem 5.4.6. Let $(X_t)_{t\geq 0}$ be a homogeneous Markov process with discrete state space $\mathcal{S}$ and for $i, j \in \mathcal{S}$ let $p_{ij} : (0, \infty) \to [0, 1]$ be such that for $s \geq 0$ with $P(X_s = i) > 0$ we have

$$p_{ij}(h) = P(X_{s+h} = j | X_s = i) \quad \text{for each } h > 0.$$

Assume that $\lim_{h\searrow 0} p_{ii}(h) = 1$ for $i \in \mathcal{S}$. Prove that for $i, j \in \mathcal{S}$ the function p_{ij} is continuous on $(0, \infty)$.

Solution 5.4.6: Let $i, j \in \mathcal{S}$ and $t > 0$. For $h > 0$, by the Chapman–Kolmogorov equation from Remark 5.1-(2), we have

$$p_{ij}(t + h) = \sum_{k\in\mathcal{S}} p_{ik}(h) p_{kj}(t).$$

Hence, for $h > 0$ we have

$$|p_{ij}(t+h) - p_{ij}(t)| \le 1 - p_{ii}(h) + \sum_{k \in \mathcal{S} \setminus \{i\}} p_{ik}(h) = 2\big(1 - p_{ii}(h)\big),$$

where we use the fact that

$$\sum_{k \in \mathcal{S} \setminus \{i\}} p_{ik}(h) = \sum_{k \in \mathcal{S} \setminus \{i\}} P(X_{s+h} = k | X_s = i)$$

$$= 1 - P(X_{s+h} = i | X_s = i) = 1 - p_{ii}$$

for some $s \ge 0$ such that $P(X_s = i) > 0$. Since $\lim_{h \searrow 0} \big(1 - p_{ii}(h)\big) = 0$, we deduce that p_{ij} is right-continuous at t. Similarly, for $h \in (0, t)$, we have

$$p_{ij}(t) = \sum_{k \in \mathcal{S}} p_{ik}(h) p_{kj}(t - h).$$

Hence, for $h \in (0, t)$ we have

$$|p_{ij}(t) - p_{ij}(t - h)| \le 1 - p_{ii}(h) + \sum_{k \in \mathcal{S} \setminus \{i\}} p_{ik}(h) = 2\big(1 - p_{ii}(h)\big).$$

We deduce as above that p_{ij} is left-continuous at t. Therefore, p_{ij} is continuous at t.

Problem 5.4.7. Let $(X_t)_{t \ge 0}$ be a homogeneous Markov process with finite discrete state space $\mathcal{S}$ and for $i, j \in \mathcal{S}$ let $p_{ij} : (0, \infty) \to [0, 1]$ be such that for $s \ge 0$ with $P(X_s = i) > 0$ we have

$$p_{ij}(h) = P(X_{s+h} = j | X_s = i) \quad \text{for each } h > 0.$$

Assume that there exits a sequence of numbers $(q_{ik})_{i,k \in \mathcal{S}}$ (called intensities) such that

$$q_{ii} = \lim_{h \searrow 0} \frac{1 - p_{ii}(h)}{h}, \quad i \in \mathcal{S},$$

and

$$q_{ij} = \lim_{h \searrow 0} \frac{p_{ij}(h)}{h}, \quad i, j \in \mathcal{S}, i \ne j.$$

Prove that for $i, j \in \mathcal{S}$ the function p_{ij} is differentiable on $(0, \infty)$ and

$$p'_{ij}(t) = -q_{ii} p_{ij}(t) + \sum_{k \in \mathcal{S} \setminus \{i\}} q_{ik} p_{kj}(t) \quad \text{for } t > 0 \qquad (5.7)$$

(this is called a Kolmogorov equation).

Hint: Use the following fact: if a continuous function defined on an open interval has a continuous right-derivative, then the function is differentiabl (see, e.g., [Anderson (1991), Lemma 7.2, p. 55]).

Solution 5.4.7: Let $i, j \in \mathcal{S}$ and $t > 0$. For $h > 0$, by the Chapman–Kolmogorov equation from Remark 5.1-(2), we have

$$p_{ij}(t + h) = \sum_{k \in \mathcal{S}} p_{ik}(h) p_{kj}(t).$$

Then for $h > 0$ we have

$$\frac{p_{ij}(t + h) - p_{ij}(t)}{h} = -\frac{1 - p_{ii}(h)}{h} p_{ij}(t) + \sum_{k \in \mathcal{S} \setminus \{i\}} \frac{p_{ik}(h)}{h} p_{kj}(t).$$

By taking the limit $h \searrow 0$ in the above equality and using the hypothesis, we get that p_{ij} is right-differentiable on $(0, \infty)$ and the right-derivative of p_{ij} at t equals $-q_{ii} p_{ij}(t) + \sum_{k \in \mathcal{S} \setminus \{i\}} q_{ik} p_{kj}(t)$ for $t > 0$. Since the function $-q_{ii} p_{ij} + \sum_{k \in \mathcal{S} \setminus \{i\}} q_{ik} p_{kj}$ is continuous on $(0, \infty)$ (by Problem 5.4.6), we deduce that p_{ij} is differentiable on $(0, \infty)$ and (5.7) holds.

Remark 5.10. The results stated in Problem 5.4.6 hold also in the case of an infinite discrete state space, see [Anderson (1991), Proposition 2.7, p. 13]. ▼

Problem 5.4.8. Let $(X_t)_{t \geq 0}$ be a birth and death process with birth rates $(\lambda_n)_{n \in \mathbb{N}}$ and death rates $(\mu_n)_{n \in \mathbb{N}^*}$. For $i, j \in \mathbb{N}$ let $p_{ij} : (0, \infty) \to [0, 1]$ be such that for $s \geq 0$ with $P(X_s = i) > 0$ we have

$$p_{ij}(h) = P(X_{s+h} = j | X_s = i) \text{ for each } h > 0.$$

Prove that for $i, j \in \mathbb{N}$ the function p_{ij} is differentiable on $(0, \infty)$ and

$$p'_{ij}(t) = \mu_i p_{i-1j}(t) - (\lambda_i + \mu_i) p_{ij}(t) + \lambda_i p_{i+1j}(t) \text{ for } t > 0, \text{ if } i \geq 1,$$

and $p'_{0j}(t) = \lambda_0(-p_{0j}(t) + p_{1j}(t))$ for $t > 0$.

Solution 5.4.8: Let $(q_{ij})_{i,j \in \mathbb{N}}$ be given as in Example 5.14-(2). Let $i, j \in \mathbb{N}$ and $t > 0$. Reasoning as in the solution of Problem 5.4.7, we have

$$\lim_{h \searrow 0} \frac{p_{ij}(t + h) - p_{ij}(t)}{h} = -q_{ii} p_{ij}(t) + \lim_{h \searrow 0} \sum_{k \in \mathbb{N} \setminus \{i\}} \frac{p_{ik}(h)}{h} p_{kj}(t).$$

Let $s \geq 0$ be such that $P(X_s = i) > 0$. Since for $h > 0$ we have

$$\sum_{\substack{k \in \mathbb{N} \\ |k-i|>1}} \frac{p_{ik}(h)}{h} p_{kj}(t) \leq \sum_{\substack{k \in \mathbb{N} \\ |k-i|>1}} \frac{p_{ik}(h)}{h} = \frac{1}{h} P(|X_{s+h} - X_s| > 1 | X_s = i),$$

we deduce, by Definition 5.13-(3), that

$$\lim_{h \searrow 0} \sum_{\substack{k \in \mathbb{N} \\ |k-i|>1}} \frac{p_{ik}(h)}{h} p_{kj}(t) = 0.$$

Moreover, since the right-derivative of p_{ij} is the function

$$-q_{ii}p_{ij} + q_{ii-1}p_{i-1j} + q_{ii+1}p_{i+1j}, \quad \text{if } i \geq 1,$$

which is continuous on $(0, \infty)$ (by Problem 5.4.6), we deduce that p_{ij} is differentiable on $(0, \infty)$ (see the hint of Problem 5.4.7) and the desired equality holds for $i \geq 1$. Similarly, one can deduce that p_{0j} is differentiable on $(0, \infty)$ and satisfies the corresponding equality.

Problem 5.4.9. An investor found a magical scheme of investment that has the following property: for each invested dollar in a week he gets independently X dollars, where X has the Poisson distribution with the mean value equal to 10 (dollars). He starts with 1\$ and invests each week all the money he gained in the previous week.

1) Simulate 1000 times the profit in the nth week, $n \in \{1, 2, 3, 4, 5\}$, using Matlab, and estimate the corresponding expected value and variance.

2) Find the expected value and the variance of his profit in the nth week for $n \in \mathbb{N}^*$.

Solution 5.4.9: 1)

```
function investment(n,N)
profit=zeros(1,N);
for i=1:N
    X=1;
    for j=1:n
        if X>0
            X=sum(poissrnd(10,1,X));
        else
            break;
        end
    end
    profit(i)=X;
end
fprintf(['The estimated expected value and variance of the\n'...
         ' profit in the %dth week are %d and %d.\n']...
         ,n,floor(mean(profit)),floor(var(profit,1)));
end
```

```
>> investment(3,1000)
The estimated expected value and variance of the
 profit in the 3th week are 997 and 112702.
```

2) We note that $X \sim Poiss(\lambda)$ and $E(X) = V(X) = \lambda = 10$. For every $n \in \mathbb{N}^*$ let Z_n be the size of the profit in the nth week and $Z_0 = 1$. Note that $(Z_n)_{n \in \mathbb{N}}$ is a branching process. In view of Remark 5.3-(1), for every $n \in \mathbb{N}$ and $i \in \mathbb{N}^*$, let $X_{n,i} \sim Poiss(10)$ be the size of the profit generated by the ith dollar invested in the nth week. So, $(X_{n,i})_{(n,i) \in \mathbb{N} \times \mathbb{N}^*}$ is a family of independent identically distributed random variables such that

$$Z_{n+1} = \sum_{i=1}^{\infty} X_{n,i} \cdot \mathbb{I}_{\{Z_n \geq i\}}, \quad n \in \mathbb{N}.$$

We deduce by Remark 5.3-(2)

$$E(Z_{n+1}) = \sum_{k=1}^{\infty} E(X_{n,1} + \ldots + X_{n,k}) P(Z_n = k)$$

$$= \sum_{k=1}^{\infty} k E(X) P(Z_n = k) = E(X) E(Z_n), \ n \in \mathbb{N}. \tag{5.8}$$

Now, it is easy to deduce that $E(Z_n) = E^n(X) = 10^n$ for $n \in \mathbb{N}$. Using again Remark 5.3-(2), as well as Theorem 3.4, we deduce that

$$E(Z_{n+1}^2) = \sum_{k=1}^{\infty} E\big((X_{n,1} + \ldots + X_{n,k})^2\big) P(Z_n = k)$$

$$= \sum_{k=1}^{\infty} \Big(V(X_{n,1} + \ldots + X_{n,k}) + E^2(X_{n,1} + \ldots + X_{n,k})\Big) P(Z_n = k)$$

$$= \sum_{k=1}^{\infty} \big(k V(X) + k^2 E^2(X)\big) P(Z_n = k) = V(X) E(Z_n) + E^2(X) E(Z_n^2)$$

$$= V(X) E^n(X) + E^2(X) E(Z_n^2), \ n \in \mathbb{N}. \tag{5.9}$$

Hence,

$$E(Z_n^2) = V(X) E^{n-1}(X) + E^2(X) V(X) E^{n-2}(X)$$
$$+ \ldots + E^{2n-2}(X) V(X) + E^{2n}(X)$$

$$= V(X) \cdot E^{n-1}(X) \cdot \frac{E^n(X) - 1}{E(X) - 1} + E^2(Z_n), \quad n \in \mathbb{N}^*,$$

and thus

$$V(Z_n^2) = V(X) \cdot E^{n-1}(X) \cdot \frac{E^n(X) - 1}{E(X) - 1} = \frac{10^n(10^n - 1)}{9}, \quad n \in \mathbb{N}.$$

Problem 5.4.10. Let X be a random variable whose range is a subset of $\mathbb{N}$ and let $p \in (0, 1)$. A certain type of cell splits into X cells. Consider a branching process that starts with one cell, which represents the 0th generation, and for every generation each existing cell splits independently of each other, according to the distribution of X. If a cell splits into 0 cells, then the cell dies; if a cell splits into 1 cell in the current generation, then the cell survives (i.e., has 1 descendant) in the next generation; if a cell splits into n cells, then the cell has n descendants in the next generation, for $n \in \mathbb{N}$, $n \geq 2$.

1) Let $X \sim Geo(p)$. Simulate in Matlab the branching process and print the number of cells in the nth generation for $n \in \{1, \ldots, 10\}$. Find the probability that the branching process dies out (i.e., some generation has no cells).

2) Let $X \sim Bino(2, p)$. Simulate the branching process and print the number of cells in the nth generation for $n \in \{1, \ldots, 10\}$. Find the probability that the branching process dies out.

Solution 5.4.10: 1)

```
function branching(p,n)
X=1;
for i=1:n
  if X~=0
    fprintf('The number of cells in generation %d is %d.\n',i,X);
    X=sum(geornd(p,1,X));
  else
    fprintf('The branching process dies out in generation %d.\n',i);
    return;
  end
end
end
```

```
>> branching(1/2,10)
The number of cells in generation 1 is 1.
The number of cells in generation 2 is 4.
The number of cells in generation 3 is 3.
The number of cells in generation 4 is 2.
The number of cells in generation 5 is 2.
The number of cells in generation 6 is 3.
The number of cells in generation 7 is 1.
The branching process dies out in generation 8.
```

For every $n \in \mathbb{N}$, let D_n be the event that the process dies out by the nth generation and let $q_n = P(D_n)$. Also, let D be the event that the process dies out and let $q = P(D)$. We observe that $D = \bigcup_{n=0}^{\infty} D_n$. Also, we note that $(q_n)_{n \in \mathbb{N}}$ is an increasing sequence of numbers in the interval $[0, 1]$ and

$$q = P(D) = P\left(\bigcup_{n=0}^{\infty} D_n\right) = \lim_{n \to \infty} q_n,$$

by Theorem 1.6-(1).

Let $n \in \mathbb{N}^*$. For every $j \in \mathbb{N}$, let C_j be the event that there are j cells in the first generation. For $i, j \in \mathbb{N}^*$, $i \le j$, let A_{ij}^n be the event that all the cells generated by the ith cell in the first generation die by the nth generation, given that there are j cells in the first generation. Note that $A_{1j}^n, \ldots, A_{jj}^n$ are independent for $j \in \mathbb{N}^*$ and $P(A_{ij}^n) = q_{n-1}$ for $i, j \in \mathbb{N}^*$, $i \le j$, because each cell in the first generation starts independently a branching process, using the same distribution of X. By the law of total probability (see Theorem 1.8), we have

$$q_n = P(D_n) = \sum_{j=0}^{\infty} P(D_n | C_j) P(C_j)$$

$$= P(X = 0) + \sum_{j=1}^{\infty} P\left(A_{1j}^n \cap \ldots \cap A_{jj}^n\right) P(X = j)$$

$$= p + \sum_{j=1}^{\infty} P(A_{1j}^n) \cdot \ldots \cdot P(A_{jj}^n) P(X = j) = p + \sum_{j=1}^{\infty} q_{n-1}^j P(X = j).$$

Then

$$q_n = p + \sum_{j=1}^{\infty} p\left(q_{n-1}(1-p)\right)^j = \frac{p}{1 - q_{n-1}(1-p)},$$

since $0 \le q_{n-1} \le 1$ and $0 < 1 - p < 1$. Letting $n \to \infty$, we deduce that q satisfies the equation

$$q\left(1 - q(1-p)\right) = p.$$

This implies that $q \in \left\{1, \dfrac{p}{1-p}\right\}$. Moreover, using the above recurrence relation for the sequence $(q_n)_{n \in \mathbb{N}}$, we can easily prove that for $n \in \mathbb{N}$ we have $q_n \le \dfrac{p}{1-p}$. So,

$$q \le \frac{p}{1-p}.$$

If $p \in \left(0, \dfrac{1}{2}\right]$, then we have $q = 1$, because in this case $\dfrac{p}{1-p} \ge 1$. If $p \in \left(\dfrac{1}{2}, 1\right)$, then $q = \dfrac{p}{1-p}$, because in this case $q \le \dfrac{p}{1-p} < 1$, in view of the above inequality for q.

2)

function branching(p,n)
X=1;
for i=1:n
  if X~=0
  fprintf('The number of cells in generation %d is %d.\n',i,X);
  X=**sum(binornd**(2,p,1,X));
  else
  fprintf('The branching process dies out in generation %d.\n',i);
  return;
  end
end
end

>> branching(1/2,10)
The number of cells in generation 1 is 1.
The number of cells in generation 2 is 2.
The number of cells in generation 3 is 3.
The number of cells in generation 4 is 2.
The number of cells in generation 5 is 3.
The number of cells in generation 6 is 1.
The branching process dies out in generation 7.

For every $n \in \mathbb{N}$ we use the same notation q_n for the probability that the process dies out by the nth generation and $q = \lim_{n \to \infty} q_n$ is the probability that process dies out. As in 1) we obtain

$$q_n = P(X = 0) + \sum_{j=1}^{\infty} q_{n-1}^j P(X = j), \ n \in \mathbb{N}^*,$$

where $X \sim Bino(2, p)$. Therefore, we obtain

$$q_n = (q_{n-1}p + 1 - p)^2, \ n \in \mathbb{N}^*.$$

This implies that q satisfies the equation

$$q = (qp + 1 - p)^2$$

and thus $q \in \left\{ 1, \left(1 - \dfrac{1}{p}\right)^2 \right\}$. Moreover, using the above recurrence relation for the sequence $(q_n)_{n \in \mathbb{N}}$, we can easily prove that for $n \in \mathbb{N}$ we have $q_n \leq \left(1 - \dfrac{1}{p}\right)^2$. So,

$$q \leq \left(1 - \frac{1}{p}\right)^2.$$

If $p \in \left(0, \frac{1}{2}\right]$, then $\left(1 - \frac{1}{p}\right)^2 \geq 1$ and thus $q = 1$. If $p \in \left(\frac{1}{2}, 1\right)$, then $q \leq \left(1 - \frac{1}{p}\right)^2 < 1$ and thus $q = \left(1 - \frac{1}{p}\right)^2$.

Problem 5.4.11. Each bacterium of a certain type generates independently 0, 1, 2 offspring with probabilities $\frac{1}{4}, \frac{1}{4}$, respectively $\frac{1}{2}$. Let Z_n be the number of bacteria in the nth generation for $n \in \mathbb{N}^*$ and assume that $Z_0 = 1$. Prove that:

1) $\left(0.8^n \cdot Z_n\right)_{n \in \mathbb{N}}$ converges in mean of order 2 to a nonconstant random variable;

2) $\left(0.8^n \cdot Z_n\right)_{n \in \mathbb{N}}$ converges a.s. to a nonconstant random variable.

Solution 5.4.11: For every $n \in \mathbb{N}^*$, let Z_n be the size of the population of bacteria in the nth generation and $Z_0 = 1$. Note that $(Z_n)_{n \in \mathbb{N}}$ is a branching process. For $n \in \mathbb{N}$ and $i \in \mathbb{N}^*$ let $X_{n,i}$ be the number of offspring generated by the ith bacterium in the nth generation. Hence,

$$P(X_{n,i} = 0) = P(X_{n,i} = 1) = \frac{1}{4}, \quad P(X_{n,i} = 2) = \frac{1}{2}$$

and thus

$$\mu = E(X_{n,i}) = \frac{5}{4}, \quad \sigma^2 = V(X_{n,i}) = \frac{11}{16} \quad \text{for } n \in \mathbb{N}, i \in \mathbb{N}^*.$$

In view of the independence and the distribution of the numbers of offspring generated by the bacteria (see also Remark 5.3-(1)), $(X_{n,i})_{(n,i) \in \mathbb{N} \times \mathbb{N}^*}$ is a family of independent identically distributed random variables such that we have

$$Z_{n+1} = \sum_{i=1}^{\infty} X_{n,i} \cdot \mathbb{I}_{\{Z_n \geq i\}}, \quad n \in \mathbb{N}.$$

Using the same arguments as in the proof of (5.8) and (5.9), we deduce that

$$E(Z_n) = \mu^n \quad \text{and} \quad E(Z_n^2) = \sigma^2 \cdot \frac{\mu^n(\mu^n - 1)}{\mu(\mu - 1)} + \mu^{2n}, \quad n \in \mathbb{N}.$$

By Theorem 3.2 and Remark 3.2-(3), we get for $n, l \in \mathbb{N}$

$$E(Z_n \cdot Z_{n+l+1}) = \sum_{\substack{k \in \mathbb{N}^* \\ P(Z_{n+l}=k)>0}} E\big(Z_n \cdot Z_{n+l+1} \big| Z_{n+l} = k\big) P(Z_{n+l} = k)$$

$$= \sum_{\substack{k \in \mathbb{N}^* \\ P(Z_{n+l}=k)>0}} E\left(Z_n \cdot \sum_{i=1}^{k} X_{n+l+1,i} \bigg| Z_{n+l} = k\right) P(Z_{n+l} = k)$$

$$= \sum_{\substack{k\in\mathbb{N}^* \\ P(Z_{n+l}=k)>0}} \sum_{i=1}^{k} E\big(Z_n \cdot X_{n+l+1,i}\big|Z_{n+l}=k\big)P(Z_{n+l}=k)$$

$$= \sum_{\substack{k\in\mathbb{N}^* \\ P(Z_{n+l}=k)>0}} \sum_{i=1}^{k} E(X_{n+l+1,i}) \cdot E\big(Z_n\big|Z_{n+l}=k\big)P(Z_{n+l}=k)$$

$$= \sum_{\substack{k\in\mathbb{N}^* \\ P(Z_{n+l}=k)>0}} k\mu E\big(Z_n\big|Z_{n+l}=k\big)P(Z_{n+l}=k)$$

$$= \sum_{\substack{k\in\mathbb{N}^* \\ P(Z_{n+l}=k)>0}} \mu E\big(kZ_n\big|Z_{n+l}=k\big)P(Z_{n+l}=k) = \mu E(Z_n \cdot Z_{n+l}),$$

where we note that the first and last expected values exist in view of The orem 3.8. Thus, we deduce that

$$E(Z_n \cdot Z_{n+l}) = \mu^l E(Z_n^2) = \mu^l \cdot \left(\sigma^2 \cdot \frac{\mu^n(\mu^n-1)}{\mu(\mu-1)} + \mu^{2n}\right)$$

for $n, l \in \mathbb{N}$. Let $U_n = \dfrac{1}{\mu^n} \cdot Z_n = 0.8^n \cdot Z_n$ for $n \in \mathbb{N}$. Then

$$E\big((U_{n+l}-U_n)^2\big) = E(U_{n+l}^2) + E(U_n^2) - 2E(U_n \cdot U_{n+l}) = \frac{\sigma^2\mu^{-n}(1-\mu^{-l})}{\mu(\mu-1)}$$

for $n, l \in \mathbb{N}$.

1) From the above relation, we get that $(U_n)_{n\in\mathbb{N}}$ is a Cauchy sequence in the L^2 sense, hence, by Theorem 4.10, there is a random variable U such that $U_n \xrightarrow{L^2} U$. By Theorem 4.12, we have $E(U_n^2) \to E(U^2)$ and $E(U_n) \to E(U)$. Since $V(U_n) = E(U_n^2) - E^2(U_n) \to \dfrac{\sigma^2}{\mu(\mu-1)} = \dfrac{11}{5}$, we have $V(U) \neq 0$, and thus U is not constant.

2) In view of the above relation and Theorem 4.12, we deduce that

$$E\big((U-U_n)^2\big) = \frac{\sigma^2}{\mu(1-\mu)} \cdot \mu^{-n}, \quad n \in \mathbb{N}.$$

Let $\varepsilon > 0$. Using Markov's inequality from Theorem 3.6, we get

$$P(|U-U_n| > \varepsilon) \leq \frac{E\big((U-U_n)^2\big)}{\varepsilon^2}, \quad n \in \mathbb{N},$$

and thus

$$\sum_{n=0}^{\infty} P(|U-U_n| > \varepsilon) \leq \frac{\sigma^2}{\varepsilon^2\mu(1-\mu)} \sum_{n=0}^{\infty} \mu^{-n} = \frac{\sigma^2}{\varepsilon^2\mu(1-\mu)^2} < \infty.$$

Now, Theorem 4.2 yields that $U_n \xrightarrow{a.s.} U$.

Problem 5.4.12. Let $(X_n)_{n \in \mathbb{N}}$ be a Markov chain with state space $\mathcal{S} = \{0, 1, \ldots, 5\}$, starting at $X_0 = 3$ with the a matrix of transition probabilities given by

$$\mathcal{P} = (p_{ij}^{(1)})_{i,j=\overline{0,5}} = \begin{pmatrix} 1 & 0 & 0 & 0 & 0 & 0 \\ 1-p & 0 & p & 0 & 0 & 0 \\ 0 & 1-p & 0 & p & 0 & 0 \\ 0 & 0 & 1-p & 0 & p & 0 \\ 0 & 0 & 0 & 1-p & 0 & p \\ 0 & 0 & 0 & 0 & 1 & 0 \end{pmatrix}, \text{ where } p \in (0,1).$$

This Markov chain describes a random walk through the points $0, 1, \ldots, 5$, having initial state $X_0 = 3$.

1) Prove that the Markov chain has a.s. two barriers: an absorption at the point 0 and a reflection at the point 5.

2) Simulate 5 values of the Markov chain for $n = 20$ time points, with each of the following probabilities: $p = 0.2$ and $p = 0.8$. What do you observe?

3) Using Matlab, compute $P(X_8 = 3)$ and $P(X_{10} = 1, X_8 = 3)$ for $p = 0.2$.

Solution 5.4.12: 1) We note that for every $n \in \mathbb{N}$ we have

$$P(X_{n+1} = 0 | X_n = 0) = p_{00}^{(1)} = 1, \text{ if } P(X_n = 0) > 0,$$

and

$$P(X_{n+1} = 4 | X_n = 5) = p_{54}^{(1)} = 1, \text{ if } P(X_n = 5) > 0.$$

Moreover, for every $n \in \mathbb{N}$ with $P(X_n = 0) > 0$ and $r \in \mathbb{N}^*$ we have $p_{00}^{(r)} = 1$ and thus $P(X_{n+r} = 0 | X_n = 0) = 1$.

2) We follow the ideas from Example 5.11.

```
function X=markov_chain_pb(p,n)
P=[1 0 0 0 0 0;1-p 0 p 0 0 0;0 1-p 0 p 0 0;0 0 1-p 0 p 0;0 0 0 1-p 0 p;0 0 0 0 1 0];
X=zeros(5,n);
for i=1:5
    j=3;
    X(i,1)=3;
    for k=2:n
        j=randsample(0:5,1,'true',P(j+1,:));
        X(i,k)=j;
    end
end
end

>> X=markov_chain_pb(0.2,20)
```

X = 3 2 3 2 1 2 3 2 3 2 1 0 0 0 0 0 0 0 0 0
    3 2 1 0 0 0 0 0 0 0 0 0 0 0 0 0 0 0 0 0
    3 4 5 4 3 2 1 0 0 0 0 0 0 0 0 0 0 0 0 0
    3 2 1 0 0 0 0 0 0 0 0 0 0 0 0 0 0 0 0 0
    3 2 3 4 3 4 3 2 3 4 3 2 1 2 1 0 0 0 0 0

>>X=markov_chain_pb(0.8,20)

X = 3 4 3 2 1 2 3 2 3 2 1 2 3 4 5 4 5 4 5 4
    3 4 5 4 5 4 5 4 5 4 5 4 5 4 5 4 3 4 5 4
    3 4 5 4 5 4 5 4 3 4 5 4 5 4 5 4 5 4 3 4
    3 2 1 2 1 0 0 0 0 0 0 0 0 0 0 0 0 0 0 0
    3 4 5 4 5 4 3 2 1 0 0 0 0 0 0 0 0 0 0 0

We observe that, in general, an absorption occurs earlier in the case $p = 0.2$ than in the case $p = 0.8$.

3) We have

$$P(X_8 = 3) = P(X_8 = 3 | X_0 = 3) P(X_0 = 3) = p_{33}^{(8)}$$

and

$$P(X_{10} = 1, X_8 = 3) = P(X_{10} = 1 | X_8 = 3) P(X_8 = 3) = p_{31}^{(2)} \cdot p_{33}^{(8)}.$$

>>P=[1 0 0 0 0 0; 0.8 0 0.2 0 0 0; 0 0.8 0 0.2 0 0; ...
    0 0 0.8 0 0.2 0; 0 0 0 0.8 0 0.2; 0 0 0 0 1 0];
>>prob=@(Q,i,j) Q(i+1,j+1);
>>fprintf('P(X_8=3)=%f\n',prob(P^8,3,3));
P(X_8=3)=0.040151
>>fprintf('P(X_10=1,X_8=3)=%f\n',prob(P^2,3,1)*prob(P^8,3,3));
P(X_10=1,X_8=3)=0.025697

Problem 5.4.13. Let $\lambda > 0$ and let $A \in \mathcal{M}_{n \times n}(\mathbb{R})$, $n \in \mathbb{N}^*$, be given by

$$A = \begin{pmatrix} -\lambda & \lambda & 0 & \dots & 0 & 0 \\ 0 & -\lambda & \lambda & \dots & 0 & 0 \\ 0 & 0 & -\lambda & \dots & 0 & 0 \\ \vdots & \vdots & \vdots & & \vdots & \vdots \\ 0 & 0 & 0 & & -\lambda & \lambda \\ 0 & 0 & 0 & \dots & 0 & -\lambda \end{pmatrix}.$$

Using some simulated sample paths of a Poisson process in Matlab, plot some trajectories that estimate the solution $\mathbf{x} = (x_0, \dots, x_{n-1})$ of the following initial value problem:

$$\begin{cases} \dfrac{d\mathbf{x}}{dt}(t) = \mathbf{x}(t) \cdot A, & t \geq 0 \\ \mathbf{x}(0) = \mathbf{e}_1, \end{cases}$$

where $\dfrac{d\mathbf{x}}{dt} = \left(\dfrac{dx_0}{dt}, \ldots, \dfrac{dx_{n-1}}{dt}\right)$ and $\mathbf{e}_1 = (1, 0, \ldots, 0) \in \mathbb{R}^n$ (the above initial value problem has a unique solution, see [Agarwal and O'Regan (2008)]).

Solution 5.4.13: Let $(N_t)_{t \geq 0}$ be a Poisson process with parameter λ. For every $k \in \mathbb{N}$ let

$$x_k(t) = P(N_t = k) = \frac{(\lambda t)^k}{k!} e^{-\lambda t}, \ t \geq 0.$$

By simple computations we deduce that the following relations hold

$$\frac{dx_0}{dt}(t) = -\lambda x_0(t), \ t \geq 0, \ x_0(0) = 1$$

$$\frac{dx_k}{dt}(t) = \lambda x_{k-1}(t) - \lambda x_k(t), \ t \geq 0, \ x_k(0) = 0, \ k \in \mathbb{N}^*.$$

Thus $\mathbf{x} = (x_0, \ldots, x_{n-1})$ satisfies the above initial value problem.

In order to estimate the trajectories of $\mathbf{x}$, we generate independently Poisson processes with parameter λ: $(N_t^{(1)})_{t \geq 0}, \ldots, (N_t^{(m)})_{t \geq 0}, \ldots$ and we note that, by Example 4.3-(2), we have for $t \geq 0$ and $k \in \{0, \ldots, n-1\}$

$$\frac{\#\{l \in \{1, \ldots, m\} : N_t^{(l)} = k\}}{m} \xrightarrow{a.s.} x_k(t) \text{ as } m \to \infty.$$

Note that for fixed $t > 0$ we choose $m \in \mathbb{N}^*$ sufficiently large to write the approximation

$$x_k(t) \approx \frac{\#\{l \in \{1, \ldots, m\} : N_t^{(l)} = k\}}{m}, \ k \in \{0, \ldots, n-1\}.$$

```
function poiss_ivp(lambda,n,t0,m)
t=linspace(0,t0,1000);
N=zeros(m,1000);
for l=1:m
    S=0;
    while S<t0
        S=S+exprnd(1/lambda);
        N(l,t>=S)=N(l,t>=S)+1;
    end
end
clf; hold on;
for k=0:n-1
    plot(t,mean(N==k,1));
end
end

>>poiss_ivp(1,4,10,10000)
```

The unique solution of the above initial value problem is given by (see [Agarwal and O'Regan (2008), Lecture 18])

$$x(t) = \mathbf{e}_1 \cdot \exp\{tA\} = \mathbf{e}_1 \cdot \left(I_n + \sum_{k=1}^{\infty} \frac{t^k}{k!} A^k \right), \ t \geq 0.$$

In the following we plot the trajectories of such a solution in Matlab, to compare them with the previously estimated ones.

```
function ivp(lambda,n,t0)
A=diag(lambda*ones(1,n-1),1)-lambda*eye(n);
e1=[1 zeros(1,n-1)];
t=linspace(0,t0,1000); x=zeros(n,1000);
for j=1:1000
    x(:,j)=e1*expm(t(j)*A);
end
plot(t,x);
end
```

>>**figure**; ivp(1,4,10)

Problem 5.4.14. Let $(W_t)_{t \geq 0}$ be a Wiener process. What distribution has the random variable $3W_s + W_t$, where $0 < s < t$?

Solution 5.4.14:
First method: We use the property that $(W_t)_{t \geq 0}$ is also a Gaussian process (see Example 5.16). Then, (W_s, W_t) has multivariate normal distribution (see Definition 5.14) and $3W_s + W_t$ is a linear combination of the components of the random vector. By Problem 3.6.1 it follows that $3W_s + W_t$ has normal distribution $N(0, 15s + t)$.
Second method: Observe that W_s and W_t are normally distributed random variables, but they are not independent. We write

$$3W_s + W_t = 4(W_s - W_0) + W_t - W_s.$$

The random variables $W_s - W_0 \sim N(0, s)$, $W_t - W_s \sim N(0, t - s)$ are independent, then by the results from Problem 3.5.4 it follows that

$$3W_s + W_t = 4(W_s - W_0) + W_t - W_s \sim N(0, 15s + t).$$

Problem 5.4.15. In some cases it is possible to describe a stock price by the process

$$X_t = \exp\{at + bW_t\}, \ t \geq 0,$$

where $(W_t)_{t \geq 0}$ is a Wiener process and $a \in \mathbb{R}, b \in \mathbb{R}^*$. Compute a density function and the expectation for X_t, $t > 0$.

Solution 5.4.15: Since the logarithm is strictly monotone and it is the inverse function of the exponential function we get for $x > 0$, $t > 0$

$$P(X_t \leq x) = P(\exp\{at + bW_t\} \leq x) = P(at + bW_t \leq \log x)$$

$$= \begin{cases} P\left(\dfrac{1}{\sqrt{t}}W_t \leq \dfrac{\log x - at}{b\sqrt{t}}\right), & \text{if } b > 0 \\ 1 - P\left(\dfrac{1}{\sqrt{t}}W_t < \dfrac{\log x - at}{b\sqrt{t}}\right), & \text{if } b < 0. \end{cases}$$

By Definition 5.15 we have that $Y_t = \frac{1}{\sqrt{t}}W_t \sim N(0,1)$ and we get

$$P(X_t \leq x) = \begin{cases} F\left(\frac{\log x - at}{b\sqrt{t}}\right), & \text{if } b > 0 \\ 1 - F\left(\frac{\log x - at}{b\sqrt{t}}\right), & \text{if } b < 0, \end{cases}$$

where F is the distribution function of the standard normal distribution. Then for $t > 0$

$$f_{X_t}(x) = \frac{1}{x|b|\sqrt{t}}F'\left(\frac{\log x - at}{b\sqrt{t}}\right) = \frac{1}{x|b|\sqrt{2\pi t}}\exp\left\{-\frac{1}{2b^2 t}(\log x - at)^2\right\},$$

if $x > 0$, and $f_{X_t}(x) = 0$, if $x \leq 0$, is a density function of X_t. By the definition of expectation

$$E(X_t) = \int_{\mathbb{R}} x f_{X_t}(x)dx = \frac{1}{|b|\sqrt{2\pi t}}\int_0^\infty \exp\left\{-\frac{1}{2b^2 t}(\log x - at)^2\right\}dx, \ t > 0.$$

We make the change of variable $\log x = -u$ and have

$$E(X_t) = \frac{1}{|b|\sqrt{2\pi t}}\int_{-\infty}^\infty \exp\left\{-\frac{1}{2b^2 t}(u + at)^2\right\}e^{-u}du$$

$$= \frac{\exp\left\{(a + \frac{b^2}{2})t\right\}}{|b|\sqrt{2\pi t}}\int_{-\infty}^\infty \exp\left\{-\frac{1}{2b^2 t}(u + (a + b^2)t)^2\right\}du$$

$$= \exp\left\{\left(a + \frac{b^2}{2}\right)t\right\}, \ t > 0.$$

Problem 5.4.16. The process $B_t = W_t - tW_1$, $t \in [0, 1]$ is called Brownian bridge. This process starts in the time point 0 at position 0 and ends up in time point 1 at the initial position 0.

1) Compute the expectation function and the covariance function of this process.

2) Are the increments of this process independent?

3) Is $(B_t)_{t \geq 0}$ a Gaussian process?

Solution 5.4.16: 1) The expectation function is $\tilde{m}(t) = 0$ for each $t \in [0,1]$ and the covariance function is given by

$$\tilde{\gamma}(s,t) = \min\{s,t\} - st, \quad s,t \in [0,1].$$

2) We choose $0 = t_0 < t_1 < t_2 < 1$ and consider the increments

$$B_{t_1} - B_{t_0} = W_{t_1} - t_1 W_1 \text{ a.s.}$$

and

$$B_{t_2} - B_{t_1} = W_{t_2} - W_{t_1} + (t_1 - t_2)W_1.$$

The increments are not independent, since

$$\mathrm{cov}(B_{t_1} - B_{t_0}, B_{t_2} - B_{t_1}) = t_1^2 - t_1 t_2 \neq 0.$$

Note that this implies that $(B_t)_{t \in [0,1]}$ is not a Wiener process.

3) Let $n \in \mathbb{N}^*$, $0 \leq t_1 < \ldots < t_n \leq 1$, $(b_1, \ldots, b_n) \in \mathbb{R}^n \setminus \{0_n\}$. We consider the linear combination

$$\sum_{i=1}^{n} b_i B_{t_i} = -W_1 \sum_{i=1}^{n} t_i b_i + \sum_{i=1}^{n} b_i W_{t_i}$$

which obviously is a linear combination of $W_{t_1}, \ldots, W_{t_n}, W_1$ and it has

$N\left(0, \displaystyle\sum_{i,j=1}^{n} b_i b_j (\min\{t_i, t_j\} - t_i t_j)\right)$ distribution (by Example 5.16 and Problem 3.6.1). By Remark 5.7-(2) it follows that $(B_t)_{t \in [0,1]}$ is a Gaussian process. Observe that in the case when $t_1 = 0$ or $t_n = 1$ the random vector follows a multivariate normal distribution included in the degenerate case (see Remark 5.7-(1)).

Problem 5.4.17. Consider the process $B_t = at + W_t$, $t \geq 0$, where $(W_t)_{t \geq}$ is a Wiener process and $a \in \mathbb{R}^*$ is a constant.
1) Compute the expectation function and the covariance function for this stochastic process.
2) Is $(B_t)_{t \geq 0}$ a Gaussian process?
3) Is $(B_t)_{t \geq 0}$ a Wiener process?

Solution 5.4.17: 1) For $(B_t)_{t \geq 0}$ the expectation function is

$$\tilde{m}(t) = E(B_t) = at, \quad t \geq 0$$

and the covariance function is

$$\tilde{\gamma}(s,t) = E\big((B_t - \tilde{m}(t))(B_s - \tilde{m}(s))\big) = E(W_t W_s) = \min\{s,t\}, \quad s,t \geq 0$$

2) Let $n \in \mathbb{N}^*$, $0 < t_1 < \ldots < t_n$, $t_0 = 0$ and denote

$$\mathbb{U}_1 = (B_{t_1}, B_{t_2}, \ldots, B_{t_n}), \quad \mathrm{m} = (t_1, t_2, \ldots, t_n),$$

$$\mathbb{U}_2 = (W_{t_1}, W_{t_2}, \ldots, W_{t_n}).$$

We also have $\mathbb{U}_1 = \mathbb{U}_2 I_n + a\mathrm{m}$. By Example 5.16 we have that $\mathbb{U}_2$ has multivariate normal distribution and by Example 2.24-(2) it follows that $\mathbb{U}_1$ has also multivariate normal distribution. Hence, $(B_t)_{t \geq 0}$ is a Gaussian process. Observe that in the case when $t_1 = 0$ the random vector follows a multivariate normal distribution included in the degenerate case (see Remark 5.7-(1)), since the corresponding covariance matrix has the first line and first column filled with zeros.

3) Since $\tilde{m}(t) - \tilde{m}(s) = a(t - s) \neq 0$ for $s \neq t$ the property (3) of the Definition 5.15 of a Wiener process is not fulfilled, hence $(B_t)_{t \geq 0}$ is not a Wiener process.

Problem 5.4.18. If $(W_t)_{t \geq 0}$ is a Wiener process and $r > 0$ is fixed, prove that $B_t = W_{t+r} - W_r$ for $t \geq 0$ is a Gaussian process. Moreover, is $(B_t)_{t \geq 0}$ a Wiener process?

Solution 5.4.18: Let $0 < t_1 < \ldots < t_n$. Denote $t_0 = 0$ and

$$\mathbb{B}_1 = (B_{t_1}, B_{t_2}, \ldots, B_{t_n}),$$

$$\mathbb{B}_2 = (W_{t_1+r} - W_r, W_{t_2+r} - W_{t_1+r}, \ldots, W_{t_n+r} - W_{t_{n-1}+r}),$$

$$A = \begin{pmatrix} 1 & 1 & 1 & \ldots & 1 & 1 \\ 0 & 1 & 1 & \ldots & 1 & 1 \\ \vdots & \vdots & \vdots & \ddots & \vdots & \vdots \\ 0 & 0 & 0 & \ldots & 1 & 1 \\ 0 & 0 & 0 & \ldots & 0 & 1 \end{pmatrix}.$$

Note that A is an invertible matrix. We have $\mathbb{B}_1 = \mathbb{B}_2 A$.
By the definition of the Wiener process

$$W_{t_1+r} - W_r, W_{t_2+r} - W_{t_1+r}, \ldots, W_{t_n+r} - W_{t_{n-1}+r}$$

are independent random variables, each of them following a normal distribution with zero mean. We have $\mathbb{B}_2 \sim MVN(0_n, C)$, where C is a diagonal matrix, with the values $c_{ii} = V(W_{t_i+r} - W_{t_{i-1}+r}) = t_i - t_{i-1}$, $i \in \{1, \ldots, n\}$ on the main diagonal. By Example 2.24-(2) it follows that $\mathbb{B}_1 \sim MVN(0_n, A^T C A)$. Observe that, in the case when $t_1 = 0$, $B_{t_1} = 0$

a.s. and $(B_{t_1}, \ldots, B_{t_n})$ follows a multivariate normal distribution included in the degenerate case (see Remark 5.7-(1)). We conclude that $(B_t)_{t \geq 0}$ is a Gaussian process.

We have $E(B_t) = 0$ for each $t \geq 0$ and

$$\text{cov}(B_s, B_t) = E((W_{s+r} - W_r)(W_{t+r} - W_r)) = \min\{s+r, t+r\} - r = \min\{s, t\}$$

for each $s, t \geq 0$. Then it follows by Example 5.16 that $(B_t)_{t \geq 0}$ is a Wiener process.

Appendix

A.1 Integration

We start with a short review about measures. Let Ω be a nonempty set, and let $\mathcal{A}$ be a σ-field of subsets of Ω.

Definition A.1. A function $\mu : \mathcal{A} \to [0, \infty]$ is called **measure** (or **countably additive measure**) if it satisfies the following conditions:

(1) $\mu(\emptyset) = 0$;

(2) $\mu\left(\bigcup_{i=1}^{\infty} A_i\right) = \sum_{i=1}^{\infty} \mu(A_i)$ whenever $A_i \in \mathcal{A}$ for all $i \in \mathbb{N}^*$, and $A_i \cap A_j = \emptyset$ for all $i \neq j$.

The triple $(\Omega, \mathcal{A}, \mu)$ is called **measure space**. If $\mu(\Omega) < \infty$, then μ is a **finite measure**. ♦

Definition A.2. A set $A \in \mathcal{A}$ for which $\mu(A) = 0$ is called **μ-null set**. If a property holds for all $\omega \in \Omega$ except for those ω in some μ-null set, then it is said that the property holds for **μ almost every** ω (abbreviated μ a.e., or a.e.). ♦

Remark A.1. If $\mu(\Omega) = 1$, then the measure μ becomes a probability and $(\Omega, \mathcal{A}, \mu)$ is a probability space (see Definition 1.6). ▼

The main properties of a measure are contained in the following theorem.

Theorem A.1. *Let $(\Omega, \mathcal{A}, \mu)$ be a measure space with $\mu(\Omega) < \infty$. Then the following properties hold:*

(1) *If $A, B \in \mathcal{A}$, then $\mu(A \setminus B) = \mu(A) - \mu(A \cap B)$.*
(2) *If $A \subseteq B$ with $A, B \in \mathcal{A}$, then $\mu(A) \leq \mu(B)$, i.e., μ is monotone.*

(3) $\mu(A \cup B) = \mu(A) + \mu(B) - \mu(A \cap B)$ *for each* $A, B \in \mathcal{A}$.

(4) μ *is subadditive, i.e.,*

$$\mu\left(\bigcup_{n=1}^{\infty} A_n \right) \leq \sum_{n=1}^{\infty} \mu(A_n)$$

for all $A_n \in \mathcal{A}, n \in \mathbb{N}^*$.

(5) *If* $(A_n)_{n \in \mathbb{N}^*}$ *is an increasing sequence of sets from* $\mathcal{A}$, *then*

$$\lim_{n \to \infty} \mu(A_n) = \mu\left(\bigcup_{i=1}^{\infty} A_i \right).$$

(6) *If* $(A_n)_{n \in \mathbb{N}^*}$ *is a decreasing sequence of sets from* $\mathcal{A}$, *then*

$$\lim_{n \to \infty} \mu(A_n) = \mu\left(\bigcap_{i=1}^{\infty} A_i \right).$$

A.1.1   *The Lebesgue Integral*

Definition A.3. Let $(\Omega, \mathcal{A}, \mu)$ be a measure space, and let $f : \Omega \to [0, \infty$
be a measurable function. Define

$$L(f) = \sup \left\{ \sum_{k=1}^{n} \mu(A_k) \inf f(A_k) : (A_i)_{i=\overline{1,n}} \text{ is a partition of } \Omega \right\},$$

where $\inf \emptyset = 0$. If $L(f) < \infty$, then $L(f)$ is called the **Lebesgue integra**
of f with respect to μ and is denoted by $\displaystyle\int_{\Omega} f d\mu$; we say that f admit
Lebesgue integral.

For a function $f : \Omega \to \mathbb{R}$ we write $f = f^+ - f^-$, where

$$f^+ = \max\{0, f\} \text{ and } f^- = -\min\{0, f\},$$

and define

$$\int_{\Omega} f d\mu = \int_{\Omega} f^+ d\mu - \int_{\Omega} f^- d\mu,$$

provided that at least one of the integrals $\displaystyle\int_{\Omega} f^+ d\mu$ and $\displaystyle\int_{\Omega} f^- d\mu$ is finite
We say that f is **Lebesgue integrable**, if both integrals are finite.
For $A \in \mathcal{A}$ we define

$$\int_{A} f d\mu = \int_{\Omega} \mathbb{I}_A \cdot f d\mu.$$

Remark A.2. If $(\Omega, \mathcal{A}, P)$ is a probability space and $f = X$ is a random variable which is Lebesgue integrable with respect to the probability measure $P,$ then the expectation of X is given by $E(X) = \int_\Omega X(\omega)dP(\omega).$ ▼

In what follows we consider $(\Omega, \mathcal{A}, \mu)$ to be a measure space. Let $(\mathbb{R}, \mathcal{B}, \lambda)$ denote the measure space of real numbers, $\mathcal{B}$ the system of Borel sets and λ the Lebesgue measure on $\mathbb{R}$.

Theorem A.2. *Let* $f, g : \Omega \to \mathbb{R}$ *be Lebesgue integrable functions. Then the following properties hold:*

(1) $\displaystyle\int_\Omega (\alpha f + \beta g)d\mu = \alpha \int_\Omega fd\mu + \beta \int_\Omega gd\mu$ *for all* $\alpha, \beta \in \mathbb{R}.$

(2) *If* $f(\omega) = g(\omega)$ *for* μ *a.e.* $\omega \in \Omega,$ *then* $\displaystyle\int_\Omega fd\mu = \int_\Omega gd\mu.$

(3) *If* $f(\omega) \le g(\omega)$ *for* μ *a.e.* $\omega \in \Omega,$ *then* $\displaystyle\int_\Omega fd\mu \le \int_\Omega gd\mu.$

(4) *If* $\displaystyle\int_\Omega |f|d\mu = 0,$ *then* $f = 0$ *for* μ *a.e.* $\omega \in \Omega.$

A.1.2 The Lebesgue–Stieltjes Integral

A special case is obtained when the measure μ is determined by a function $F : \mathbb{R} \to \mathbb{R}$ having the following properties:
(LS1) F is increasing;
(LS2) F is right-continuous;
(LS3) $\displaystyle\lim_{x \to -\infty} F(x) = 0$ and $\displaystyle\lim_{x \to \infty} F(x) = 1.$
Then there exists a unique probability measure $\mathbb{P}$ on $(\mathbb{R}, \mathcal{B})$ such that

$$\mathbb{P}((a, b]) = F(b) - F(a) \quad \text{for each } a < b.$$

This measure is called **Lebesgue–Stieltjes measure** and one can define the corresponding **Lebesgue–Stieltjes integral** $\displaystyle\int_\mathbb{R} g(t)dF(t)$ for a function $g : \mathbb{R} \to \mathbb{R}$, which is $\mathcal{B}/\mathcal{B}$ measurable, by

$$\int_\mathbb{R} g(t)dF(t) = \int_\mathbb{R} g(\omega)d\mathbb{P}(\omega).$$

Let X be a random variable on a probability space $(\Omega, \mathcal{F}, P)$, and let F_X be its distribution function. This random variable induces a probability measure P_X, called distribution of X (see Definition 2.5), on $\mathcal{B}$ via

$$P_X(B) = P(\{\omega \in \Omega : X(\omega) \in B\}), \quad B \in \mathcal{B}.$$

In particular for $a, b \in \mathbb{R}, a < b$, we can take $B = (a, b]$ and get

$$P_X((a, b]) = P(a < X \le b) = F_X(b) - F_X(a).$$

But F_X satisfies the properties (LS1), (LS2) and (LS3) from above (see Theorem 2.4), then the distribution P_X coincides with the (uniquely determined) Lebesgue–Stieltjes measure induced by F_X. Therefore we can write

$$\int_\Omega g(X(\omega)) dP(\omega) = \int_\mathbb{R} g(t) dF_X(t),$$

where $g : \mathbb{R} \to \mathbb{R}$ is a $\mathcal{B}/\mathcal{B}$ measurable function and $g(X)$ is an integrable function with respect to P.

All these results can be formulated also on $\mathbb{R}^n$. For more details and properties see Sections 6.1, 6.2 and 6.3 in [Chow and Teicher (1997)].

Remark A.3. Let $(\Omega, \mathcal{F}, P)$ be a probability space and let X be an integrable random variable, then the expectation of X is given by

$$E(X) = \int_\Omega X(\omega) dP(\omega) = \int_\mathbb{R} t dF_X(t),$$

where F_X is the distribution function of X.
If X is a continuous random variable with density function f_X, then

$$E(X) = \int_\mathbb{R} t dF_X(t) = \int_\mathbb{R} t f_X(t) dt.$$

If X is a discrete random variable with the representation

$$X(\omega) = \sum_{i \in I} x_i \mathbb{I}_{A_i}(\omega) \quad \text{for all } \omega \in \Omega,$$

where $A_i = \{\omega \in \Omega : X_i(\omega) = x_i\} \in \mathcal{F}, i \in I$, forms a partition of Ω, then

$$E(X) = \int_\Omega X(\omega) dP(\omega) = \sum_{i \in I} \int_{A_i} X(\omega) dP(\omega) = \sum_{i \in I} x_i P(X = x_i).$$

A.2 Review on Combinatorics

We recall some basic counting formulas from combinatorics:
- number of permutations of n elements: $n!$, where $n \in \mathbb{N}$ and $0!=1$;
- number of k-permutations of n elements without repetitions:

$$P(n, k) = \frac{n!}{(n-k)!}, \text{ where } k \in \mathbb{N}, n \in \mathbb{N}^*, n \geq k;$$

- number of k-permutations of n elements with repetitions: n^k, where $k, n \in \mathbb{N}^*$;
- number of k-combinations from n elements without repetitions:

$$C(n, k) = \binom{n}{k} = \frac{n!}{k!(n-k)!}, \text{ where } k \in \mathbb{N}, n \in \mathbb{N}^*, n \geq k;$$

- number of k-combinations from n elements with repetitions:

$$C(n+k-1, k) = \frac{(n+k-1)!}{k!(n-1)!}, \text{ where } k, n \in \mathbb{N}^*.$$

A.3 Euler Functions

Euler's **Gamma function** $\Gamma : (0, \infty) \to (0, \infty)$ is defined by

$$\Gamma(a) = \int_0^\infty x^{a-1}e^{-x}dx, \quad a > 0.$$

Proposition A.1. *The Gamma function has the following properties:*

(1) $\Gamma(a+1) = a\Gamma(a)$ *for each* $a > 0$;

(2) $\Gamma(n+1) = n!$ *for* $n \in \mathbb{N}$;

(3) $\Gamma\left(\dfrac{1}{2}\right) = \sqrt{2}\displaystyle\int_0^\infty e^{-\frac{t^2}{2}}dt = \int_{\mathbb{R}} e^{-t^2}dt = \sqrt{\pi}$;

(4) $\Gamma\left(n+\dfrac{1}{2}\right) = \sqrt{\pi}\dfrac{(2n-1)!!}{2^n} = \sqrt{\pi}\displaystyle\prod_{j=1}^n \dfrac{2j-1}{2}$ *for* $n \in \mathbb{N}^*$.

Euler's **Beta function** $B : (0, \infty) \times (0, \infty) \to (0, \infty)$ is defined by

$$B(a, b) = \int_0^1 x^{a-1}(1-x)^{b-1}dx, \quad a, b > 0.$$

Proposition A.2. *The Beta function has the following properties:*

(1) $B(a, 1) = \dfrac{1}{a}$ *for* $a > 0$;

(2) $B(a, b) = B(b, a)$ *for* $a > 0, b > 0$;

(3) $B(a, b) = \dfrac{a-1}{b} B(a-1, b+1)$ *for* $a > 1, b > 0$;

(4) $B(a, b) = \dfrac{b-1}{a+b-1} B(a, b-1) = \dfrac{a-1}{a+b-1} B(a-1, b)$ *for* $a > 1,\ b > 1$;

(5) $B(a, b) = \dfrac{\Gamma(a)\Gamma(b)}{\Gamma(a+b)}$ *for* $a > 0, b > 0$;

(6) $B(k+1, n-k+1) = \dfrac{1}{(n+1)C(n, k)}$ *for* $k, n \in \mathbb{N}$ *with* $k \le n$.

A.4   Review on Matrix Theory

In the following we recall some results regarding positive (semi)definite symmetric matrices. For basic notions and results in linear algebra, see [Horn and Johnson (1990)].

Definition A.4. Let $C \in \mathcal{M}_{n \times n}(\mathbb{R})$ be a symmetric matrix. It is **positive semidefinite**, if $xCx^T \ge 0$ for all $x \in \mathbb{R}^n$. It is a **positive definite matrix**, if $xCx^T > 0$ for all $x \in \mathbb{R}^n \setminus \{0_n\}$. ◀

Proposition A.3. (1) *If* $C \in \mathcal{M}_{n \times n}(\mathbb{R})$ *is a positive definite matrix and* $B \in \mathcal{M}_{n \times n}(\mathbb{R})$ *is an invertible matrix, then* $B^T C B$ *is a positive definite matrix.*
(2) *If* $B \in \mathcal{M}_{n \times n}(\mathbb{R})$ *is an invertible matrix, then* $B^T B$ *is a positive definite matrix.*

Proposition A.4. *Let* $C \in \mathcal{M}_{n \times n}(\mathbb{R})$ *be a positive semidefinite matrix. Then the following properties hold:*
(1) *C has positive eigenvalues* $\lambda_1, \ldots, \lambda_n$ *(counting multiplicities) and corresponding n orthogonal unit-length eigenvectors* $\mathbb{b}_1, \ldots, \mathbb{b}_n$, *i.e.,*

$$C \mathbb{b}_i^T = \lambda_i \mathbb{b}_i^T, \quad i \in \{1, \ldots, n\},$$

where $\mathbb{b}_i \mathbb{b}_i^T = 1$, $i \in \{1, \ldots, n\}$, *and* $\mathbb{b}_i \mathbb{b}_j^T = 0$ $i, j \in \{1, \ldots, n\}, i \ne j$.
(2) *C can be written as* $C = BDB^T$, *where* $D = \mathrm{diag}(\lambda_1, \ldots, \lambda_n)$ *is the diagonal matrix of the eigenvalues of* C, *and* B *is the orthogonal matrix whose columns are the eigenvectors* $\mathbb{b}_1^T, \ldots, \mathbb{b}_n^T$.
(3) *C is positive definite* $\iff \lambda_i > 0$ *for each* $i \in \{1, \ldots, n\} \iff C$ *invertible.*
(4) *The square root of the matrix* C *is* $C^{\frac{1}{2}} = BD^{\frac{1}{2}}B^T$, *where* $D^{\frac{1}{2}} = \mathrm{diag}(\lambda_1^{\frac{1}{2}}, \ldots, \lambda_n^{\frac{1}{2}})$ *and* B *is the orthogonal matrix mentioned in statement (2).*

List of Abbreviations and Symbols

♦	end of a definition
▲	end of an example
▼	end of a remark
a.e.	almost everywhere
a.s.	almost surely
WLLN	weak law of large numbers
SLLN	strong law of large numbers
$\#A$	number of elements of the set A
$\mathbb{I}_A$	indicator function of the event A
$P(A)$	probability of the event A
$P(A|B)$	conditional probability of A given B
$r_n(A)$	relative frequency of the occurrence of the event A after repeating the experiment n times
$\mathbb{N}$	set of positive integers $\{0, 1, 2 \ldots\}$
$\mathbb{N}^*$	set of strictly positive integers $\{1, 2 \ldots\}$
$\mathbb{Z}$	set of integers
$\mathbb{Q}$	set of rational numbers
$\mathbb{R}$	set of real numbers
$\mathbb{R}^n$	Euclidean n-dimensional space
$\mathcal{M}_{n \times n}(\mathbb{R})$	set of $n \times n$ real matrices
0_n	zero vector in $\mathbb{R}^n$
I_n	$n \times n$ identity matrix
$\mathcal{B}$	Borel σ-field generated by the open sets in $\mathbb{R}$
$\mathcal{B}^n$	Borel σ-field generated by the open sets in $\mathbb{R}^n$
$\lambda_{\mathbb{R}^n}(M)$	Lebesgue measure of a measurable set $M \subset \mathbb{R}^n$
log	natural logarithm, ln
$G(a - 0)$	$\lim_{x \nearrow a} G(x)$, limit to the left of a function G at a
$G(a + 0)$	$\lim_{x \searrow a} G(x)$, limit to the right of a function G at a

f	density function
F	(cumulative) distribution function
$E(X)$	expectation of X
$E(X\|B)$	conditional expectation of X given the event B
$E(X\|\mathcal{G})$	conditional expectation of X given the σ-field $\mathcal{G}$
$E(X\|Y)$	conditional expectation of X given the random variable Y
$V(X)$	variance of X
$\mathrm{cov}(X, Y)$	covariance of X and Y
$\rho(X, Y)$	correlation coefficient of X and Y
M_X	moment generating function of X
$x_n \to x$	convergence of real numbers
$X_n \xrightarrow{P} X$	convergence in probability
$X_n \xrightarrow{a.s.} X$	convergence almost surely
$X_n \xrightarrow{L^r} X$	convergence in mean of order r
$(N_t)_{t \geq 0}$	Poisson process
$(W_t)_{t \geq 0}$	Wiener process (Brownian motion)
Γ	Euler's Gamma function
B	Euler's Beta function
$P(n, k)$	k-permutations of n elements without repetitions (variations)
$C(n, k)$	combinations of k elements from n elements without repetitions
$Ber(p)$	Bernoulli distribution
$Bino(n, p)$	binomial distribution
$Geo(p)$	geometric distribution
$NBin(n, p)$	negative binomial distribution
$Poiss(\lambda)$	Poisson distribution
$Unid(n)$	discrete uniform distribution
$Beta(a, b)$	Beta distribution

$Cauchy(a, b)$	Cauchy distribution
$\chi^2(n, \sigma)$	χ^2 distribution
$Exp(\lambda)$	exponential distribution
$F(m, n)$	Snedecor–Fischer distribution
$Gamma(a, b)$	Gamma distribution
$N(m, \sigma^2)$	normal distribution
$T(n)$	Student distribution
$Unif[a, b]$	continuous uniform distribution
$Multino(n, p_1, ..., p_k)$	multinomial distribution
$Hyge(n_1, n_2)$	hypergeometric distribution
$MVHyge(n, n_1, ..., n_k)$	multivariate hypergeometric distribution
$MVN(\mathrm{m}, A)$	multivariate normal distribution
$MVUnif([a_1, b_1] \times \cdots \times [a_n, b_n])$	multivariate uniform distribution

Bibliography

Agarwal, R. P. and O'Regan, D. (2008). *An Introduction to Ordinary Differential Equations* (Springer, New York)

Anderson, W. J. (1991). *Continuous-Time Markov Chains: An Applications-Oriented Approach* (Springer–Verlag, New York)

Bauer, H. (1996). *Probability Theory* (Walter de Gruyter, New York)

Baldi, P. (2017). *Stochastic Calculus: An Introduction Through Theory and Exercises* (Springer International Publishing, Cham)

Baron, M. (2007). *Probability and Statistics for Computer Scientists* (Chapman & Hall, Boca Raton)

Beekman, R. M. (2017). *The Puzzles of Mars: A Mathematical Journey* (Lulu.com, USA)

Billingsley, P. (1986). *Convergence of Probability Measures* (John Wiley & Sons, New York)

Billingsley, P. (1986). *Probability and Measure* (2nd edn., John Wiley & Sons, New York)

Box, G. E. P. and Muller, M. E. (1958). *A Note on the Generation of Random Normal Deviates* (The Annals of Mathematical Statistics, 29 (2): 610–611)

Borodin, A. N. (2017). *Stochastic Processes* (e-book, Springer, Cham)

Breiman, L. (1992). *Probability* (SIAM, Philadelphia)

Choudary, A. D. R. and Niculescu, C. P. (2014) *Real Analysis on Intervals* (Springer, New Delhi)

Chow, Y. S. and Teicher, H. (1997). *Probability Theory: Independence, Interchangeability, Martingales* (Springer–Verlag, New York)

DasGupta, A. (2010). *Fundamentals of Probability: A First Course* (Springer, New York)

Davis, M. H. A. and Lleo, S. W. (2015). *Risk-Sensitive Investment Management* (World Scientific Pub., Singapore)

DeGroot, M. H. (1989). *Probability and Statistics* (Addison–Wesley Publishing Company, Reading, Massachusetts)

Dehling, H. and Haupt, B. (2004). *Einführung in die Wahrscheinlichkeitstheorie und Statistik* (Springer–Verlag, Berlin)

Doob, J. L. (1953). *Stochastic Processes* (John Wiley & Sons, New York)

Durrett, R. (2010). *Probability — Theory and Examples* (Cambridge University Press, Cambridge)

Durrett, R. (2016). *Essentials of Stochastic Processes* (e-book, Springer, Cham)

Feller, W. (1971). *An Introduction to Probability Theory and Its Applications, Volume I*, (3rd edn., John Wiley & Sons, New York)

Feller, W. (1971). *An Introduction to Probability Theory and Its Applications, Volume II* (2nd edn., John Wiley & Sons, New York)

Flajolet, P. (1985). *Approximate Counting: A Detailed Analysis* (BIT, 25: 113–134)

Ghahramani, S. (2005). *Fundamentals of Probability with Stochastic Processes* (3rd edn., Pearson Education LTD., London)

Gikhman, I. I. and Skorokhod, A. V. (2004). *Theory of Stochastic Processes, Part I* (Springer, Berlin)

Gikhman, I. I. and Skorokhod, A. V. (2004). *Theory of Stochastic Processes, Part II* (Springer, Berlin)

Girardin, V. and Limnios, N. (2018). *Applied Probability. From Random Sequences to Stochastic Processes* (e-book, Springer, Cham)

Grimmett, G. and Stirzaker, D. (2001). *Probability and Random Processes* (3rd edn., Oxford University Press, Oxford)

Grinstead, Ch. M. and Snell, J. L. (1997). *Introduction to Probability* (American Mathematical Soc., Rhode Island)

Gut, A. (2005). *Probability: A Graduate Course* (Springer, New York)

Gut, A. (2009). *An Intermediate Course in Probability* (Springer, New York)

Hajek, B. (2015). *Random Processes for Engineers* (Cambridge University Press Cambridge)

Harris, T. E. (1963). *The Theory of Branching Processes* (3rd edn., Springer-Verlag, Berlin)

Henze, N. (2013). *Irrfahrten und verwandte Zufälle — Ein elementarer Einstieg in die stochastischen Prozesse* (Springer Spektrum, Wiesbaden)

Hesse, Ch. (2003). *Angewandte Wahrscheinlichkeitstheorie* (Friedr. Vieweg & Sohn Verlagsgesellschaft mbH, Braunschweig, Wiesbaden)

Hoeffding, W. (1963). *Probability Inequalities for Sums of Bounded Random Variables* (Journal of the American Statistical Association, 58 (301): 13–30)

Holcman, D. (Editor) (2017). *Stochastic Processes, Multiscale Modeling and Numerical Methods for Computational Celluar Biology* (e-book, Springer Cham)

Horn, R. A. and Johnson, C. R. (1990). *Matrix Analysis* (Cambridge University Press, Cambridge)

Iancu, M. and Lisei, H. (2017). *Properties of Random Walks in Dimension One* (Didactica Mathematica, 35: 45–58)

Karr, A. F. (1993). *Probability* (Springer–Verlag, New York)

Kay, S. M. (2006). *Intuitive Probability and Random Processes Using MATLAB* (Springer, New York)

Kirkwood, J. R. (2015). *Markov Processes* (CRC Press, Boca Raton, FL.)

Knill, O. (2009). *Probability and Stochastic Processes with Applications* (Oversea Press, New Delhi)

Knuth, D. (1998). *The Art of Computer Programming, Volume 2/Seminumerical Algorithms* (3rd edn., Addison–Wesley, Boston)

Kolokoltsov, V. N. (2011). *Markov Processes, Semigroups and Generators* (Walter de Gruyter, Berlin)

Lisei, H. (2004). *Probability Theory* (Casa Cărţii de Ştiinţă, Cluj-Napoca)

Lisei, H., Micula, S. and Soos, A. (2006). *Probability Theory Through Problems and Applications* (Presa Universitară Clujeană, Cluj-Napoca)

Mitzenmacher, M. and Upfal, E. (2005). *Probability and Computing: Randomized Algorithms and Probabilistic Analysis* (Cambridge University Press, Cambridge)

Mörters, P. and Peres, Y. (2010). *Brownian Motion* (Cambridge University Press, Cambridge)

Neapolitan, R. E. (2003). *Learning Bayesian Networks* (Prentice-Hall, Inc. Upper Saddle River, NJ, USA)

Nelson, G. S. (2015). *A User-Friendly Introduction to Lebesgue Measure and Integration* (American Mathematical Society, Providence, Rhode Island)

Pfeiffer, P. E. (1990). *Probability for Applications* (Springer–Verlag, New York)

Protter, P. E. (2004). *Stochastic Integration and Differential Equations* (Springer–Verlag, Berlin)

Ross, S. (2007). *Introduction to Probability Models* (9th edn., Elsevier, Amsterdam)

Ross, S. (2013). *Simulation* (5th edn., Elsevier, Amsterdam)

Ross, S. (2014). *A First Course in Probability* (9th edn., Pearson Education, New Jersey)

Rudin, W. (1987). *Real and Complex Analysis* (3rd edn., McGraw-Hill, New York)

Schilling, R. E. and Partzsch, L. (2014). *Brownian Motion: An Introduction to Stochastic Processes* (e-book, Walter De Gruyter, München)

Schinazi, R. B. (2012). *From Calculus to Analysis* (Springer, New York)

Sericola, B. (2013). *Markov Chains. Theory, Algorithms and Applications* (ISTE Ltd, London and John Wiley & Sons Inc., Hoboken, New Jersey)

Skorokhod, A. V. (2005). *Basic Principles and Applications of Probability Theory* (Springer–Verlag, Berlin)

Stoyanov, J. M. (2013). *Counterexamples in Probability* (3rd edn., Dover Publications, New York)

Vaart, A. W. van der (1998). *Asymptotic Statistics* (Cambridge University Press, New York)

Vitali, E., Motta, M. and Galli, D. E. (2018). *Theory and Simulation of Random Phenomena: Mathematical Foundations and Physical Applications* (e-book, Springer, Cham)

Weiss, G. H. (1983). *Random Walks and Their Applications: Widely Used as Mathematical Models, Random Walks Play an Important Role in Several Areas of Physics, Chemistry, and Biology* (American Scientist, 71: 65–71)

Yaglom, A. M. and Yaglom, I. M. (1964). *Challenging Mathematical Problems with Elementary Solutions, Volume 1: Combinatorial Analysis and Probability Theory* (Dover Publications, New York)

Index

Printed in the United States
By Bookmasters